Remediation and Management

of

Degraded Lands

Remediation
and Management
— of —
Degraded
Lands

Edited by

M.H. Wong, Ph.D.
Hong Kong Baptist University

J.W.C. Wong, Ph.D.
Hong Kong Baptist University

A.J.M. Baker, Ph.D.
University of Sheffield

LEWIS PUBLISHERS

Boca Raton London New York Washington, D.C.

Library of Congress Cataloging-in-Publication Data

Remediation and management of degraded lands / edited by M.H. Wong.
       p.   cm.
       Includes bibliographical references and index.
       ISBN 1-57504-109-X
       1. Abandoned mined lands reclamation. 2. Land degradation—Management.
   3. Mines and mineral sources—Environmental aspects.
   I. Wong, M.H.
   TD195.M5R46 1998
   631.6′4—dc21
                                   98-24752
                                      CIP

© 1999 by CRC Press LLC
Lewis Publishers is an imprint of CRC Press LLC

No claim to original U.S. Government works
International Standard Book Number 1-57504-109-X
Library of Congress Card Number 98-24752
Printed in the United States of America 1 2 3 4 5 6 7 8 9 0
Printed on acid-free paper

Preface

The rapid industrialization and economic development in some countries have resulted in the achievement of greatly improved standards of living, and life expectancy has risen with gross domestic product per capita. However, one of the prices to be paid for such rapid growth is environmental deterioration, and our natural resources are under severe threat constantly. Land is one of our most precious resources. Many countries have broadly similar types of damaged land due to the exploitation of materials for mineral and energy resources, through our utilization of land for waste disposal, through agricultural abuse or misuse, and through indiscriminate exploitation of natural resources. Degradation can arise from municipal refuse, sewage sludge, toxic industrial wastes, pulverized fuel ash, sand and gravel, metalliferous and nonmetalliferous ores, smelter wastes, coal spoil, and acid mine drainage, as well as overgrazing and extraction of natural resources such as logging for firewood or lumber.

The International Conference on the Remediation and Management of Degraded Lands provided an excellent opportunity for scientists from a variety of disciplines to exchange and update information on issues related to "Remediation and Management of Degraded Lands;" in particular, to review the extent of degradation of land resources, the ecological requirements for reclamation or restoration of derelict and disturbed lands and their beneficial end-uses. The collaborative efforts involving scientists, managers, and administrators from academia, business, and government will no doubt lead to progressive and enlightened approaches to addressing derelict land issues. During the Closing Section, it was decided to hold follow-up meetings regularly, and the 2nd meeting will be held in Perth, Australia in 1998/99.

Thirty-five papers presented at the conference have been included in this volume, and they are grouped under three subheadings: I. Mine Management and Rehabilitation; II. Management of Derelict Lands; and III. Soil Contamination and Remediation. It is intended that the collection of these papers should act as a framework for discussion of the major processes, issues, and concepts involved in the process of land remediation. It is also hoped that the volume will serve as a reference book for undergraduate and postgraduate students, practitioners, professional or community members, having a common goal of ensuring the long-term sustainable reuse of land.

We would like to use this opportunity to thank authors for their contribution and the members of the Guest Editorial Board for their hard work. We would also like to express our sincere thanks for the advice given by the members of the Advisory Committee, and the effort contributed by the members of the Organizing Committee. The financial support from various funding agents is gratefully acknowledged.

M.H. Wong
J.W.C. Wong
A.J.M. Baker

The Editors

Prof. M.H. Wong, PhD, DSc (Durham University); FIBiol; CFIWEM

Ming H. Wong is Director of the Institute for Natural Resources and Waste Management and Chair Professor of the Department of Biology, Hong Kong Baptist University, and also Visiting Professor of South China Agricultural University, Zhongshan University, and Yangzhou University, PRC and University of Middlesex, UK. Professor Wong's research interests included ecotoxicological assessments of heavy metals and restoration of derelict land.

Dr. J.W.C. Wong, PhD (Murdoch University)

Jonathan Wong is Associate Professor in Environmental Science of Hong Kong Baptist University and also Visiting Professor of China Agricultural University and Nanjing Agricultural University, PRC. His research interests included soil contamination and remediation, and composting of organic wastes.

Dr. A.J.M. Baker, PhD (London University); FIBiol; FLS; MIEEM

Alan Baker is Reader in Environmental Science in the Department of Animal and Plant Sciences, University of Sheffield, and also Visiting Professor at Hong Kong Baptist University. Dr. Baker is an international authority on ecological and evolutionary aspects of the impact of heavy metal pollution on plants. His research focus over the last six years has been on the development of phytoremediation technology for the decontamination of metal-polluted soils and effluents.

Guest Editorial Board

Dr. S.G. McRae
Wye College
University of London, Ashford
Kent, TN25 5AH, United Kingdom

Dr. D.W. Merrilees
Environmental Sciences Department,
   SAC
Auchincruive
Ayr. KA6 5HW, Scotland

Dr. D.R. Mulligan
Centre for Mined Land Rehabilitation
Department of Agriculture
The University of Queensland
Brisbane, Qld 4072, Australia

Prof. N.F.G. Rethman
Department of Plant Production and
   Soil Science
University of Pretoria
Pretoria 0002, Republic of South Africa

Dr. G.E. Schuman
High Plains Grasslands Research
   Station
8408 Hildreth Road
Cheyenne, WY 82009

Prof. R.B.E. Shutes
Urban Pollution Research Centre
Middlesex University
Bounds Green Road
London N11 2NQ, United Kingdom

Dr. B.A. Whitton
Department of Biological Sciences
University of Durham
Durham, DH1 3LE, United Kingdom

Contributors

D.A. Angers
Soils and Crops Research Centre
Agriculture and Agri-Food Canada
Ste-Foy, Qc., Canada G1V 2J3

C.J. Beauchamp
Département de Phytologie
Centre de Recherche en Horticulture,
   FSAA
Université Laval
Ste-Foy, Qc., Canada G1K 7P4

L.C. Bell
Australian Centre for Minesite
   Rehabilitation Research
Kenmore, Qld 4069, Australia

R.W. Bell
Institute for Environmental Science
Murdoch University
Murdoch, WA 6150, Australia

R.W. Bell
School of Biological and
   Environmental Sciences
Murdoch University
Murdoch, WA 6150, Australia

S.M. Bellairs
Centre for Mined Land Rehabilitation
   and Department of Agriculture
The University of Queensland
St. Lucia, QLD-4072, Australia

A.D. Bradshaw
School of Biological Sciences
University of Liverpool
Liverpool L69 3BX, United Kingdom

T.M. Chaudhry
Department of Biological Sciences
Faculty of Business and Technology
University of Western Sydney,
   Macarthur
Campbelltown, NSW 2560, Australia

R. Chen
Department of Ecology & Biodiversity
University of Hong Kong
Hong Kong

K.W. Cheung
Agriculture and Fisheries Department
Government of the Hong Kong Special
   Administrative Region
Hong Kong

R.Y.H. Cheung
Department of Biology and Chemistry
City University of Hong Kong
Kowloon, Hong Kong

K.K. Chiu
Civil and Structural Engineering
   Department
The Hong Kong Polytechnic
   University
Hung Hum, Hong Kong

P.-T. Chiueh
Graduate Institute of Environmental
   Engineering
National Taiwan University
Taipei, Taiwan 10617

S.L. Chong
Territory Development Department
Hong Kong

R.T. Corlett
Department of Ecology & Biodiversity
University of Hong Kong
Hong Kong

N.M. Dickinson
School of Biological and Earth Sciences
Liverpool John Moores University
Liverpool, United Kingdom

A. Duncan
Scottish Agricultural College
Auchincruive
Ayr KA6 5HW, Scotland

C. Fallon
Urban Pollution Research Centre
Middlesex University
London N11 2NQ, United Kingdom

A. Fierro
Département de Phytologie
Centre de Recherche en Horticulture,
  FSAA
Université Laval
Ste-Foy, Qc., Canada G1K 7P4

C.H. Fung
Agriculture & Fisheries Department
Hong Kong Government
Canton Road
Hong Kong

A.H. Grigg
Centre for Mined Land Rehabilitation
The University of Queensland
Brisbane, QLD 4072, Australia

L. Hill
Department of Biological Sciences
Faculty of Business and Technology
University of Western Sydney, Macarthur
Campbelltown, NSW 2560, Australia

R.D. Hill
Department of Ecology & Biodiversity
University of Hong Kong
Hong Kong

G.E. Ho
Institute for Environmental Science
Murdoch University
Murdoch, WA 6150, Australia

Y.B. Ho
Department of Botany
The University of Hong Kong
Hong Kong

L.R. Hossner
Soil and Crop Sciences Department
College of Agriculture and Life Sciences
Texas A&M University
College Station, Texas 77843-2474

M.A. House
Urban Pollution Research Centre
Middlesex University
London N11 2NQ, United Kingdom

T. Isuta
Department of Environmental Science
  and Resources
Tokyo University of Agriculture and
  Technology, Fuchu
Tokyo 183, Japan

C.Y. Jim
Department of Geography and Geology
University of Hong Kong
Hong Kong

A.G. Khan
Department of Biological Sciences
Faculty of Business and Technology
University of Western Sydney, Macarthur
Campbelltown, NSW 2560, Australia

C. Kuek
Department of Biological Sciences
Faculty of Business and Technology
University of Western Sydney, Macarthur
Campbelltown, NSW 2560, Australia

P.-S. Kwan
Ocean Man Consultant Ltd.
Hong Kong

C.M.L. Lai
Department of Biology and Chemistry
City University of Hong Kong
Kowloon, Hong Kong

S.K.S. Lam
Department of Biological Sciences
University of Western Sydney
Kingswood, NSW 2747, Australia

S.P. Lau
Agriculture & Fisheries Department
Hong Kong Government
Canton Road
Hong Kong

C.-D. Lee
Department of Civil Engineering
Chung-Yuan University
Chung-Li, Taiwan 32023

S.C. Lee
Civil and Structural Engineering
    Department
The Hong Kong Polytechnic
    University
Hung Hum, Hong Kong

S.-C. Lee
Environmental Engineering Unit
Department of Civil and Structural
    Engineering
Hong Kong Polytechnic
    University
Hong Kong

N.W. Lepp
School of Biological and
    Earth Sciences
Liverpool John Moores University
Liverpool, United Kingdom

J. Liang
Department of Agronomy
Agricultural College
Yangzhou University
Jiangsu, P.R. China

J.P. Lindeque
Department of Plant Production & Soil
    Science
University of Pretoria
Pretoria 0002, South Africa

M. Liu
Environmental Protection Agency of
    Pingyuan County
Guangdong Province
Guangdong 514600, P.R. China

S.L. Lo
Graduate Institute of Environmental
    Engineering
National Taiwan University
71 Chou Shan Road
Taipei 106, Taiwan

T.A. Madsen
Centre for Mined Land Rehabilitation
The University of Queensland
Brisbane, QLD 4072, Australia

J.A. Manakov
Kuzbas Botanical Gardens
Kemerovo Scientific Centre
650060 Kemerovo, Russia

S.G. McRae
Wye College
University of London, Ashford
Kent, TN25 5AH, United Kingdom

D.W. Merrilees
Scottish Agricultural College
Auchincruive
Ayr KA6 5HW, Scotland

S.D. Miller
Environmental Geochemistry
    International Pty Ltd.
Balmain NSW 2041, Australia

D.R. Morrey
Golder Associates, Inc.
Lakewood, CO

D.R. Mulligan
Centre for Mined Land Rehabilitation
The University of Queensland
Brisbane, QLD 4072, Australia

A.S. Mungur
Urban Pollution Research Centre
Middlesex University
London N11 2NQ, United Kingdom

A.B. Pearce
Centre for Mined Land Rehabilitation
The University of Queensland
Brisbane, QLD 4072, Australia

M.R. Peart
Department of Geography and Geology
University of Hong Kong
Hong Kong

C. Penny
School of Biological and
    Earth Sciences
Liverpool John Moores University
Liverpool, United Kingdom

C.S. Poon
Civil and Structural Engineering
    Department
The Hong Kong Polytechnic
    University
Hung Hum, Hong Kong

N.F. Protopopov
Department of Ecology and Safety
Tomsk Polytechnic University
634045 Tomsk-45, Russia

N.F.G. Rethman
Department of Plant Production &
    Soil Science
University of Pretoria
Pretoria 0002, South Africa

D.M. Revitt
Urban Pollution Research Centre
Middlesex University
London N11 2NQ, United Kingdom

P.A. Roe
Centre for Mined Land Rehabilitation
The University of Queensland
Brisbane, QLD 4072, Australia

M.K.S.A. Samaraweera
Institute for Environmental Science
Murdoch University
Murdoch, WA 6150, Australia

G.E. Schuman
Rangeland Resources Research
High Plains Grasslands Research Station
Agricultural Research Service
United States Department of Agriculture
Cheyenne, WY

Y. Shan
The United Graduate School
Tokyo University of Agriculture and
    Technology, Fuchu
Tokyo 183, Japan

R.B.E. Shutes
Urban Pollution Research Centre
Middlesex University
London N11 2NQ, United Kingdom

B.R. Striganova
Laboratory of Soil Zoology
Institute of Ecology and Evolution
117071 Moscow, Russia

G.E.D. Tiley
Scottish Agricultural College
Auchincruive
Ayr KA6 5HW, Scotland

T. Totsuka
Department of Environmental Science
    and Resources
Tokyo University of Agriculture and
    Technology, Fuchu
Tokyo 183, Japan

R.H.M. van de Graaff
van de Graaff & Associates Pty Ltd.
Park Orchards
Victoria 3114, Australia

L. Wall
Woodward Clyde Pty Ltd.
China-Australia Research Institute for
  Mine Waste Management
Xizhimenwai, Beijing 100044, China

D.H. Wen
China-Australia Research Institute for
  Mine Waste Management
Beijing General Research Institute of
  Mining and Metallurgy
Xizhimenwai
Beijing 100044, P.R. China

M.H. Wong
Department of Biology
Hong Kong Baptist University
Kowloon Tong, Hong Kong

Y.J. Wu
China-Australia Research Institute for
  Mine Waste Management
Beijing General Research Institute of
  Mining and Metallurgy
Xizhimenwai
Beijing 100044, P.R. China

L. Xu
South China Institute of Environmental
  Sciences
NEPA
Guangzhou 510655, P.R. China

C. Yang
Department of Natural Conservation
National Environmental Protection
  Agency
Beijing 100035, P.R. China

M.-L. Yau
Ecosystems Ltd.
Aberdeen, Hong Kong

J. Zhang
Department of Biology
Hong Kong Baptist University
Kowloon Tong, Hong Kong

L.B. Zhou
China-Australia Research Institute for
  Mine Waste Management
Beijing General Research Institute of
  Mining and Metallurgy
Xizhimenwai
Beijing 100044, P.R. China

X. Zhuang
Ecosystems Ltd.
Aberdeen, Hong Kong

Acknowledgments

The organizers of the International Conference on the Remediation and Management of Degraded Lands are grateful for the support of our sponsors.

- Croucher Foundation, Hong Kong
- Environment and Conservation Fund, Hong Kong
- Hong Kong Baptist University
- National Natural Science Foundation of China
- Asia-Oceania Network of Biological Science
- Annals of Botany, United Kingdom
- British Council, Hong Kong
- Guyline Technologies Ltd., Hong Kong
- Perkin Elmer Hong Kong Ltd.
- Tegent Technology Ltd., Hong Kong
- Schmidt & Co., (Hong Kong) Ltd.

Contents

II. Management of Derelict Lands

III. Soil Contamination and Remediation

MINE MANAGEMENT AND REHABILITATION

Chapter 1

A Multidisciplinary Approach to Producing Solutions for Sustainable Mine Rehabilitation— The Role of the Australian Centre for Minesite Rehabilitation Research

L.C. Bell

INTRODUCTION

Although data on the proportion of land disturbed by mining globally is not available, extrapolation from data available for major mining countries such as the United States and Australia would suggest that this percentage is small (<1%) in relation to that devoted to other land uses. For example, in Australia it is estimated that land disturbed by mining is less than 0.02% (Farrell and Kratzing, 1996), whereas 54% is devoted to grazing and 6% to cropping (Alexander, 1996).

Notwithstanding the relatively small proportion of land disturbed by mining globally, there is the potential for severe local environmental impact, particularly with surface mining. In the latter case, areas of existing vegetation and fauna habitat are destroyed, and the removal of the overburden covering the mineral resource may result in a marked change in topography, hydrology, and stability of the landscape. In the absence of effective environmental management, these on-site effects may also cause further impacts off-site, resulting from the water and wind erosion of unstabilized overburden spoil or of tailings resulting from mineral processing. These effects may include sedimentation of surrounding streams, and an additional decrease in water quality through the increase in salinity, acidity, and toxic element load in these streams, these latter effects often being associated with acid mine drainage (AMD) resulting from oxidation of sulfidic minerals (Mulligan, 1996).

Given the potential for mining to create severe environmental impact, it is not surprising that communities now expect companies to put in place effective environmental management programs which ensure that mined land is rehabilitated. The immediate objective of rehabilitation is to ensure the postmining landscape is stable to the erosive forces of wind and water. A second objective is to return the land to a condition suitable for other forms of land use (Bell, 1996).

Important elements of a rehabilitation plan for a mine site are: (1) construction of a landform which satisfies land use, drainage, and erosion requirements; (2) development of a suitable plant growth medium through selective handling of soil and overburden; and (3) establishment of a vegetative cover. Thus the technologies required to achieve effective rehabilitation are based upon input from disciplines spanning engineering and the physical and biological sciences. Additional challenges for mine-site environmental managers arise with the need to apply these technologies to components such as waste rock dumps, open pits and, in those cases where mineral processing is undertaken, tailings disposal facilities.

Although the mining industry worldwide has made significant advances in the development of technologies for rehabilitation in the past decade, there are a number of major issues for which best-practice solutions are still being sought. For example, in 1994 the Australian Mining Industry Council (AMIC) (now the Minerals Council of Australia), in a review of priorities and needs for environmental research in the Australian mining industry, identified the main strategic issues as ecosystem establishment and resilience, final voids, biodiversity, acid drainage, greenhouse effects, tailings, and waste rock and overburden dumps (AMIC, 1994).

The value of undertaking research and development to underpin environmental management is exemplified by the bauxite and mineral sand mining industries in Australia which have achieved international recognition for their standard of rehabilitation (Ward et al., 1996; Brooks, 1989). However, there has been recognition that, generally, rehabilitation research in Australia in the past has tended to be: (1) site-specific with less emphasis on strategic issues; (2) lacking interaction between researchers in the engineering and biological science disciplines; and (3) undertaken by research groups acting independently. In 1993 the Australian Centre for Minesite Rehabilitation Research was established as an initiative of the mining industry to overcome these deficiencies. This chapter describes the objectives, structure, research, and technology transfer programs of this Centre which may serve as a model for similar centres in other countries.

CENTRE VISION AND GOALS

The vision of the Centre is the development, with industry, of sustainable systems for land affected by exploration, mining, and mineral processing activities, acceptable to government and the community. The challenge of the Centre is to pursue the vision by undertaking research which:

- promotes the development of rehabilitation technologies for mines sited across a diverse range of ecosystems;
- integrates the spectrum of disciplinary skills represented by engineering, and the physical and biological sciences; and
- harnesses the skills of world-class research scientists in a number of organizations in a cooperative manner to produce synergistic benefits for the industry and the community.

The vision is being pursued, and the challenge accepted, by working toward the following goals:

- to conduct strategic research into mining rehabilitation to provide ecologically sustainable environmental solutions;
- to attain international recognition as a centre of excellence undertaking commissioned research on mining rehabilitation in an independent and thorough manner;
- to provide scientific and technological foundations to facilitate industry and government in setting acceptable rehabilitation standards;
- to act as a networking and communications focus of rehabilitation practice, and to enhance education and training in mining rehabilitation.

CENTRE PARTNERS AND EXPERTISE

The ACMRR is an unincorporated joint venture between five of the major groups in Australia carrying out minesite rehabilitation research:

- Centre for Land Rehabilitation, University of Western Australia
- Centre for Mined Land Rehabilitation, University of Queensland
- Environmental Science Program, Australian Nuclear Science and Technology Organisation (ANSTO)
- Minesite Rehabilitation Research Program, CSIRO
- Mulga Research Centre and Mine Rehabilitation Group, Curtin University of Technology

and the Australian minerals industry, through the Australian Mineral Industries Research Association Limited (AMIRA).

The Centre brings together an impressive grouping of multidisciplinary research expertise to focus on mining rehabilitation and is unequaled in the major mining countries of the world. Its aim is to use this expertise effectively in collaborative research projects, so that a significant multidisciplinary synergy is developed.

The Centre for Land Rehabilitation at the University of Western Australia, in Perth, combines the expertise of staff from the faculty of Agriculture, Department of Civil Engineering, and the Department of Environmental Engineering, and has particular expertise in consolidation and stability of constructed landforms (tailings, waste dumps, and pits), contaminant transport processes, waste characterization, and ecosystem reconstruction processes.

The Centre for Mined Land Rehabilitation (CMLR) at the University of Queensland, in Brisbane, is comprised of the Departments of Agriculture, Botany, Chemical Engineering, Civil Engineering, Earth Sciences, Economics, Geographical Sciences and Planning, Management Studies, Mining, Minerals and Materials Engineering, Plant Production and Zoology, and has particular expertise in erosion and water quality control on constructed landforms, leachate and groundwater dynamics, tailings disposal techniques, and ecosystem reconstruction.

The Environmental Science Program of ANSTO, south of Sydney, is multidisciplinary incorporating Environmental Physics, Environmental Chemistry, Ecological Impacts, Chemical and Waste Engineering, and Radioanalytical Applications sections. The Centre has specific expertise in the processes of leaching of heavy metals from mine wastes and tailings, acid mine drainage, treatment of mine and waste waters, bioremediation and physicochemical rehabilitation of mining and contaminated industrial sites, and the design of mine waste and tailings structures to minimize environmental impacts.

The Minesite Rehabilitation Research Program within CSIRO is a collaboration between the Divisions of Soils, Exploration and Mining, Water Resources, Coal and Energy Technology, Wildlife and Ecology, Tropical Crops and Pastures, and Entomology with staff located across Australia. Major research themes in the CSIRO program are the characterization of mine wastes, long-term behavior and processes of evolution of landforms constructed of mine wastes (waste rock/spoil dumps, tailings dams, final voids), and ecosystem establishment and long-term sustainability.

The Mulga Research Centre and Mine Rehabilitation Group at the Curtin University of Technology, in Perth, are organized around the School of Environmental Biology with links which include the Schools of Applied Geology, Mines, and Agriculture. The Centre has strong expertise in the biology of flora and fauna and a particular interest in revegetation strategies for degraded landscapes, seed biology, faunal recolonization, and monitoring of ecosystem reconstruction.

The Australian Mineral Industries Research Association Ltd. (AMIRA), in Melbourne, was established in 1959 by the mineral industries in Australia to initiate and coordinate jointly sponsored research and development contracts on behalf of its member companies. The organization is an important partner in the ACMRR because of its long experience in identification of problems in the minerals industry, the development of research proposals to solve those problems, and the management of research contracts.

The ACMRR can also involve other research groups as necessary, when additional skills or resources are required. This involvement may be as a subcontractor for specific projects, depending upon the Centre's ability to meet industry requirements.

STRATEGIC RESEARCH PROGRAM

The research program areas of the Centre have been established on the basis of considerable market research involving industry and government input. The Ecologically Sustainable Development Working Group (Mining) (1991), which reported to the Prime Minister, identified rehabilitation of mined land as a priority area for research and development to ensure "ecologically sustainable development for the mining and mineral processing industry" and listed specific aspects requiring attention.

Subsequently AMIC further defined research needs in the five priority areas of waste rock dumps and overburden, final voids, acid mine drainage, tailings management, and ecosystem establishment and resilience (AMIC 1994). The research programs of the ACMRR closely follow the priority areas defined in the AMIC (1994) report. Views of government departments were also sought in establishing the research focus.

The strategic research efforts of the Centre are directed in four main areas:

- Landform Stability—Key processes controlling the long-term behavior and stability of constructed landforms
- Water Systems—Key processes controlling downstream surface and groundwater quality
- Ecosystem Reconstruction—Key processes controlling the long-term sustainability of constructed ecosystems
- Waste Treatment and Disposal—Key processes controlling long-term treatment and disposal of waste products.

Many specific rehabilitation issues facing the mining industry involve one or more of the above areas, and thus it has been found useful to classify the research programs and subprograms of the Centre as shown in Figure 1.1. The multidisciplinary nature of the research programs is shown in Table 1.1. A brief description of each program area follows.

Waste Rock Dump and Final Void Stability

Mining produces waste rock dumps and voids such as abandoned pits and strips, which must be rehabilitated to uses that are sustainable and compatible with normal land management practices. Waste rock is expensive to move, and final landforms should be constructed as an integral part of mine design to optimize operations throughout the life of the mine. The critical values for topographic and substrate parameters necessary to achieve site stability are not well defined, and the aim of this program is to provide a sound techni-

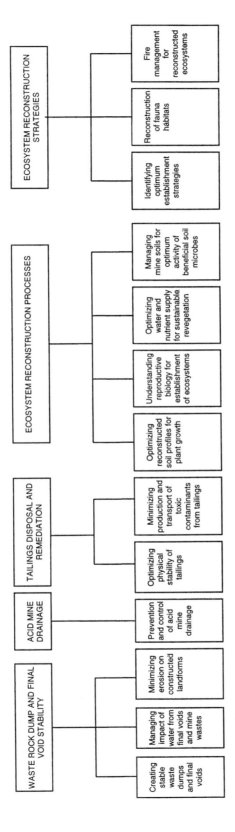

FIGURE 1.1. Research program structure for the ACMRR.

Table 1.1. Multidisciplinary Nature of the Research Programs of ACMRR.

Disciplinary Area / Research Programs	Mining Engineering and Mineral Processing	Landscape Evolution	Contaminant Hydrology	Disturbance Ecology
Waste Rock Dump and Final Void Stability	●	●		•
Acid Mine Drainage	●		●	•
Tailings Disposal and Remediation	●	•	●	•
Ecosystem Reconstruction		•	•	●

● Major association;

• Minor association

cal and cost-effective basis to select and design the optimum final rehabilitated landform for mine sites situated in diverse climates and geomorphologies.

One of the first projects of the Centre reviewed the management and impact of final mining voids. Through reviews of worldwide practice and visits to open-cut mines throughout Australia, a decision process was developed to evaluate final completion options for any mining void. A key feature of the study was the classification of mining void types in Australia into eight major groups depending on the interaction of climate, geology, and social setting.

Prediction of final void water quality is an important consideration in mine planning, and, at industry's request, the Centre has established a team representing the disciplines of geochemistry, hydrology, limnology, and ecology to undertake case studies with a view to improving water quality prediction models.

The optimum design of waste rock dumps for acceptable erosion control will depend on many factors including the nature of the materials and climate. As a basis for identifying best practice in landform design, the Centre is conducting a national survey of waste rock dump design and stability.

Acid Mine Drainage

The prevention and control of acid mine drainage (AMD) is a major environmental concern of many metalliferous mines. Exposure of sulfides to air and water can produce sulfuric acid and acid-leached heavy metals which can be transported through waste rock dumps, tailings dams, or abandoned mines and enter surface and groundwaters with potentially damaging environmental effects.

The aim of this program is to develop cost-effective techniques to prevent and remediate AMD through investigations of existing and potential sources of AMD, laboratory experimentation and modeling of the processes. The ultimate aim is to develop a set of tools that can be used by both mine operators and regulatory authorities to design the disposal of sulfide wastes so that they have a planned long-term, low-level environmental impact. Currently the Centre is undertaking a major study of the nature and extent of acid drainage associated with mining and other forms of land use in Australia and the associated financial liability.

Tailings Disposal and Remediation

Most mineral processing plants produce, in slurry form, fine-grained waste tailings which must be disposed of close to the plant. The tailings, which contain residual quantities of chemicals and potential contaminants from the plant, are generally confined to a specific area for safety and environmental reasons, and government guidelines and regulations require that this material is left in an environmentally acceptable state upon mine closure. The long-term effects of contaminant transport via wind, surface drainage, and seepage are pressing concerns for the mining industry. Many tailings problems arise from the low strength of tailings deposits, their ongoing consolidation, erosion by wind and water, spillages, and tailings water seepage.

The aim of this program is to develop economic design, management, and closure strategies for environmentally acceptable tailings disposal facilities by researching the critical processes governing the geotechnical, chemical, and hydrological behavior of tailings deposits.

A current scoping project involves a review of the research needs for the management and rehabilitation of tailings disposal facilities, and a project being developed involves an assessment of the benefits of codisposal of tailings with waste rock.

Ecosystem Reconstruction Processes

One objective of best mine rehabilitation practice is to create a stable, near-natural ecosystem that requires no management inputs additional to those in operation prior to mining. This program investigates the physical, chemical, and biological processes and their interactions that create robust sustainable ecosystems. Although the main focus is on native ecosystems, research to develop sustainable productive systems such as agriculture and forestry are also undertaken.

An important consideration in ecosystem reconstruction is to be able to assess, in the short to medium term, whether the new ecosystems will be sustainable in the long term. A current project involves a review of potential indicators of ecosystem rehabilitation success in order to delineate those indicators which the mining industry can use to effectively manage its rehabilitation programs.

A project being developed involves assessment of the tolerance of native plant species to heavy metals and of tests for bioavailability.

Ecosystem Reconstruction Strategies

A first step in the reconstruction of an ecosystem is the reestablishment of the vegetation; the functioning of other life forms, both micro and macro, depend upon this component. This program addresses issues relating to establishment procedures for all structural levels appropriate to a diverse native species community. A suitable floristic, structural, and physiographic diversity encourages immigration of fauna from surrounding areas, or indeed provides the opportunity to reintroduce threatened or vulnerable species to the newly-constructed habitats. Appropriate management techniques are also addressed in this program to ensure that fire does not destroy the newly-created systems, while at the same time recognizing the crucial role that fire plays in maintaining the structure and integrity of many Australian ecosystems.

Among the projects the Centre is researching in this program is an investigation of the limitations to the reestablishment of spinifex grass on mined land and of the procedures required to achieve successful recolonization. This plant species covers approximately 25% of the semiarid and arid land surface of Australia, where many mining companies are experiencing difficulty in reestablishing the plant.

The technologies developed from the research in this program have the potential to be applied to the repair of land degraded by other forms of land use such as grazing.

TECHNOLOGY TRANSFER PROGRAM

An important activity of the Centre is the Technology Transfer Program which complements the Strategic Research Program. Thus one of the roles of the Centre is to organize training and extension courses dealing with best practice for environmental officers in the mining industry, and to enhance technology transfer to project sponsors through workshops and training courses. Recent examples of short courses include Mine Rehabilitation in Tropical Environments and Cyanide Management in Mining. Recent ACMRR workshops such as Acid Mine Drainage, Post-mining Landform Design and Stability and Native Seed Biology for Revegetation not only provide a form for dissemination of research results from existing projects but also serve to identify future research and training needs.

The Centre also plays a part in the eduction of future environmental personnel by provision of postgraduate scholarships for research through its joint venture research partners.

MANAGEMENT OF THE CENTRE

The ACMRR is run by a Board of Management which is comprised of nominees of each of the joint venture partners plus five additional industry nominees selected by AMIRA. The Board's role is to establish and monitor the policies and overall strategies of the Centre. It meets four times a year in various locations around Australia and uses these opportunities to interact with personnel from government, industry, and research institutions.

The Director is responsible for carrying out the policies and strategies of the Board and for management of the Centre's affairs, including the research and technology transfer programs.

The Management Committee is responsible for identifying appropriate research projects for the Centre, and is comprised of the Directors of each of the joint venture research groups, an AMIRA nominee, and the Director.

The Strategic Research Program Advisory Committee is an advisory group consisting of environmental professionals representing mining companies covering a wide range of resource categories and climatic zones in addition to government representatives. The

Committee assists the Director by identifying research and training needs of industry and assisting the research teams in project development.

The Centre was initially funded by start-up contributions from industry and joint venture partners but now derives its operational income from management of its core activities and from state and federal governments.

CONCLUSIONS

In Australia, the development of strategic rehabilitation research through coordinated multidisciplinary research programs represents the opening of a new research market. This presents a challenge to both industry and the researchers to develop new approaches to research, planning, and marketing involving partnership building. The ACMRR has been addressing these challenges and is developing new ways to relate to its research customers in industry. However, the stakeholders extend beyond industry. Government agencies need the same research outcomes as industry in order to set effective regulations, standards, and completion criteria. They also benefit from the education and training services provided by ACMRR.

The demand for solutions to environmental management problems that are economically viable and acceptable to the community extends beyond Australia. Australian and other companies involved in overseas mining ventures, particularly in the Asia and Pacific regions, must have access to the best available rehabilitation expertise, and there are exciting opportunities for the ACMRR to market new rehabilitation research and training services in this area.

REFERENCES

Alexander, N., Ed. *Australia: State of the Environment 1996. Executive Summary.* CSIRO Publishing, Collingwood, Victoria, 1996.

AMIC. *Environmental Research in the Australian Mining Industry—Priorities and Needs.* Australian Mining Industry Council, Canberra, 1994.

Bell, L.C. Rehabilitation of Disturbed Land, in *Environmental Management in the Australian Minerals and Energy Industries—Principles and Practices*, Mulligan, D.R., Ed., University of New South Wales Press, Sydney, 1996, pp. 227–261.

Brooks, D.R. Reclamation in Australia's Mineral Sands Industry, in *Proceedings of the Conference: Reclamation, A Global Perspective*, Calgary, Canada, Walker, D.G., C.B. Powter, and M.W. Pole, Eds., Report No. RRTAC 89-2, Alberta Land Conservation and Reclamation Council, Edmonton, Vol. I, 1989, pp. 11–26.

Ecologically Sustainable Development Working Group (Mining). *Final Report—Mining.* Australian Government Publishing Service, Canberra, 1991.

Farrell, T.P. and D.C. Kratzing. Environmental Effects, in *Environmental Management in the Australian Minerals and Energy Industries—Principles and Practices*, Mulligan, D.R., Ed., University of New South Wales Press, Sydney, 1996, pp. 14–45.

Mulligan, D.R., Ed., *Environmental Management in the Australian Minerals and Energy Industries—Principles and Practices.* University of New South Wales Press, Sydney, 1996.

Ward, S.C., G.C. Slessar, D.J. Glenister, and P.S. Coffey. Environmental Resource Management Practices of Alcoa in Southwest Western Australia, in *Environmental Management in the Australian Minerals and Energy Industries—Principles and Practices*, Mulligan, D.R., Ed., University of New South Wales Press, Sydney, 1996, pp. 383–402.

Chapter 2

Development of Success Criteria for Reestablishment of Native Flora Habitats on Coal Mine Rehabilitation Areas in Australia

S.M. Bellairs

INTRODUCTION

Rehabilitation success after mining is difficult to assess. It has often been judged by a superficial resemblance to a local vegetation type, whether that be pasture, forest, native woodland, or wetland. Where agricultural production is desired it can be assessed by the relative productivity of the land compared to similar unmined land with similar inputs. In Australia there is an increasing desire to create sustainable native vegetation communities after mining. These are seen to provide an option which requires minimal ongoing maintenance and allows flexibility for subsequent land uses. In setting completion criteria, regulatory authorities tend to set vegetation composition, richness, density, and cover values; however, they also expect that the ecosystem will be functional and often stipulate that it will be sustainable and require minimal maintenance. There is a desire to assess rehabilitation success more objectively, and increasingly the focus is on functional aspects of the rehabilitated ecosystems.

The work presented here is largely the result of research sponsored by the Blair Athol Coal Project and Tarong Coal Pty Ltd, two open-cut coal mines in Queensland, Australia which are operated by Pacific Coal Pty Ltd, a CRA-RTZ subsidiary. For several years both operations have involved the Centre for Mined Land Rehabilitation at the University of Queensland in developing criteria to assess the success of rehabilitation of mined areas. In this chapter the process that we have undertaken to develop and refine these success criteria is discussed.

BACKGROUND

The Blair Athol mine and the Tarong Meandu Creek mine are both large open-cut coal mines which commenced production in the 1980s. Both are reestablishing native ecosystems on the areas being rehabilitated after mining.

The Blair Athol mine is located in central Queensland about 250 km from the coast and 900 km NNW of Brisbane (Figure 2.1). The Tarong Meandu mine site is located 180 km northwest of Brisbane. Rainfall at both sites is summer dominated (Figure 2.2). Blair Athol has an average rainfall of 676 mm. Temperatures exceed 40°C in summer and may occasionally drop below 0°C in winter. Rainfall is slightly higher at Tarong with an annual rainfall of 805 mm, and temperatures are lower.

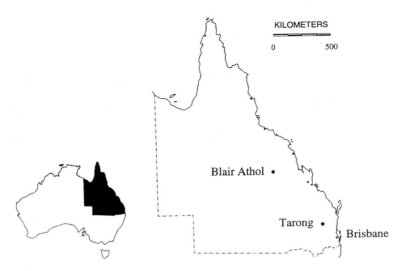

FIGURE 2.1. Location of the Blair Athol and Tarong coal mines in Queensland, Australia.

Blair Athol has a lease of 3000 ha and extracts approximately 10 M tonnes of steaming coal per annum. The deposit of 270 M tonnes has been worked since 1890, originally by underground methods. First attempts at open-cut mining began in 1923 and the current operation commenced production in 1984. The first rehabilitation of old workings was also carried out in 1984, and the anticipated mine life is until the year 2010. The native vegetation is *Eucalyptus* woodland with an *Acacia* midstorey, scattered shrubs, and a grass and herbaceous understorey. On stony soils the woodlands are dominated by *Acacia rhodoxylon* with scattered *Eucalyptus crebra*, while on sandy soils several *Eucalyptus* species dominate.

The Meandu Creek mine operated by Tarong Coal commenced coal production in 1983 with reserves of 180 M tonnes of bituminous steaming coal and an anticipated mine life of 30 years. The coal is used to fuel the nearby Tarong Power Station which supplies about 40% of the power needs of the state of Queensland. Research into rehabilitation techniques commenced in 1981, with the first mined area being rehabilitated in 1985. At Tarong the native vegetation communities are more varied. *Eucalyptus* forest with a range of *Acacia* spp., shrubs, herbs, and grasses would have been the major vegetation type prior to European settlement but some dry vineforest communities also occur on the ridgetops. These communities have a diverse range of rainforest tree and vine species. Since early this century, both sites have been disturbed by grazing and logging. At least one-third of the vegetation on the Tarong mining lease had been cleared for cattle grazing prior to mining.

At the Blair Athol and Tarong open-cut coal mines in Queensland the rehabilitation objectives are to:

1. create a stable landform,
2. protect downstream water quality, and
3. reestablish sustainable flora communities.

These flora communities should be compatible with the local vegetation, capable of supporting the local wildlife, and have the potential for future timber production and other

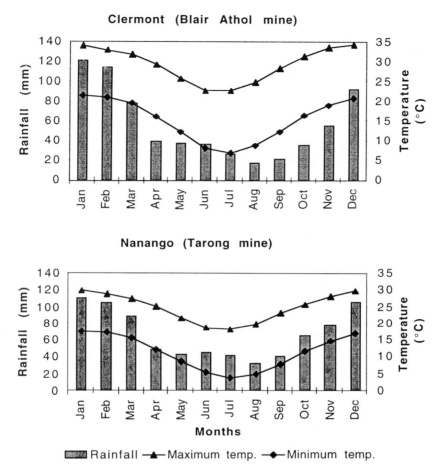

FIGURE 2.2. Ombrothermic diagrams for the closest towns to the Blair Athol and Tarong coal mines.

land uses. A proportion of the Tarong rehabilitation will include introduced pasture species, but generally local species are used. Blair Athol is particularly interested in establishing native vegetation communities on the rehabilitation areas that are suitable for supporting the local koala population.

For both operations, rehabilitation methods are similar. Vegetation ahead of the mining front is cleared and the topsoil is removed, generally to a depth of about 30 cm. The overburden is drilled and blasted and then removed using a dragline. The coal is blasted and removed using truck and shovel methods. The spoil piles deposited by the dragline are recontoured to a maximum slope of generally less than 15% and topsoil is spread over the surface of the spoil using scrapers. The respreading depth is about 10 cm at Blair Athol and about 30 cm at Tarong. After the topsoil is spread, the rehabilitation areas are ripped, fertilized, and seeded. Ripping is carried out along the contour to reduce compaction and assist with erosion control. The seed mix includes native tree and shrub species and at Tarong a cover crop of introduced grasses are included to assist with soil stabilization. *Acacia* species are an important component of the seed mix, as they grow rapidly and are legumes with the ability to fix nitrogen.

The aim of research carried out by the Centre for Mined Land Rehabilitation at the mines is to determine objective and justifiable criteria that the rehabilitation should achieve to be considered successful, and show that the rehabilitation does meet those criteria, or develop techniques to improve the rehabilitation to achieve those criteria.

DEVELOPMENT OF SUCCESS CRITERIA

To determine success criteria and indicators of those criteria involves several stages. The first is to identify what is required to be achieved for the rehabilitation to be successful. The next is to determine what structural, compositional, and functional processes are required to achieve those success criteria. Thirdly, how to assess whether those processes are occurring or going to be achieved needs to be determined. From an assessment of these processes the critical steps in these processes can be determined and indicators that these processes are occurring can be chosen.

Successful Rehabilitation

Success of rehabilitation is determined by sociopolitical considerations and from these, acceptability judgments or completion criteria are derived by regulators and the company. They include site-specific criteria that may be related to the end land use and, increasingly, ecosystem functioning requirements such as sustainability and requiring minimal maintenance. Tacey et al. (1993) provide the following definition of rehabilitation success. Success is indicated by stable productive ecosystems with self-sustaining biophysical processes similar to their surroundings. No unusual management inputs should be required and the site should be manageable either as part of its surroundings or for a designated land use. Rehabilitation is defined by Bell (1986) as the process of stabilizing land against erosion and returning the land to a form and productivity that conforms with a predetermined postmining land use.

Rehabilitation performance objectives are generated by regulatory authorities on a site-specific basis to denote milestones in biophysical processes that predict with reasonable surety that the site will achieve its desired end land use. Such criteria include fauna and flora composition, vegetation density, cover, water quality, erosion rates, visual quality, and land capability (Tacey et al., 1993).

Often such criteria are developed with a stable community as the target and without regard to ecosystem functioning. However, Australian ecosystems tend not to be stable but rather tend to be regulated by disturbance regimes and these are often vital for recruitment to occur (Noble and Slatyer, 1980). Where native vegetation communities and landscapes are the aim, assessing the ability of the ecosystems to respond adequately to disturbance is more likely to result in success than attempting to create an ecosystem that is not affected by disturbance. However, irrespective of the lack of a stable target and acknowledging that the rehabilitation community is dynamic, certain species and community types would generally be expected to persist in the landscape if the rehabilitation were to meet its designated end-use criteria and be successful.

The Commonwealth EPA recognizes three generic success criteria.

- The site can be managed for its designated land use without any greater management inputs than other land in the area being used for a similar purpose.

- Restored native ecosystems may be different in structure to the surrounding native ecosystems, but there should be confidence that they will change with time along with or toward the makeup of the surrounding area.
- The rehabilitated land should be capable of withstanding normal disturbances such as fire or flood.

In Queensland, environmental management of mining is regulated through the Mineral Resources Act, the Environmental Management Policy for Mining, and the Environmental Protection Act (DME, 1995; Welsh, 1992). When mining is proposed, the company provides an Initial Advice Statement. This is assessed against trigger criteria to determine if an Environmental Impact Statement (EIS) is required. The proposal is then assessed through an Environmental Management Overview Strategy (EMOS) with or without an EIS, depending on the scale of the operation and the likely environmental impact. The EMOS provides a life-of-mine strategy for the management of environmental effects. It should contain clear commitments, framed in a way that shall allow later assessment of the extent to which the commitment has been met. The EMOS has three basic objectives: (1) achievement of acceptable postdisturbance land use capability, (2) achievement of a stable postdisturbance landform (it should be self-sustaining or in a condition where the maintenance requirements are consistent with the agreed postmining land use) and (3) preservation of present and future downstream water quality. It should address environmental impacts and control of land resources, water resources, noise, air quality, conservation values, heritage and cultural values, and social impacts. A security deposit is lodged which is calculated on the cost of rehabilitating the site. As the mine achieves greater levels of environmental compliance, so the size of the security deposit held is reduced. At the end of mine life, a Final Rehabilitation Report is submitted. This addresses commitments regarding protection and rehabilitation of the environment and assesses the extent to which these commitments have been met. During the mining period the company has to periodically submit a Plan of Operations which provides a detailed account of how the company is going to carry out the management programs necessary to achieve the EMOS objectives in the next one to five years.

To achieve the appropriate vegetation community for the desired end land use at the Tarong and Blair Athol mine sites; the conservation of the local fauna or koala population and timber production, the vegetation community must have certain compositional and structural characteristics. To meet the sustainability and stability criteria in particular, it must also be a functional ecosystem. For a successfully functioning ecosystem there should be conservation of nutrients, moisture, and desired biota in the landscape through time. There should not be greater leakage of these resources than in the native communities (Ewel, 1987).

Functional Requirements to Achieve Success

For conservation of water through time, the following ecosystem processes need to be considered.

- infiltration and runoff
- water storage capacity
- transpiration and evaporation
- the maintenance of landscape and site features that affect water movement

- the distribution pattern of water in the landscape
- the maintenance of vegetation density, cover, and rooting structure as this will affect water flows, transpiration, and evaporation

For conservation of nutrients through time, the following processes have been identified as being important.

- soil nutrient status to provide sufficient nutrient capital to establish the desired community
- erosion and maintenance of stable landforms as this will affect the rate of loss of nutrients
- soil organic matter and cation exchange capacity to retain nutrients
- distribution and movement of nutrients in the landscape
- soil microbial activity (OM breakdown, mycorrhiza and nitrogen fixation) to capture and transform nutrients
- the vegetation biomass. In an ecosystem containing a mature perennial vegetation community, most of the nutrients are typically in the biomass
- ability to recover nutrients after disturbance including fire and drought. This will be affected by the types of species present as well as landform characteristics

Processes for the conservation of desired biota through time include the following.

- propagule availability, including production and/or dispersal, and the appropriate phenology for the disturbance regime of the site
- invasion and maintenance of desired fauna, particularly species involved in pollination or dispersal of desired flora, or those important for maintaining foliage breakdown and community structure
- maintenance of an appropriate vegetation density, structure, cover, and patchwork or mosaic of communities.

The vegetation structure and composition is important for allowing species recruitment and coexistence, for providing food and shelter for fauna, and for water and nutrient conservation.

Selection of Criteria Required to Indicate Success

There are many processes to be taken into account to determine that ecosystem establishment at the Blair Athol and Tarong mine sites is successful. These processes can be categorized to make the task manageable.

- Some processes can be visually observed or predetermined during the planning process. For example, landscape drainage patterns can be predetermined by contouring the replaced spoil.
- Some processes can be inferred by the occurrence of other dependent processes. For example, if tree recruitment is occurring then propagules and establishment microsites must be present.
- Some need to be shown to occur but they may occur across the range of sites that are successful according to other criteria. If research shows that when slopes are less than 20% and grass or litter cover is greater than 30% then erosion is controlled, then, provided that the slope and cover criteria are shown to occur, it would not be necessary to measure the erosion rates.
- Finally, a selection of critical processes or features will be left that show considerable variation across the rehabilitation areas, and these are most critical to monitor. Measure-

ments of these processes can be used as indicators to provide support of the occurrence of other processes that indicate a successfully functioning community.

EXAMPLES OF CRITERIA SELECTED TO INDICATE SUCCESS AT TARONG AND BLAIR ATHOL

Two examples of success criteria developed for rehabilitation of the Blair Athol and Tarong mine sites will be discussed in more detail. Two important functions for successful rehabilitation of mined lands at Tarong and Blair Athol are the occurrence of nutrient cycling on the rehabilitation areas at Tarong and the creation of appropriate habitat for koalas at Blair Athol.

Nutrient Cycling at the Tarong Mine Site

The following key points were identified as being critical for nutrient cycling at the Tarong mine site: ground stability, initial nutrient capital, presence of nitrogen fixing species, storage and capture of nutrients in the soil, and soil microbial activity.

Initially, reasonable ground stability must be maintained to prevent rapid loss of nutrients from the site. Erosion needs to be controlled by maintaining slope angle and length, along with the establishment of adequate vegetation and litter cover to reduce surface water flow. Transpiration by plants also reduces subsurface water flow and loss of nutrients to the groundwater (Begon et al., 1990). Slope angle, slope length, and ground cover parameters for erosion control are being investigated by a large multimine project, "Postmining Landscape Parameters for Erosion and Water Quality Control" (So et al., 1995). For plots with a 15% slope at Tarong, vegetation cover of greater than 47% results in minimal soil loss, even from a simulated 1:100 year storm event (Loch and Bourke, 1996).

Sufficient nutrients must be present at the site to allow initial establishment and growth of the grass cover crop and the trees, and eventually to provide the nutrient capital to construct the desired community (Bradshaw, 1983; Marrs, 1989). The fertilizer application rate is determined from soil analyses. Sufficient fertilizer is applied to permit reasonable grass growth but not too much to prevent the grasses out-competing the trees. Research has been carried out over several years, and is continuing, to determine the most appropriate grass cover crop species, sowing rates, and fertilizer rates to achieve this balance.

Another important factor for the long-term sustainability of the native woodland and forest communities at Tarong is the presence of nitrogen-fixing species. Nitrogen is particularly important as it is required in larger amounts than any other mineral nutrient for healthy plant growth (Allen et al., 1974). Nitrogen-fixing *Acacia* shrubs and trees are an important component of the unmined *Eucalyptus* forest ecosystem, and guidelines have been established to return these species to the rehabilitation areas at similar densities to that occurring in the surrounding *Eucalyptus* forest. Few studies have determined the contribution of nitrogen fixation by legumes to Australian native ecosystems. Hamilton et al. (1993) determined nitrogen fixation in a mixed *Eucalyptus* forest following fire to be very low for the first 12 months and was only 121 g ha^{-1} after 27 months, however the *Acacia* density of their study site was fairly low. For an *Acacia holosericea* plantation in Senegal, Cornet et al. (1985) calculated that about 4 kg N ha^{-1} yr^{-1} would be fixed from a density of 2000 trees ha^{-1}. This is a similar *Acacia* density to that occurring in the native forest communities near Tarong. For a 50 year old *Eucalyptus* forest, nitrogen fixation from asymbiotic and symbiotic sources has been estimated to be about 7 t ha^{-1} (Attiwill and Adams, 1993).

Typically, the inputs and outputs of nutrients are very low in comparison to the amounts that are held in biomass and recycled within an ecosystem (Begon et al., 1990). For a high productivity *Eucalyptus* forest with a standing biomass of 500 t ha^{-1} and 500 kg N ha^{-1} and a litter layer of 150 kg N ha^{-1}, the loss through stream flow may only be 5 kg ha^{-1} (Attiwill and Adams, 1993). However, in disturbed communities loss of nitrogen from the system can be extremely high. Likens and Borman (1975) found that deforestation of an experimental forest catchment area resulted in a sixtyfold increase in the loss of nitrate in stream flow.

Nitrate retention can be increased by the presence of organic matter. The carbon to nitrogen ratio of litter material tends to be much higher (70:1 for *Eucalyptus* forest, Attiwill and Adams, 1993; 44:1 for Brigalow woodland, Bligh, 1990) than that of decomposer organisms (8–10:1, Begon et al., 1990). As more litter is available and the material is broken down by the decomposers, the nitrate is rapidly consumed rather than being leached (Begon et al., 1990). A mean litterfall value for a 50 year old *Eucalyptus* forest would be 8 t ha^{-1} (Attiwill and Adams, 1993). Litter production at a 20 year old rehabilitation site in central Queensland is similar at about 8.5 t ha^{-1} year (Grigg, personal communication).

A limiting step in the nutrient cycling can be microbial breakdown of litter. However, litter also needs to be broken down to allow release and recycling of the nutrients contained in the litter. The transformation of organic matter to nitrate predominantly occurs through the action of soil microbial organisms (Russell, 1973). At some coal mine rehabilitation areas in Queensland, soil surface conditions are harsh with extremes of pH, high sodicity, and high salinity and this has led to concern about the survival and functioning of soil microbial organisms. While soil conditions are more favorable on the Tarong rehabilitation areas, this is still an aspect that is being investigated.

At Tarong, litter breakdown, soil organic matter buildup, and microbial biomass have been assessed. Soil organic matter is less than for the two native vegetation communities on the mining lease but it is similar to the levels in the Tarong State Forest (Figure 2.3).

The microbial biomass has been measured (Figure 2.4) for five rehabilitation sites as well as two unmined forest sites and a site that had been cleared for grazing prior to mining but had then been invaded by *Acacia* and other species. As long as vegetation cover is established it appears that the microbial biomass returns to levels of the abandoned pasture within five years. The vegetation diversity of the site or whether the litter was of grass or legume shrubs did not affect this relationship.

To assess the functioning of nutrient cycling at the Tarong mine site, these results suggest that it is not necessary to routinely measure microbial biomass, but an assessment of the soil conditions and litter ground coverage should be made to ensure the sites are within the range of the sites assessed above. Thus, simple criteria that indicate success include soil chemical characteristics, litter and grass cover, visual evidence of litter breakdown, and the density of nitrogen-fixing legume species. If sites have less litter cover or more extreme soil conditions, then those sites would also need to be assessed to determine if microbial breakdown of litter is likely to be affected.

Establishment of Koala Habitat at the Blair Athol Mine Site

The second example is the reestablishment of koala (*Phascolarctos cinereus*) habitat at the Blair Athol mine site. This medium-sized (4–10 kg) arboreal marsupial has a fairly restricted range of food trees, which are predominantly several species of *Eucalyptus*. It will also move on the ground for some distance between food and shelter trees (Strahan, 1983).

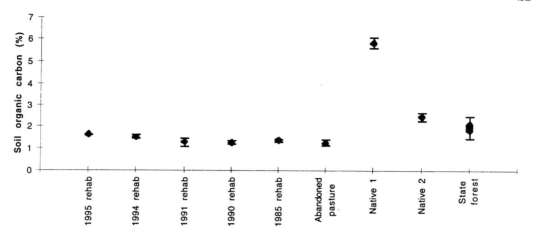

FIGURE 2.3. Soil organic carbon content of rehabilitation areas and native vegetation sites. Sites Native 1 and 2 are unmined native vegetation communities on the mine lease. The State Forest sites are from the adjacent Tarong State Forest.

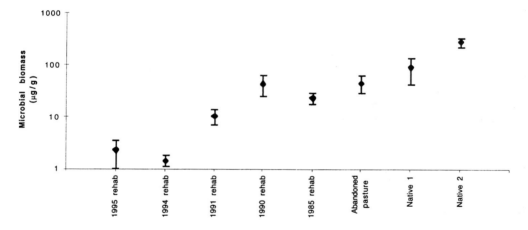

FIGURE 2.4. Microbial biomass of surface soil from rehabilitation areas, abandoned pasture, and native vegetation sites on the Tarong mining lease.

Considerable research has been carried out on the local koala population at the Blair Athol mine site including radio tracking of all individuals on the mining lease, analysis of food and water intake, and analysis of the social and genetic structure of the population (Ellis, 1997). Information on the use of food and shelter tree species is then used to determine tree species to be reestablished as tubestock or broadcast seed onto the mine site. The tree species favored for shelter are often different from the tree species the koalas eat. *Eucalyptus populnea* and *E. crebra* are the favored food species but a range of species are used for shelter, often in preference to these species (Ellis, 1997). Non-*Eucalyptus* species may be used for shelter in hot weather, as they provide better shade cover. We also need to take into account the vegetation structure requirements of the koalas. Their habitat needs to allow them freedom to move between trees and provide them with some cover from predators. Thus, success criteria also need to include criteria on the composition and density of shrubs and grasses as well as for the food and shelter tree species.

For the maintenance of the desired flora species, conditions on the rehabilitation areas must also be created that allow the continued recruitment of these species, and particularly the koala food and shelter species that are introduced. Research has indicated that tree establishment may be more successful if some grass cover is present (Forster, 1995) and certainly moderate grass cover does not result in greater mortality of perennial dicotyle-don species (Bellairs and Clowes, 1995). However, where there is dense cover of intro-duced perennial grasses, tree establishment does not occur. Thus, success criteria also include maximum values for exotic grass cover and soil surface conditions to promote continued establishment of the trees. As long as the tree density is adequate then the intro-duced grass species can maintain only a sparse cover while the less competitive native grass species can establish successfully (Bellairs, unpublished data).

CONCLUSIONS

These two examples emphasize the complexity of recreating native ecosystems on mined lands. Rehabilitation needs to be carried out with regard to the functional processes that need to occur and needs to aid natural developmental processes. There is a desire from the industry for as few as possible, and very simple indicators. While indicators need to be simple in order to be used operationally, it may be optimistic (and risky) to expect to tie too much into one indicator of successful rehabilitation. Ecological indicators serve a similar function in evaluating the health of an ecosystem, because economic indicators serve in evaluating the economic health of a nation (Salwerowicz, 1994). Numerous economic in-dicators are required to evaluate aspects of the economy, and so different ecological indi-cators assess different functions of the rehabilitated ecosystem. The challenge is to understand the system sufficiently to choose the most critical indicators that will give the most discriminatory information to assess the direction of the rehabilitation and maximize the likelihood that it will achieve and maintain the desired community.

ACKNOWLEDGMENTS

Research carried out has been funded by Pacific Coal Pty Ltd, Blair Athol Coal Project, and Tarong Coal. Research assistance has been provided by Celeste Beavis, Paulette Clowes, and Mary-Anne Murray. Additionally, input from and discussions with Randal Hinz, James Sullivan, Peter Baker, David Esser, David Mulligan, and Clive Bell were invaluable and appreciated.

REFERENCES

Allen, S.E., H.M. Grimshaw, J.A. Parkinson, and C. Quarmby. *Chemical Analysis of Ecological Materials.* Blackwell Scientific Publications, Cambridge, MA, 1974, p. 565.
Attiwill, P.M. and M.A. Adams. Tansley Review No. 50 Nutrient Cycling in Forests. *New Phytol.*, 124, pp. 561–582, 1993.
Begon, M., J.L. Harper, and C.R. Townsend. *Ecology: Individuals, Populations and Communities*, 2nd ed. Blackwell Scientific Publishers, Cambridge, MA, p. 945, 1990.
Bell, L.C. Mining, in *Australian Soils: The Human Impact*, Russell, J.S. and R.F. Isbell, Eds., St. Lucia, Australia, The University of Queensland Press, p. 522, 1986.

Bellairs, S.M. and P. Clowes. *Seedling Establishment at the Ecolodge Rehabilitation Area at the Blair Athol Coal Mine*. Unpublished report to Blair Athol Coal Project. April 1996. Centre for Mined Land Rehabilitation, The University of Queensland, St. Lucia, Australia.

Bradshaw, A.D. The Reconstruction of Ecosystems. *J. Appl. Ecol.*, 20, pp. 1–18, 1983.

Cornet, F., C. Otto, G. Rinaudo, G.D. Hoang, and Y. Dommergues. Nitrogen Fixation by *Acacia holosericea* Grown in Field-Simulating Conditions. *Acta Oecologica/Oecologia Plantarum* 6, pp. 211–218, 1985.

DME. *Environmental Impact Assessment and Management for Mining in Queensland*. Department of Minerals and Energy, Brisbane, Australia, 1995.

Ellis, B. *Koalas at Blair Athol*. Report to Blair Athol Coal Project, December 1996. The Koala Study Program, The University of Queensland, St. Lucia, 1997.

Ewel, J.J. Restoration is the Ultimate Test of Ecological Theory, in *Restoration Ecology: A Synthetic Approach to Ecological Research*, Jordan, W.R., III, M.E. Gilpin, and J.D. Aber, Eds., Cambridge University Press, Cambridge, MA, pp. 31–33, 1987.

Forster, J.D. Competitive Effects Arising from the Simultaneous Establishment of Pasture Species with Woody Perennial Species in the Initial Stages of Revegetating Rehabilitation Sites at Tarong Coal Mine. Thesis presented to The University of Queensland, St. Lucia, Australia for Bachelor of Applied Science with Honours, 1995.

Hamilton, S.D., P. Hopmans, P.M. Chalk, and C.J. Smith. Field Estimation of N_2 Fixation by *Acacia* spp. Using ^{15}N Isotope Dilution and Labelling with ^{35}S. *Forest Ecol. Manage.*, 56, pp. 297–313, 1993.

Likens, G.E. and F.G. Borman. An Experimental Approach to New England Landscapes, in *Coupling of Land and Water Systems*, Hasler, A.D., Ed., London, Chapman and Hall, 1975, pp. 7–30.

Loch, R.J. and J.J. Bourke. Effects of Vegetation on Runoff and Erosion: Results of Rainfall Simulator Studies on 12m Long Plots, Meandu Mine, Tarong, April 1995. Unpublished report to Tarong Coal, 1996.

Marrs, R.H. Nitrogen Accumulation, Cycling and the Restoration of Ecosystems on Derelict Land. *Soil Use Manage.*, 5, pp. 127–134, 1989.

Noble, I.R. and R.O. Slatyer. The Use of Vital Attributes to Predict Successional Changes in Plant Communities Subject to Recurrent Disturbances. *Vegetatio*, 43, pp. 5–21, 1980.

Russell, E.W. *Soil Conditions and Plant Growth*. London, Longman Scientific, 1973, 849 p.

Salwerowicz, F. Mineral Development and Ecosystem Management, in *Proceedings of the 3rd International Conference on Environmental Issues and Waste Management in Energy and Mineral Production*. Perth, Australia, Curtin University of Technology, 1994, pp. 3–14.

So, H.B., B. Kirsch, L.C. Bell, R.J. Loch, C. Carroll, and G. Willgoose. Post-Mining Landscape Parameters for Erosion and Water Quality Control, in *Proceedings of the 20th Annual Environmental Workshop*, Darwin, 2–6 October 1995, Canberra, Australia, Minerals Council of Australia, 1995, p. 427.

Strahan, R., Ed. *The Australian Museum Complete Book of Australian Mammals*. London, Angus and Robertson, 1983, p. 530.

Tacey, W., J. Treloar, and R. Gordine. Completion Criteria for Mine Site Rehabilitation in the Arid Zone of Western Australia, in *Proceedings Green and Gold: Goldfields International Conference on Arid Landcare*. 29 October–1 November 1993. Kalgoorlie, Goldfields Arts Centre, 1993, pp. 151–162.

Welsh, D.R. Overview of Environmental Practices in the Queensland Coal Mining Industry—The Regulatory Framework, in *Proceedings of the 17th Annual Australian Mining Industry Council Environmental Workshop*, Yeppoon, 5–9 October 1992, Canberra, Australian Mining Industry Council, 1992, pp. 27–48.

Chapter 3

Research Initiatives for the Remediation of Land Following Open-Cut Coal Mining in Central Queensland

D.R. Mulligan, A.H. Grigg, T.A. Madsen, A.B. Pearce, and P.A. Roe

INTRODUCTION

In 1992, BHP Coal Pty Ltd, the operator of eight open-cut coal mines in the Bowen Basin of central Queensland, embarked on a major research and development program which still continues today. The Centre for Mined Land Rehabilitation at the University of Queensland was one of the research groups that became involved with the program. A series of linked, multidisciplinary projects was developed, focusing on developing procedures to overcome, or at least minimize, the major impediments to the establishment and persistence of self-sustaining ecosystems on postmined land in the central Queensland coalfields. The research addressed both control and/or amelioration of factors that affect plant establishment, as well as control and/or management of those factors limiting growth and ongoing sustainability. Surface crusting, salinity, competition during establishment, and water availability had been identified as the major limitations to successful establishment, while among the sustainability issues, water availability, nutrient availability through the reestablishment of effective nutrient cycling, compaction, and erosion were identified as key issues. Linking both programs is the role of species selection, both plant and microbial, in fostering establishment and sustainability success (Figure 3.1).

This chapter presents an outline and some of the results from five component projects that contribute to the overall program.

BACKGROUND

Location and Size

In central Queensland on the east coast of Australia, the Bowen Basin covers some 32,000 km^2 extending 550 km north to south at latitudes between about 20 and 25° south. It is approximately 100 km east to west at its broadest section, and about 175–300 km inland from the coast (Figure 3.2). Coal reserves in the Bowen Basin and the outlying basins of Blair Athol and Callide are estimated at 26,000 Mt, and these support 23 mining operations with a total rated production of about 87 Mt yr^{-1}, most of which is won by open-cut operations (Queensland Coal Board, 1995).

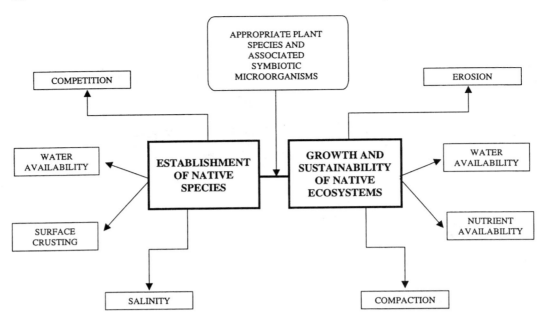

FIGURE 3.1. An overview of the major limitations to effective establishment and sustainability of ecosystems established on postmined land in the Bowen Basin of central Queensland.

Biophysical Environment

The climate of the region is subtropical with a concentration of rainfall in the period November to March when 75% of the annual average of 625 mm is received. Early wet season rain is usually associated with thunderstorms, leading to high spatial variability and high rainfall intensities. Heat-wave conditions of 35–40°C are frequently experienced in the warmer months. Evaporation exceeds rainfall in every month of the year and annually by a factor of three.

The patterns of vegetation in the Basin are strongly related to the supporting soil types (Kelly, 1979a). Mining operations are generally on the undulating valley bottoms and side slopes where two vegetation types predominate. Open forests of *Acacia harpophylla* (brigalow), alone or with other tree species generally occur on grey and brown cracking clays, while grassy woodlands dominated by *Eucalyptus populnea* (poplar box) occur on shallow duplex soils, often with highly dispersible subsoils. A major proportion of the vegetation in the region has been cleared to improve agricultural value. Extensive cattle grazing was the main use of the land prior to mining activity, which expanded rapidly during the 1970s.

Nature of the Operations

The impact of surface mining is generally much greater than that of underground mining as much of the existing vegetation and fauna habitat are destroyed, and the removal of overburden and its subsequent placement creates major changes in the topography, hydrology, and stability of the landscape (Bell, 1991). Although soil stripping and replacement were not common in the early years of mining development in the region, mines now routinely recover surface soil prior to overburden blasting and removal. Since the soils are

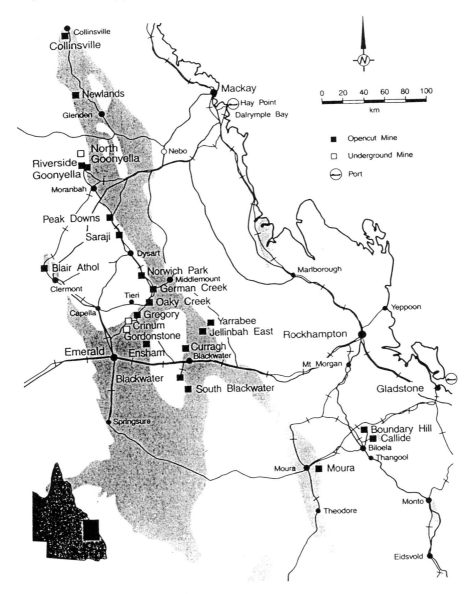

FIGURE 3.2. Location of coal mines in the Bowen Basin, central Queensland. (Source: BHP Coal Pty Ltd).

commonly duplex with dispersible saline subsoils, only surface (upper 10–40 cm) soil is stripped, and this is then normally stockpiled for some years prior to respreading and subsequent revegetation (Roe et al., 1996).

All the open-cut mines in the Bowen Basin have draglines as the primary machines to expose the coal. Draglines tend to cast spoil in parallel linear ridges with sawtooth crests (Kelly, 1979a, b). For most spoil materials, major earthworks reshaping is required to reform this spoil to a relatively stable landform. Progressively, as the depth to coal has increased, mines are employing shovel and truck systems to prestrip spoil ahead of the draglines, thus keeping the draglines operating at optimal digging depths. The use of shovel and truck prestripping has a major advantage over straight dragline stripping, as the prestrip spoil can be placed to a design form and hence reduce the amount of earthworks required to achieve a measure of stability. The usual practice has been to truck the prestrip spoil into the dragline spoil and fill the valleys between the rows of spoil, essentially eliminating the need to regrade the spoil to a more stable landform (Roe et al., 1996).

Where past practices have resulted in topsoil deficits, spoil remains as the surface media awaiting revegetation. Many of these spoils have properties that are adverse to plant establishment and if left unmanaged, not only fail to support an acceptable level of appropriate vegetative cover, but also have a high potential to erode and result in serious adverse environmental effects off-site.

The outcome from all this type of activity is therefore a land that has been drastically disturbed and degraded, and ready for rehabilitation.

REHABILITATION RESEARCH

Surface Crusting

The spoils of the region range from very acidic to very alkaline; are typically extremely nutrient-deficient, and often saline and very sodic (Table 3.1). The sodic materials are highly dispersive and form crusts upon wetting and subsequent drying. A lack of quality topsoil available in many areas has seen research efforts concentrate on surface management practices through the use of organic mulches, both with and without the additional use of gypsum. Extensive studies in the United States on bentonite spoil which has similar chemical and physical characteristics, have proven the value of such ameliorants (Schuman and DePuit, 1991).

Organic mulches with the potential for widespread availability in the area included straw, sawdust and wood residue (from nearby timber mills), bagasse (a by-product of the sugar refinery industries near the coast), and cotton trash (a waste product from the cotton processing plants). Carbon:nitrogen ratios of these materials were determined in the laboratory (Table 3.2) to assist in determining the rate of nitrogen application required to ensure the mulched system did not leave the establishing plant community grossly nitrogen-deficient. As a prelude to the treatments and rates of ameliorants to be tested in the field, greenhouse trials demonstrated the value of mulches in reducing crust strength, increasing water-holding capacity and improving emergence. On unamended spoil, there was only limited emergence. On Blackwater spoil, the least hostile of the three spoils examined (Table 3.1), application and incorporation of 20 t ha^{-1} of sawdust resulted in an emergence rate of 16% for *Eucalyptus camaldulensis*; the same treatment on Goonyella spoil resulted in only 1% emergence. The experiments also provided useful information regarding how long seedlings would be likely to survive between 'rain' events. Laboratory experiments on Goonyella spoil using a rainfall simulator further demonstrated the advantages of mulch applications with respect to increasing infiltration (Figure 3.3) and hence reducing runoff.

A field trial was set up on prestrip spoil on the bucketwheel-excavator dump at Goonyella mine in mid-1994, using straw and sawdust mulches as well as a treatment that replaced a 30 cm capping of stockpiled topsoil onto the spoil. The treatments used are presented in Table 3.3. A smaller trial using only the sawdust and spoil was established at Saraji mine at the same time. At Goonyella, there were in excess of 600 5 m x 5 m plots with each treatment being replicated four times. Species sown were predominantly Australian native tree, shrub, and grass species (namely, *Acacia salicina, A. harpophylla, A. holosericea, Casuarina*

Table 3.1. Chemical Characteristics of Tertiary Spoils from Three Mines in Central Queensland.

Parameter	Units	Extractant	Blackwater	Goonyella	Saraji
pH		1:5 H_2O	8.7	8.4	4.2
EC	dS m^{-1}	1:5 H_2O	0.67	1.09	1.56
Organic C	%	K_2CrO_7/H_2SO_4	1.8	0.2	0.2
NO_3-N	mg kg^{-1}	1:5 H_2O	7.9	3.1	3.8
SO_4-S	mg kg^{-1}	0.01M $Ca(H_2PO_4)_2$	550	21	205
P	mg kg^{-1}	0.5 M $NaHCO_3$	5	5	5
K	cmol$_c$ kg^{-1}	1 M NH_4OAc	0.45	0.17	0.22
Ca	cmol$_c$ kg^{-1}	1 M NH_4OAc	11.24	4.35	0.32
Mg	cmol$_c$ kg^{-1}	1 M NH_4OAc	9.32	10.39	6.58
Na	cmol$_c$ kg^{-1}	1 M NH_4OAc	3.99	15.52	6.20
Cl	mg kg^{-1}	1:5 H_2O	215	4850	1150
Cu	mg kg^{-1}	0.005 M DTPA	2.6	0.3	0.3
Zn	mg kg^{-1}	0.005 M DTPA	3.7	0.6	1.3
Mn	mg kg^{-1}	0.005 M DTPA	4	1	1
Fe	mg kg^{-1}	0.005 M DTPA	14	2	13
B	mg kg^{-1}	0.01 M $CaCl_2$	0.55	1.53	0.16
CEC [a]	cmol$_c$ kg^{-1}		25.0	30.4	14.4
Ca:Mg			1.21	0.42	0.05
ESP [b]	%		16.0	51.0	43.1

[a] CEC = cation exchange capacity.
[b] ESP = exchangeable sodium percentage.

Table 3.2. Carbon and Nitrogen Contents of Spoils and Organic Mulches Used in Field Trials.

Material	Carbon (%)	Nitrogen (%)	C:N
Goonyella spoil	0.12	n.d.[a]	
Saraji spoil	0.10	n.d.	
Straw	40.17	1.08	37.2
Sawdust	43.08	0.11	391.6
Bagasse	43.88	0.28	156.7
Cotton trash	41.31	1.93	21.4

[a] n.d. = not detectable.

cristata, Eucalyptus camaldulensis, E. citriodora, E. crebra, E. populnea, Astrebla lappacea, Dicanthium sericeum, and *Heteropogan contortus*). These were typical of the range of species that have been and are being used in the central Queensland coalfields region. Since the initiation of the trial, sequential drought years have resulted in limited seedling emergence, although soil physical and chemical characteristics have improved over time. Results from planned resowing of the field trial during the wet season of 1997 are expected to confirm the positive outcomes from the laboratory and greenhouse trials, and reflect the observed improvements in the properties of the ameliorated spoil in the field.

Microbial Associations and Salinity—Significance and Interaction

Intact cores from the upper 10 cm of 15 mine media across three mines were collected and used in greenhouse trials for microbiological assessment. Media included different

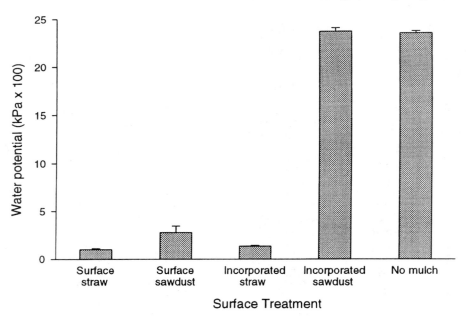

FIGURE 3.3. Effect of surface treatment on the water potential of Goonyella mine spoil at 5 cm depth following three wetting and drying cycles using a rainfall simulator. Bars are standard errors of the means of six replicates.

Table 3.3. Treatments Used in the Field Trial at Goonyella Mine.

Parameter	Treatment/Rate
Growth media	Tertiary spoil (as per Table 3.1)
	Stockpiled soil (30 cm) over spoil
Mulch type	Straw
	Sawdust
Application method	Surface applied
	Incorporated to 30 cm
Mulch rate (surface spoil and soil)	Straw: 0, 2.5, 5, 10 t ha^{-1}
	Sawdust: 0, 5, 10, 20 t ha^{-1}
Mulch rate (incorporated into spoil)	Straw: 0, 5, 10, 20 t ha^{-1}
	Sawdust: 0, 20, 40, 80 t ha^{-1}
Gypsum rate	5, 20 t ha^{-1}
Nitrogen rate	Straw: 0, 17.5, 35, 70 kg t^{-1} mulch
	Sawdust: 0, 22.5, 45, 90 kg t^{-1} mulch
Other nutrient rates	P: 100 kg ha^{-1}
	Cu: 4 kg ha^{-1}
	Mn: 4 kg ha^{-1}
	Zn: 4 kg ha^{-1}

spoil types, soil in stockpiles, soil replaced on spoil from stockpiles, and coarse coal reject used as a surface capping on spoil, as well as native unmined soils adjacent to the Goonyella, Saraji and Blackwater mines. Four native species, *Acacia salicina*, *Casuarina cunninghamiana*, *Eucalyptus citriodora*, and the grass *Dicanthium sericeum*, were used as 'bait' plants on the unamended media to compare the growth, mycorrhizal activity, and level of nodulation.

Higher rates of development of microbial symbionts on the roots of species growing in the three unmined soils were clearly evident, and these media also supported the highest rates of plant growth (Table 3.4).

A subsequent experiment in which fresh native inoculum was added to cores of two contrasting spoil types was undertaken to determine whether the low rates of infectivity of mine spoils were due to the absence of the appropriate microbes or a chemical/physical limitation of the media to support microbial development. In the absence of the soil inoculum, nodules either did not form or showed only limited development on the roots of *Acacia salicina* (Table 3.5), and no infection by vesicular-arbuscular mycorrhizas (VAM) or ectomycorrhizas (ECM) was recorded for either the acacia or *Eucalyptus citriodora*. With the addition of the inoculum, nodule weight per plant increased and ectomycorrhizal infection levels increased for one spoil only (Table 3.5). No improvement in microbial infection was recorded for the harsher of the two spoils. The results suggested that the chemical and/or physical characteristics of the spoils were a major limitation to the formation of plant-microbe associations.

Having identified surface salinity as a major impediment to direct-seeding success in several of the Bowen Basin spoil materials, a follow-up study was initiated to determine whether salt-tolerant strains of mycorrhizas would confer a survival and/or competitive advantage to particular species. This study was conducted in parallel with laboratory and greenhouse experiments examining the effects of high salt levels on the germination and growth of three *Eucalyptus* species (*E. camaldulensis*, *E. citriodora*, *E. populnea*) and *Acacia salicina*. Apart from the local provenances of the eucalypts, other provenances of known salt-tolerance from around Australia (a total of 15 between the three species) were used in this study. However, the local provenances were shown to be the equal of other more tested seed lines in their abilities to tolerate saline conditions. With respect to impacts on seedling emergence, reductions of 80–100% occurred in the eucalypts at 100 mM NaCl (10 dS m^{-1}), and no emergence occurred at 200 mM (Figure 3.4). Emergence of *A. salicina* was 45% that of the controls at 100 mM NaCl, but only 3% at 200 mM. In contrast, when established seedlings (nine weeks old) were transplanted into saline media, survival and growth four weeks later were much higher. Again, the local *A. salicina* was the most tolerant, with shoot weights reduced by 30% and 50% at 300 and 400 mM NaCl, respectively. For *E. citriodora*, shoot weights were reduced by 70–85% at 300 mM, while at 400 mM, most plants died. No deaths occurred among the *E. camaldulensis* and *E. populnea* provenances at the highest salt level tested.

Subsequent microbial/salinity interaction experiments concentrated on ectomycorrhizas (in particular, known salt-tolerant strains of *Hebeloma westraliense*, *Laccaria laccata*, and *Pisolithus tinctorius*) and a local provenance of *E. citriodora*. The three fungi differed in their tolerance of high NaCl, with *P. tinctorius* showing the highest tolerance. Laboratory-based experiments examining the influence of salt-tolerant *P. tinctorius* in promoting salt tolerance in eucalypt seedlings have indicated that at moderate NaCl concentrations (100–200 mM), there can be greater growth in inoculated seedlings than in uninoculated seedlings. Inoculated and uninoculated seedlings have been outplanted into the field recently and survival, growth rates, levels of infection, and nutrient status will be monitored over time.

Competition

Since much of the rainfall in central Queensland falls as high intensity storms during summer, it becomes important to try to establish a vegetative cover as quickly as possible

Table 3.4. Effect of Media Type (Unmined "Native" Soils and Spoils) on the Growth and Levels of Nodulation and Mycorrhizal Activity in Two Species Commonly Used in Revegetation Programs.

Species	Media	Shoot Dry Weight (g)	Nodule Dry Weight (mg)	% VAM[a]	% ECM[a]
Acacia	Saraji – Native 1	0.240 (0.045)[c]	14.9 (3.3)	82a	
salicina	Saraji – Native 2	0.530 (0.090)	4.3 (3.9)	88a	
	Saraji – Spoil 1	0.065 (0.005)	0.0 (0.0)	0c	
	Saraji – Spoil 2	0.095 (0.015)	7.0 (1.8)	0c	
	Goonyella – Native	0.240 (0.030)	13.6 (4.2)	43b	
	Goonyella – Spoil 1	0.085 (0.007)	0.7 (0.4)	2c	
	Goonyella – Spoil 2	0.035 (0.014)	0.8 (0.7)	4c	
	Blackwater – Native	0.410 (0.160)	13.5 (5.9)	54ab	
	Blackwater – Spoil 1	0.130 (0.035)	0.7 (0.6)	0c	
	Blackwater – Spoil 2	0.100 (0.020)	6.8 (1.9)	4c	
Eucalyptus	Saraji – Native 1	0.14 (0.05)		13ab	51a
citriodora	Saraji – Native 2	0.25 (0.07)		15ab	7b
	Saraji – Spoil 1	0.07 (0.02)		0c	0b
	Saraji – Spoil 2	0.04 (0.02)		0c	0.5b
	Goonyella – Native	0.17 (0.03)		20ab	40a
	Goonyella – Spoil 1	0.00 (0.00)		b	b
	Goonyella – Spoil 2	0.00 (0.00)		b	b
	Blackwater – Native	0.19 (0.06)		40a	0.25b
	Blackwater – Spoil 1	0.04 (0.01)		4bc	0b
	Blackwater – Spoil 2	0.03 (0.01)		36a	0b

[a] Analysis run on arc sin transformed data. Means followed by different letters are significantly different at $P<0.05$.
[b] Plants died and therefore could not be assessed.
[c] Mean and standard error (in parentheses) of four replicates.

Table 3.5. Effect of Soil Inoculum on the Development of Symbioses in Two Species Growing in Two Saraji Spoils of Contrasting Chemical Characteristics.

Spoil	Species	Parameter	Soil Inoculum Absent	Soil Inoculum Present
Ebony pit	*A. salicina*	Nodule dry weight (mg)	1.0a[a]	2.8a
(high pH,		% VAM	0a	0.8a
moderate	*E. citriodora*	% VAM	0a	1.5a
salinity)		% ECM	0a	19.7b
Coolibah pit	*A. salicina*	Nodule dry weight (mg)	0a	0a
(low pH,		% VAM	0a	1.6a
high salinity)	*E. citriodora*	% VAM	0a	0.3a
		% ECM	0a	0a

[a] For each parameter, means followed by a different letter are significantly different at $P<0.05$.

on the reshaped, though often steep, landforms. To this end, and also because the original post-mining land use for much of the area was designated as a return to grazing, rapid-growing exotic pasture grasses such as *Cenchrus ciliaris* (buffel grass) and *Chloris gayana* (Rhodes grass) have been widely used. An increasing interest in native species (but tem-

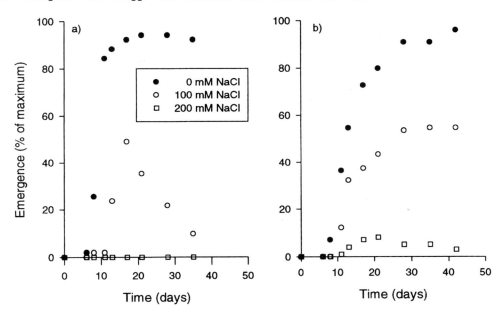

FIGURE 3.4. Percentage emergence for (a) *Eucalyptus citriodora* and (b) *Acacia salicina* at three concentrations of NaCl.

pered by the high costs of planting seedlings) has resulted in the simultaneous sowing of these species with the grasses; unfortunately, such a practice has exposed the slower growing natives to intense competition for water, nutrients, and light from the faster growing pastures. An example of the impacts of such competition on native species is presented in Figure 3.5, where there are dramatic initial and persistent effects of the presence of a cover crop of *C. ciliaris* and *C. gayana* on the density of sown native trees and shrubs. Effects on the growth rates of those natives that do survive can be similarly marked. The lack of response to the cover crop addition on the bare spoil medium was due to the fact that only *A. salicina*, a successful colonizer species elsewhere on rehabilitated areas, was able to establish.

The aim of this component of the program has been to examine options in which the exotic pastures that have already been established are removed through combinations of herbicide application, fire, and cultivation. An added advantage of this approach is that the seedbed is a media with a much higher organic matter content and vastly improved physical characteristics than would have existed if the pasture phase had not been initially established. At the Gregory mine site, a native species community was sown into such pretreated field plots. Trials were established on different replaced soil types and with different ages of existing pasture, and early results are promising.

Nutrient Cycling

In an earlier survey of existing tree trials throughout the Bowen Basin (Mulligan and Bell, 1991, 1992), it was observed that for many trials established directly on spoil (i.e., where topsoil was not replaced prior to planting), there was very little humus development, even after 10–15 years of litter accumulation. Such a situation clearly has implica-

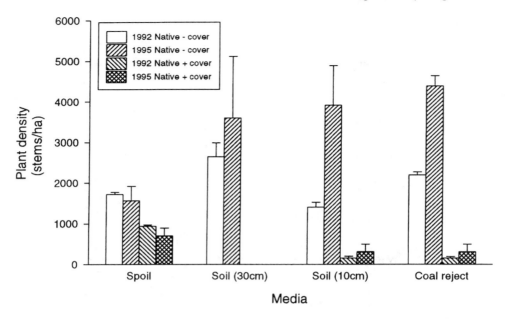

FIGURE 3.5. Total densities of native seedlings on amended and unamended spoil media in the presence or absence of a cover crop at a trial site at Saraji mine, one and four years after establishment. Bar graphs show the means and standard errors of four replicate plots.

tions for the long-term sustainability of such sites, because without effective nutrient cycling, the established vegetation in the rehabilitated communities could eventually become severely nutrient-deficient. As a part of a research program to examine the sustainability of reconstructed ecosystems, litter decomposition experiments have been carried out to identify if a bottleneck in the cycle exists and if so, what the cause of such a throttle is and what management inputs could assist in alleviating it.

Sets of litterbags were placed at 28 sites across six of the mines in the Bowen Basin. The sites encompassed the range of surface media that occur in the rehabilitated landscapes (spoils, replaced soils, coarse coal reject), the type of revegetation community (improved pasture grasslands, grasslands with a savanna-type frequency of acacias, eucalypt-dominated trial areas) and the age of those communities. Three unmined sites were included as reference areas. The decomposition bags, containing freshly fallen leaf material of either the exotic pasture grass commonly used in the area, *Cenchrus ciliaris*, or the trees *Acacia salicina* or *Eucalyptus maculata*, were retrieved from the field after one and two years and analyzed for weight loss, changes in nutrient, tannin, and lignin contents and microbial respiration activity. Chemical, physical, and microbiological properties of each of the sites were measured, as were the local topographic and climatic conditions. An assessment of the soil and litter invertebrate populations was also conducted as a part of the strategy to identify the contributing factors to differing rates of litter breakdown.

Data presented by Grigg (1995) showed that after 12 months, litter weight loss varied from 4 to 77%, with losses above 58% confined to a limited number of samples with observed termite activity. The contributions of leaching and microbial activity/respiration to the weight losses were also estimated. Averaged across all sites and substrates, weight loss differed significantly between litter types with *Acacia* > *Eucalyptus* > *Cenchrus*. The most important treatment effect on weight loss was the type of vegetation in the rehabilitated community, and it is postulated that a lack of plant diversity in many areas may be contributing to low rates of decomposition. Greater species diversity and a consequent wider variety in litter qualities may have contributed to the better decomposition rates in unmined native communities, areas where there is also a higher richness of soil and litter invertebrates.

In reestablishing an ecosystem on drastically disturbed land such as that which follows open-cut coal mining, the question arises as to whether there is enough nutrient capital in the rehabilitation to support the desired type of end-use community. As a step toward answering that question, the biomass and nutrient contents of two study sites were obtained from biometric relationships derived either directly at the site or from previous studies on similar communities. Nutrient budgets for both a 20-year-old rehabilitated *C. ciliaris – A. salicina* savanna at Blackwater mine, and an unmined *E. populnea* woodland at Saraji mine, were constructed from the results of these nutrient distribution studies, together with additional data on soil and litter nutrient concentrations, and from litter production and the decomposition experiments (Grigg, unpublished data). The outcomes from these compilations have shown that in the rehabilitated community, >90% of the aboveground biomass is in the litter pool, a pool which also contains 80% of the aboveground nitrogen and phosphorus. In contrast, in the unmined woodland, >90% of the aboveground biomass, nitrogen and phosphorus, is held in the overstorey species. Rates of litter production and disappearance are similar in this mature community, while at the rehabilitated site, litter production is 70% higher than disappearance.

Such findings clearly have implications for the resilience and sustainability of some of the rehabilitation areas, and future revegetation research and strategies will investigate ways to alter the compositional balance so as to provide greater surety of long-term success.

Community Structure

In the past, the emphasis of many of the rehabilitation programs in the coal mining industry in eastern Australia has been on the establishment of improved pastures for possible future grazing, with some attention given to the provision of an intermittent upperstorey of shade trees. With an increasing interest by the industry in returning some of the postmined land to native species communities which, in turn, will support native fauna, there was a need to develop establishment procedures for native understorey and ground cover species as well. This stratum is critical to ecosystem stability and functioning because it supports soil microbial and mesofaunal communities, assists in erosion control and improving soil structure, and provides food, shelter, and habitat for fauna.

Because very limited work had been undertaken on these species, an assessment of seed viability and germination characteristics of understorey species, and determination of conditions for optimizing germination and establishment, was required. One particular focus for this project has been those species that provide food and/or shelter for the bridled nailtail wallaby (*Onychogalea fraenata*), a rare species endangered due to habitat destruction caused by massive land clearing by agricultural practices of the past. Its present range has contracted to an area in central Queensland where the once-abundant *Acacia harpophylla* forms the overstorey, below which a fairly diverse understorey exists. Beyond the seed biology issues being addressed, this research is screening the native understorey species for tolerance to the new set of environmental conditions to be encountered in the rehabilitated landscape. Field trials involve both establishment into newly replaced media as well as into previously established areas where a degree of canopy protection already exists.

CONCLUSIONS

Land disturbed by open-cut coal mining clearly falls into the category of severely degraded, and yet with the research and technologies now available, significant advances into the repair of this land have been, and will continue to be, achievable. This chapter has provided an overview of some of the rehabilitation research being undertaken and sponsored by one of the major coal producers in Australia that will no doubt lead to an improvement in both the establishment and sustainability of reconstructed ecosystems. The approaches and techniques developed through such industry-sponsored research have implications beyond the mining industry and can in many cases be applied to the remediation of degraded lands in general.

REFERENCES

Bell, L.C. Assisting the Return of the Living Environment After Mining—An Australian Perspective, in *Proceedings of the Workshop: Issues in the Restoration of Disturbed Land, Palmerston North, New Zealand*, Gregg, P.E.H., R.B. Stewart, and L.D. Currie, Eds., Massey University, Palmerston North, New Zealand, 1991, pp 7–21.

Grigg, A.H. Self Sustainability of Rehabilitated Ecosystems Following Open-Cut Coal Mining in Central Queensland, in *Proceedings of the 20th Annual Minerals Council of Australia Environmental Workshop, Darwin, Northern Territory*, Minerals Council of Australia, Canberra, ACT, 1995, pp. 144–165.

Kelly, R.E. A Description of Natural Environment Features of the Isaac River Catchment and of the Coal Strip-Mining Operations, in *Proceedings of a Workshop on Management of Lands Affected by Mining, Kalgoorlie, Western Australia*, Rummery, R.A. and K.M.W. Howes, Eds., CSIRO Division of Land Resources Management, Perth, Western Australia, 1979a, pp. 33–54.

Kelly, R.E. Rehabilitation of Mined Land in the Central Queensland Coalfields, in *Mining Rehabilitation—'79*, Hore-Lacy, I., Ed., Canberra, ACT, Australian Mining Industry Council, Canberra, ACT, 1979b, pp. 120–126.

Mulligan, D.R. and L.C. Bell. Native Tree and Shrub Regeneration on Coal Mines in Queensland—A Review, in *Proceedings of the 16th Annual Australian Mining Industry Council Environmental Workshop, Perth, Western Australia*, Australian Mining Industry Council, Canberra, ACT, 1991, pp. 38–61.

Mulligan, D.R. and L.C. Bell. An Assessment of Tree and Shrub Growth on Rehabilitated Land Following Open-Cut Coal Mining in Queensland. Report to the Queensland Department of Resource Industries, March 1992, University of Queensland, Brisbane.

Queensland Coal Board (1995). 44th Queensland Coal Industry Review 1994–95. Queensland Coal Board, Brisbane.

Roe, P.A., D.R. Mulligan, and L.C. Bell. Environmental Management of Coal Mines in the Bowen Basin, Central Queensland, in *Environmental Management in the Australian Minerals and Energy Industries—Principles and Practices*, Mulligan, D.R., Ed., University of New South Wales Press, Sydney, New South Wales, 1996, pp. 290–315.

Schuman, G.E. and E.J. DePuit. Innovations in Abandoned Bentonite Spoil Reclamation, in *Proceedings of the 16th Annual Australian Mining Industry Council Environmental Workshop, Perth, Western Australia*, Australian Mining Industry Council, Canberra, ACT, 1991, pp. 221–239.

Chapter 4

Restoration of Opencast Coal Sites to Agriculture, Forestry, or Native Heathland

D.W. Merrilees, G.E.D. Tiley, and A. Duncan

INTRODUCTION

Opencast mining for the low-cost extraction of shallow coal deposits is a major industry in the United Kingdom with an annual output of 18 million tonnes. In Scotland, 70 opencast sites produce 5 million tonnes per annum. Sites are located on the Carboniferous sediments of the Central Lowlands in land of variable agricultural quality (Figure 4.1). The size of opencast sites varies between 50 and 1,000 hectares.

Restoration of opencast sites is now generally to a high standard with the potential of restoring land to a wide range of end uses including agriculture, forestry, and nature conservation. In the past, restoration has primarily been to agriculture (Procter, 1989) but, in line with current land management policy, opencast sites on lower quality marginal land will be reinstated with a vegetation cover similar to the vegetation of the surrounding area. The restored land use is likely to include a mix of agriculture, forestry, wetland, and heathland to enhance the ecological value of the site.

OPENCAST RESTORATION STRATEGY

Restoration and aftercare should be regarded as an integral part of the working of the site, and the protection of the soil resource during the life of the site is of paramount importance to the success of the reclamation.

One of the requirements for planning consent is to present a reclamation and aftercare strategy covering the following objectives (Department of the Environment, 1996):

- to minimize the area of land stripped at any one site by implementing a rolling program of restoration
- to protect and fully reinstate prime quality land to agriculture
- to improve the ecological value of the site
- to enhance the landscape value of the site
- to protect major watercourses and water quality
- to rehabilitate any derelict land within the site
- to provide final restoration which is compatible with the surrounding countryside.

The planning submission must also include an environmental assessment detailing all likely impacts of the opencast development on the environment and the necessary mea-

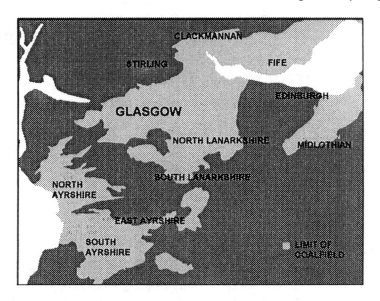

FIGURE 4.1. Location of opencast coal deposits in the Central Lowlands of Scotland. (Scale 1:1,000,000)

sures required to minimize these impacts (The Environmental Assessment [Scotland] Regulations 1988).

On large sites where coaling will occur over a number of years, the land use during each phase of the development must also be projected.

LAND RESTORATION SPECIFICATIONS

Once coaling has been completed and following overburden and subsoil replacement the soil is loosened to a minimum depth of 0.6 meters and stones greater than 0.3 meter diameter are removed. Topsoil is then replaced and the land enters a two- to five-year restoration and aftercare program which is dependent upon the final land use:

Agricultural Restoration

Restoration to agriculture is generally the most intensive and most expensive specification with a cost of up to £5000 per hectare. A major part of this cost is allocated to the installation of an intensive underdrainage system to alleviate surface waterlogging problems in these slowly permeable gley soils. The outline restoration specification is detailed below:

Year 1   – Topsoil placement
            – Cultivation
            – Stone picking (> 100 mm Ø)
            – Lime and fertilizer application
            – Temporary grass establishment
            – Structure drainage installation.
Years 2/3 – Pipe drainage installation
            – Soil loosening (x 2) to alleviate compaction
            – Stone picking (>75 mm Ø)
            – Cultivation
            – Hedges/fencing/shelter belts planted
            – Water supplies installed
            – Permanent grass establishment *or* arable cropping
            – Fertilizer application
Years 4/5 – Herbicide application
            – Fertilizer application

Digested sewage sludge cake and liquids are applied to improve soil quality and to supplement soil fertility during the aftercare period (Hall and Wolstenholme, 1997).

Forestry Restoration

Forestry planting generally occurs on the poorer quality mineral soils, deep peat, or subsoils amended with sewage sludge cake or peat. Opencast forestry soils have poor rooting potential due to soil compaction and surface wetness, necessitating deep soil loosening and mounding to improve tree establishment (Moffat and McNeill, 1994). Cultivation equipment, The Maclarty Mounder, has been designed specifically for soil loosening and soil mounding in a one-pass operation as an alternative to forestry plowing. Forestry restoration and aftercare covers three years as detailed below:

Year 1 – Topsoil placement *or* Amended soil preparation
       – Cultivation
       – Stone picking (> 300 mm Ø)
       – Structure drainage installation.

Year 2 – Soil loosening
       – Plowing *or* Mounding
       – Sewage sludge or Fertilizer application
       – Tree planting

Year 3 – Maintenance replanting
       – Herbicide application
       – Fertilizer application

Conservation Restoration

Nature conservation is a primary end-use in the restoration of upland sites with poorly-drained acid organic soils and a natural wet heathland vegetation (*Molinia, Juncus, Calluna* spp.). Construction of wetlands and areas of open water are also an integral part of the conservation specification.

HEATHLAND RESTORATION

As the large-scale restoration of wet heathland is a relatively new requirement on opencast sites in Scotland, a study was commenced in 1991 on the Headlesscross opencast site on the Lanarkshire coalfield (OS Ref NS 905580) to assess different methods of heathland reinstatement (Merrilees et al., 1995). A second site was later established on the Ayrshire coalfield (OS Ref 465075) in 1992 to supplement the initial study.

Study Area

Headlesscross opencast site is on the drumlin topography of the Carboniferous shales and sandstones at an elevation of approximately 240 m a s l. Soil cover varied from blanket peat, 50–100 cm of amorphous peat, to stagnohumic gley soils with <50 cm of peat topsoil on slowly permeable clay subsoil. The peat was very poorly drained, strongly acid (pH 3.9) in the upper horizon, and nutrient deficient (NPK status very low and Mg status moderate).

The native vegetation is a wet heath/acidic grassland mosaic, the predominant species being *C. vulgaris*, *E. vaginatum*, *J. acutiflorus*, and *Agrostis* spp. (Nature Conservancy Council, 1990). Prior to opencast mining, the land had been managed for intensive and extensive grazing after drainage improvement by shallow ditches and introduction of a *Lolium perenne/ Trifolium repens* sward on the dried areas. The climate is classified as fairly warm, rather wet foothill with an average annual rainfall of 1,050 mm, a potential summer soil moisture deficit of 0–25 mm, and an accumulated temperature (Day degrees C) of 1,100 (Birse and Dry, 1970). The Land Capability Classification for agriculture is 5.2 (Dry et al., 1986).

TREATMENT AND METHODS

Once the overburden and subsoil had been replaced to the agreed contour (3% slope), peat that had been stored for five years was spread on the experimental area to a depth of 0.5 m. The trial layout consisted of five main unreplicated treatments in large plots of 400 m² and eight subtreatments with plot sizes of 50 m². A buffer area, 30 m wide, was left untreated around the site (Figure 4.2).

The main treatments established in August 1991 were:

A. *Turves:* The surface vegetation (L F and H horizons) was stripped to a depth of 200 mm by excavator bucket from the next section to be opencast and spread directly onto the reinstatement plot.

B. *Litter:* Coarse vegetation was removed from the next section to be opencast and the litter layer (L and F horizons) rotavated to a depth of 40 mm, collected and spread onto the reinstatement plot at the rate of 3 kg m^{-2}.

C. *Calluna capsules and nurse grass:* A nurse grass of the same species mix as used in Treatment D was sown at a rate of 20 kg ha^{-1} and later overseeded in November 1991 with heather capsules at a rate of 1 kg ha^{-1}. *Calluna* seed rate was based on a germination rate of 10% achieved under mist conditions in the greenhouse. The capsules were collected from native heather on the site.

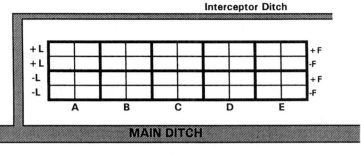

Main Treatments:
A- Turves
B- Litter
C- Heather Seed & Native Grass
D- Native Grass Mixture
E- Control (Stored Peat)

Sub-Treatments
Lime (+L) v No Lime (-L)
Fertiliser (+F) v No Fertiliser (-F)

FIGURE 4.2. Layout of main treatments and subtreatments established in August 1991. (Scale 1:1,000)

D. *Native grasses:* The following seed mixture, similar to the indigenous acidic grassland, was sown in August 1991 at the rate of 50 kg ha^{-1}.

Agrostis capillaris	25%	*Festuca longifolia*	15%
Cynosurus cristatus	10%	*Deschampsia flexuosa*	5%
Festuca rubra	40%	*Anthoxanthum odoratum*	5%

E. *Control:* Stored peat spread and left untreated to assess the seedbank survival from the storage heaps.

The following subtreatments were superimposed on each of the main treatments:

1. *Lime:* 10 t ha^{-1} of magnesian limestone, to raise the pH to 4.8, was applied to half of each cultivation subtreatment.
2. *Fertilizer:* A 5:22:22 N:P:K base fertilizer was applied at 400 kg ha^{-1} to half of each subplot. As ground cover establishment was slow in Year 2, the fertilized plots were further split, half receiving a maintenance fertilizer (5:22:22 NPK) in Years 2 and 3 at 250 kg ha^{-1}.

All treatments were lightly rolled after establishment to conserve surface moisture.

MONITORING

A botanical assessment of the total vegetation cover and percentage species establishment was carried out twice a year (April and August) for each subtreatment, using visual scores and randomly placed quadrats (5 x 0.25 m^2 quadrats/subtreatment). Soil pH and nutrient status were assessed each April to monitor any treatment differences, and a photographic record was kept of the progress of establishment.

RESULTS

Rates of vegetation establishment and diversity of species were used as indicators of the reinstatement success and its conservation value. Scottish Coal and Scottish Natural Heritage were particularly interested in the rate of *Calluna* regeneration. Results for each subtreatment (mean values of August quadrat assessment) are described below and rate of total vegetation establishment (1991–1996) is given as a mean of August subtreatment values in Figure 4.3. Variability within each subtreatment was low, with variability between subtreatments being shown in Figure 4.4: Effect of seedbed fertilizer and maintenance fertilizer, Figure 4.5: Effect of pH increase, and Table 4.1: Effect of seedbed lime and fertilizer on species diversity and numbers of species.

A. *Turves:* The transplanted vegetation gave an initial ground cover of 100% but subsequently died back during the immediate postestablishment period to 15%. In Year 2 there was a rapid recovery resulting in 68 and 95% vegetation cover in Years 2 and 3, respectively. *Calluna* cover increased from 0 to 45% cover over a 6-year period on the unlimed plots.

B. *Litter:* 16% ground cover was achieved in the initial postestablishment period, but this was reduced to 5% by winterkill. Total vegetation cover increased to over 60% in Year 3 and 90% in Year 6. *Calluna* was slow to establish from the litter seed bank, giving a ground cover of only 1% after two years but increasing markedly to over 30% in Year 6 on the unlimed plots.

C. *Heather capsules: Calluna* establishment from overseeding into the nurse grass was unsuccessful, producing <1% ground cover after two years and 10% in Year 6. A more successful establishment may be possible by direct drilling of capsules into the nurse grass rather than by surface broadcasting.

D. *Native grasses:* Application of base fertilizer was a significant factor in grass establishment. After three months the fertilized plots had 62% cover compared with 22% cover where no fertilizer was applied. Total vegetation cover on the main plot increased from 48% in Year 1 to 80% in Year 3 and 94% in Year 6. *C. cristatus* and *A. odoratum* were the dominant species in Year 1 but were later replaced by *Agrostis* spp./*Festuca* spp. in Year 3.

E. *Control:* Vegetation cover, mainly *A. stolonifera*, on the untreated stored peat was 6% after nine months, increasing to 35% during the following three summer months. In Year 3, cover was 50% rising to 90% in Year 6 and dominated by *D. flexuosa, A. stolonifera* and *Juncus* spp.

Fertilizer Effect

Application of a base fertilizer increased ground cover by approximately 50% in all treatments except A, Turves, in Year 1 (Figure 4.4).

Fertilizer increased yield as well as ground cover. This effect was particularly visible in Treatment E where it enhanced the rate of establishment from the seed bank. Current assessments confirm that a seedbed fertilizer is essential to provide rapid establishment of both sown and native species. Application of a low-rate maintenance fertilizer further enhances ground cover. Fertilizer application has encouraged heather establishment (Treatments A and B) but suppressed colonization by other species where grass growth has been enhanced (Treatments C and D).

Liming Effect

Lime application has encouraged more rapid ground cover on Treatments C, D, and E (Figure 4.5) and has also favored an increase in species diversity (Table 4.1). The inhibiting effect of lime on heather establishment, even in a strongly acid soil, is evident on all main treatments.

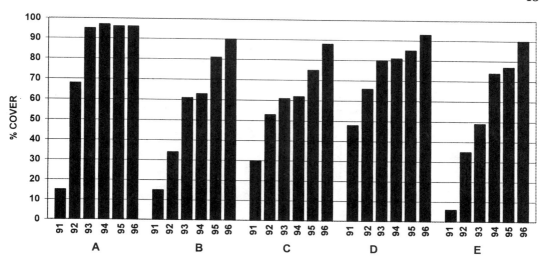

Main Treatments:
A- Turves
B- Litter
C- Heather Seed & Native Grass
D- Native Grass Mixture
E- Control (Stored Peat)

FIGURE 4.3. Changes in mean percentage of vegetation establishment (1991–1996) on main treatments A (Turves), B (Litter), C (*Calluna* capsules and Nurse grass), D (Native grasses) and E (Control).

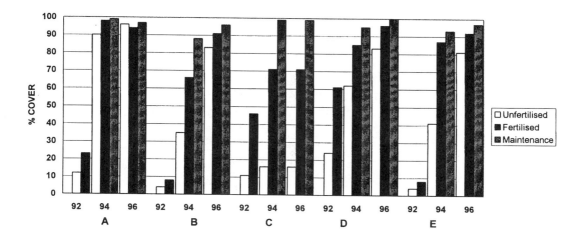

FIGURE 4.4. Effect of seedbed fertilizer (1991) and maintenance fertilizer (1992 and 1993) on mean vegetation cover (1992, 1994, and 1996) on main treatments A (Turves), B (Litter), C (*Calluna* capsules and Nurse grass), D (Native grasses), and E (Control).

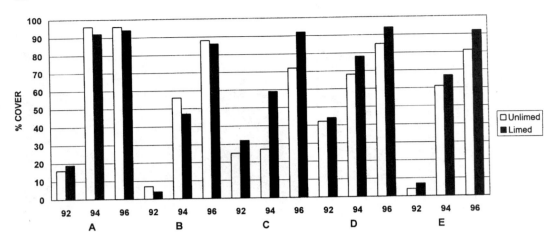

FIGURE 4.5. Effect of pH increase from 3.8 to 4.8 on mean vegetation cover (1992, 1994, and 1996) on main treatments A (Turves), B (Litter), C (*Calluna* capsules and Nurse grass), D (Native grasses) and E (Control).

Table 4.1. Effect of Seedbed (1991) Lime (L) and Fertilizer (F) on Species Diversity and Numbers of Species (1966) on Main Treatments A (Turves), B (Litter), C (*Calluna* Capsules and Nurse Grass), D (Native Grasses) and E (Control).

Lime	−L				+L			
Fertilizer	−F		+F		−F		+F	
A	H G C	13	H G	12	H G J C	19	H G	18
B	H JQ G	15	H JQ G	12	G J,JQ O	19	G J,JQ	18
C	O G	17	H G M	14	G C	17	G	16
D	G J	19	G J	11	G	16	G	14
E	O G J	17	G J	14	O G J	18	G J	16

Note: H – *Calluna*; G – Grass spp.; C – *E. angustifolium*; JQ – *J. squarrosus*; J – *J. effusus*; M – *Spagnum* spp.; O – Bare ground

Species Diversity

Recreation of the original complement of heathland species is a measure of the conservation value of the reinstated vegetation. Number of species are now broadly similar across main treatments (Table 4.1) with an annual increase of approximately one species per treatment between 1991 and 1996. Highest species numbers are on the unfertilized plots with lime applied.

J. effusus has been conspicuous in all main treatments since Year 1 but now appears to have reached equilibrium with no increase in juvenile plants. Established clumps are, however, continuing to expand. Grass species, both sown and indigenous, occupy a major part of the vegetation cover on all treatments. *J. squarrosus* and *Spagnum* spp. are prominent in Treatments A and B. A small progressive increase in *Calluna* establishment is evident on most treatments with successful establishment on Treatments A, B, and C (unlimed). A significant increase in the plant/shoot density of *Calluna* was recorded between 1995 and 1996 in Treatments A, B, and C (unlimed).

DISCUSSION

Opencast mining operations have a major impact on the rural environment and land use. The impact of such operations is both predictable and assessable, particularly when planning permission is subject to the provision of an environmental statement and restoration plan. Provision of such information, at an early stage, allows the mining company to more efficiently manage the site restoration and also provides the land users with an input into the restoration strategy and some prediction on the final end-use of the site.

Provided the land restoration specification is followed and efficiently managed, there is no reason why restored land should not be returned to its precoaling land-use potential. Provided the necessary soil husbandry inputs are carried out during the aftercare period, soil quality should also be reinstated. Major earthworks also provide an opportunity to recontour the site to different slopes and drainage status, allowing enhancement of existing land-use or a change in land-use to meet the requirements of landowners, planners, and the local community.

Specifications for agricultural reinstatement, developed particularly over the last 20 years, provide mining companies with a reasonably successful prescription for land restoration, provided the management input is satisfactory.

Major improvements have also been made in the specifications for woodland planting, successful establishment being largely dependent on compaction alleviation and creation of adequate rooting potential in reinstated soils of high density and low permeability.

This study on heathland reinstatement on the Scottish coalfields indicates that, although it is not possible to replace exactly the previously existing vegetation, reinstatement techniques are available for developing a substantially similar vegetation. Indeed, there is potential to create, on large sites, a mosaic of revegetated surface by using a mix of reinstatement techniques on the same site.

Reinstatement using stripped peat turves (Treatment A) results in the closest match to the original indigenous heathland vegetation and also provides the most rapid vegetation establishment. This method, however, requires a degree of sympathetic handling by the coal contractors during stripping, storage, transport, and replacement of the turves. With this proviso, the turves method offers a practical choice with fairly predictable results.

Use of litter (Treatment B) has, after an initial period of slow vegetation establishment, resulted in a close match to the original vegetation. The litter method involves handling a smaller bulk of material which, however, retains an adequate mix of seeds and propagules for ground cover establishment. Damp conditions are desirable immediately following litter application to ensure good ultimate establishment.

Sowing native grasses (Treatments C and D) is a simple and reliable method of replacing vegetation cover. Choosing an appropriate species mix can ensure a good match to the original vegetation, particularly if a final grazing management is intended. Seedbed fertilizer and possibly lime are essential for successful sown grass establishment. A small application of maintenance fertilizer in Years 2 and 3 is also desirable.

Calluna establishment from capsules broadcast into a nurse grass (Treatment C) has been slow but is now evident after three years on the unlimed subtreatments. Drilling of capsules into the nurse grass may have provided a more rapid establishment.

The control (Treatment E) provides a relatively uncontrolled development of vegetation with the slow ground cover in the initial years, resulting in an unacceptable risk of erosion. Ground cover and species diversity, by Year 6, on the fertilized subtreatments are on a par with other treatments, but the high percentage of bare ground, particularly on the unfertilized plots, results in an invasion of unwanted species such as *Juncus effusus*.

After six years, the heathland restoration treatments have reached a degree of stabilization in terms of ground cover and species diversity, and it is now appropriate to select the most appropriate of these treatments and implement them on a larger-site scale.

Using our combined knowledge of agriculture, forestry, and heathland restoration we can, at the precoaling planning stage, design a reinstatement specification to meet the end-use requirements of landowners, planners, and the local community.

ACKNOWLEDGMENTS

We are grateful to the Scottish Coal Company Ltd for funding the Heathland Restoration Study and the Scottish Office Agriculture, Environment and Fisheries Department for part funding SAC.

REFERENCES

Burse, E.L. and F.T. Dry. *Assessment of Climatic Conditions in Scotland, No. 1.* Macaulay Land Use Research Institute, Aberdeen, 1970.
Department of the Environment. *The Reclamation of Mineral Workings: Revision of Mineral Planning Guidance 7.* HMSO, London, 1996.
Dry, F.T., J.H. Gauld, and J.S. Bell. *Land Capability for Agriculture—Sheet 65.* Macaulay Land Use Research Institute, Aberdeen, 1986.
Hall, J.E. and R. Wolstenholme. *Manual of Good Practice for the Use of Sewage Sludge in Land Reclamation.* WRc Ref: C04014 (Draft), 1997.
Merrilees, D.W., G.E.D. Tiley, and D.C. Gwynne. Restoration of Wet Heathland after Opencast Mining, in *Proceedings of Restoration of Temperate Wetlands.* Sheffield, J. Wiley & Sons, Chichester, 1995, pp. 523–532.
Moffat, A. and J. McNeill. *Reclaiming Disturbed Land for Forestry.* Forestry Commission Bulletin 110. HMSO, London, 1994.
Nature Conservancy Council. *Handbook for Phase 1 Habitat Survey.* Peterborough, 1990.
Procter, R. Surface Mining—Future Concepts, in *Proceedings of Symposium on Surface Mining,* Nottingham, Marylebone Press Ltd., Manchester, 1989, pp. 3–4.
Scottish Office. *The Environmental Assessment (Scotland) Regulations 1988.* HMSO, 1988.

Chapter 5

Land Reclamation After Open-Pit Mineral Extraction in Britain

S.G. McRae

INTRODUCTION

Mineral extraction, particularly in a densely populated country such as Britain, is unpopular with the general population both because of its environmental impact during working (Roy Waller Associates, 1992) and the long-term effects on the land. Many environmental campaigners see mineral extraction as the ultimate form of land degradation (e.g., Adams, 1991). However, well-planned and executed reclamation of mineral workings can do much to overcome local opposition. The government position is that "restoration and aftercare should provide the means to maintain or in some circumstances even enhance the long term quality of land and landscapes taken for mineral extraction, so that there is no net loss of land for use by future generations. At the same time local communities are provided with an asset of equal or added value" (Department of the Environment, 1996a).

Guidance on how to achieve successful reclamation of mineral workings and landfill sites to a variety of afteruses is widely available, much of it sponsored by government agencies (e.g., Dobson and Moffat, 1993; Ministry of Agriculture, Fisheries and Food, 1993; Department of the Environment, 1996a; Environment Agency, 1998) or the mineral extraction industry (e.g., RMC Group, 1987; British Coal Opencast Executive, 1988; Andrews and Kinsman, 1990; Giles, 1992). Other relevant publications include those by Coppin and Bradshaw (1982) and Harris, Birch, and Palmer (1996).

Many of the traditional mining activities in Britain such as for ferrous and nonferrous metals have now largely ceased. The deep-mining of coal, with its attendant pitheaps, has declined sharply in recent years but opencast extraction has been maintained at historic levels, even though very unpopular with the general public. In terms of areal extent, the main mineral extraction activities are for constructional and related materials such as sand and gravel, limestone (for cement and construction aggregates), sandstone and igneous rock (mostly crushed for constructional aggregates and road-making materials), and clay (for bricks and cement manufacture). All these are worked by open-pit methods. Extraction of peat, mainly for horticultural purposes, and china clay, with its associated tip heaps, also affects substantial areas. Table 5.1 (Department of the Environment, 1996b) gives the most recent statistics, but for England only. The relative values for Wales, Scotland, and Northern Ireland would be generally similar.

Table 5.1. Areas Affected in 1994 by Mineral Workings (Excavation Areas and Surface Tips) in England and Areas Restored Between 1988 and 1994.[a]

	Area Being Worked in 1994[b] (ha)	Area Used for Surface Tips in 1994 (ha)	Total Area Affected in 1994 (ha)	Area Restored Between 1988 and 1994 (ha)
Sand and Gravel (Constructional)	17313	742	18055	8626
Limestone, Dolomite, and Chalk	8872	588	9460	1126
Opencast Coal	5853	185	6038	5241
Deep-Mined Coal	—	5190	5190	1692
Clay and Shale	4767	288	5055	852
Peat	4917	1	4918	108
China Clay	1863	2230	4093	184
Sandstone	1970	94	2064	226
Igneous Rock	1173	279	1452	176
Silica and Industrial Sand	1345	27	1372	250
Ironstone	775	38	813	65
Gypsum	244	1	245	137
Slate	168	58	226	17
Onshore Oil and Gas	109	—	109	49
Vein and Other Minerals	780	173	953	383
Totals	50149	9894	60043	19132

[a] Data from Department of the Environment (1996b).
[b] Including areas still open and not yet restored, though working may have ceased.

LEGISLATION

Mineral extraction in Britain is governed by various pieces of legislation designed to reduce the short- and long-term environmental impacts. Those relating to reclamation are summarized in a Mineral Planning Guidance Note issued by the Department of the Environment (1996a). All new sites, including extensions to existing workings, have to obtain formal planning permission. Most applications are accompanied by an environmental statement or similar description of the proposals and their likely environmental impacts. These may include conflicts with other recognized interests such as areas of high landscape value, high quality agricultural land, ancient woodland, nature conservation sites, sites with important archaeological remains, land or buildings of national value, water resources and general amenity (noise, dust, and traffic generation). Matters such as these are embodied in the overall planning policies set out in various development plans. These usually indicate preferred areas for mineral extraction (i.e., areas free from major constraints) or may list specific sites where mineral extraction would normally be permitted because significant constraints are absent or where the impacts could be reduced to an acceptable level.

Planning applications are considered first by professional planners, employed by the relevant Mineral Planning Authority, who have to notify or consult other interested parties. The final decision is taken by elected local government representatives who may or may not follow the recommendations of their planning officers. If permission is granted, this is always subject to Conditions, intended to ensure an orderly working of the site and to minimize the environmental impact (Whitbread and Tunnell, 1993). If permission is refused, the applicant has the right of appeal to the Department of the Environment or Scottish Office, and a Public Inquiry takes place, presided over by an Inspector or Reporter, during which all the issues are debated at length in a quasi-legal forum. Sometimes the Inspector or Reporter has the right of decision, but more commonly the matter is decided by the Secretary of State for the Environment (or Secretary of State for Scotland), who may or may not follow the recommendation of the Inspector or Reporter.

Legislation was introduced in 1981 which obliged mineral operators to undertake a formal period of aftercare, normally five years, following restoration to an agricultural, forestry, or amenity afteruse. During this time the land has to be managed sympathetically and any necessary remedial operations carried out. Aftercare Conditions normally require the preparation of an outline strategy for the aftercare period and detailed programs for particular stages. In British nomenclature, "reclamation" now includes both "restoration," which is essentially the preparation of the final landform and the replacement of soils, and "aftercare," the steps subsequently needed to bring the land to the required final standard for the agreed afteruse.

INTEGRATED SCHEMES OF WORKING AND PROGRESSIVE RECLAMATION

In the past, reclamation has been considered as a matter to be addressed only at the end of the operations, with the result that many sites were either inadequately restored or not restored at all. The main reasons were that materials with which to carry out the reclamation had not been conserved or other working practices had rendered the site impossible to restore to the intended afteruse; e.g., digging below the water table on a site intended for restoration to agriculture. In order to overcome this, all permissions within the last 30 years or so have had to provide detailed reclamation proposals and put forward plans showing how these would be achieved. Such reclamation schemes have become increasingly detailed and are based on a comprehensive preinvestigation of the site. This will include information about the existing landscape, drainage, and anticipated hydrogeological conditions, soils and other overburden, volumes of minerals to be removed, anticipated generation of mine- or quarry-waste materials, and the likely configuration of the excavation void and of any tipheaps which may be retained.

It is now generally accepted that reclamation is part of the overall operational scheme, leading to the production of integrated schemes of working and reclamation. Most open-pit mines and quarries in Britain are worked in a series of phases, so that extraction proceeds in an orderly fashion through the site, with progressive or "rolling" reclamation of worked-out phases taking place concurrently with the opening up of new phases. The schemes are usually prepared by a team of experts, including specialists in landscape design, geology, soil science, hydrology, noise, and any other disciplines required to deal with site-specific issues. Representatives of the mineral company will also be involved to make sure the final proposals are workable.

These schemes indicate the proposed end-use or uses, the final landform and landscape, the soil resources and methods of soil handing, and the seeding, planting, and other operations required to produce the final end-product. They also seek to demonstrate how environmental impacts are taken into account; for example, by appropriate siting of soil and overburden mounds to act as visual or acoustic screens. The most modern schemes provide a series of drawings and plans (examples are given in Figures 5.1, 5.2, and 5.3) showing the progressive nature of the operations, and are accompanied by tables quantifying

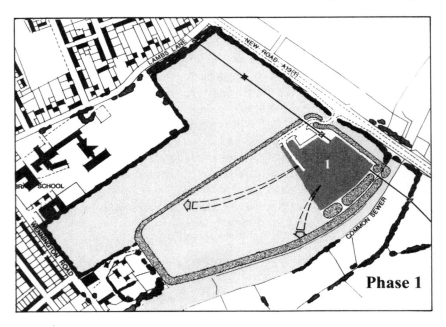

FIGURE 5.1. Integrated scheme of working and reclamation—Phase 1 of an 8-phase sand and gravel extraction site. The soils have been stripped from phase 1 and are placed in storage heaps adjacent to the phase and, as shown by the arrows, as screening mounds round what are to be future working phases.

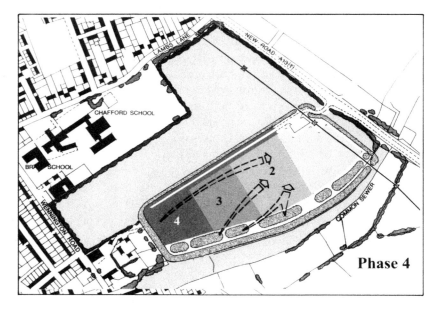

FIGURE 5.2. Integrated scheme of working and reclamation—Phase 4 of an 8-phase sand and gravel extraction site. Phase 1 has already been restored (with soils from Phase 3) and is in the aftercare stage, while Phase 2 is undergoing restoration with soils directly moved from Phase 4 (which will then be excavated) plus additional soils from store since Phase 2 is larger in area than Phase 4. Phase 3 is being infilled with imported wastes in preparation for restoration, while the rest of the site remains in agricultural use while awaiting extraction.

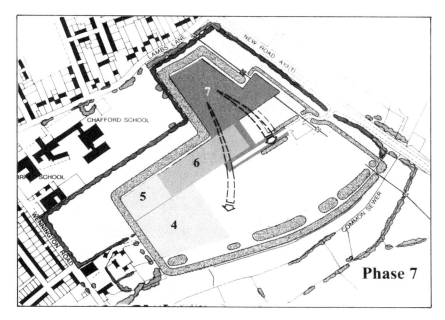

FIGURE 5.3. Integrated Scheme of Working and Reclamation—Phase 7 of an 8-phase sand and gravel extraction site. Phases 1, 2, and 3 are restored, the last with soils directly moved from Phase 7, with some surplus soils going into temporary store. They are now in the aftercare stage. Phases 4 and 5 are at various stages of soil spreading, Phase 6 is being infilled and Phase 7 is ready for extraction.

the anticipated volumes of soils and other materials which will be handled at each stage of the operations.

There is a preference for schemes in which most soils are directly moved from that part of the site being stripped to that part being restored. In this way there is less chance for soils to become lost, and interim storage is avoided. It is now generally accepted that soil deterioration during storage is relatively unimportant if the storage heaps are well constructed and storage is for only a few years. However, soils are often lost during storage by inappropriate siting of storage heaps and poor management; for example, loss of material into the excavation void. If storage is required; for example, for soils stripped from early phases when there is no place being restored ready to receive them, then additional space is required for this purpose. Unfortunately, the provision made for this is often insufficient.

FINAL TOPOGRAPHY AND LANDFILLING

The design of the final topography for a reclaimed mineral extraction site has to take into account both site drainage requirements and broader aesthetic issues of compatibility with surrounding landscapes. Particular attention is paid to gradients. On the one hand, minimum gradients; e.g., 1 in 60 over landfill sites, are often specified to assist in site drainage. On the other hand, slopes that are too steep can make the reclaimed land more susceptible to erosion and cause cultivation difficulties which may preclude an agricultural afteruse. In the case of opencast coal, there is usually enough material to allow for restoration back to approximately original levels. Sand and gravel, clay, chalk, or limestone extraction sites can be reclaimed by respreading soils on the excavation floor in a so-called "low-level" reclamation. With low-level reclamation it is vital to ensure the site will drain satisfactorily. This is easiest where the quarry floor is permeable and well above the local water table. Elsewhere it requires careful grading of the quarry floor to suitable gravity outfalls. There is increasing interest in permanent pump-drainage of sites that would otherwise have to be left as lakes (M.J. Carter Associates, 1989). Alternatively, the void can be infilled to approximately original levels with imported wastes including pulverized fuel ash or municipal and industrial wastes.

Landfilling with municipal and other putrescible wastes raises other environmental issues and often results in much more local objection than the actual mineral extraction, especially if the mineral void has remained inactive for several years previously. The main concerns are with the proper monitoring and control of leachates and landfill gases (Department of the Environment, 1991; 1994; 1995; 1996a; Environment Agency, 1998). Most new landfills in Britain are operated on the containment principle, with natural or synthetic liners and capping layers to control ingress and egress of water, and are fitted with systems to collect and treat leachates and gas. The latter is usually actively pumped from a series of gas wells at intervals over the site and either flared or used as an energy source; e.g., for electricity generation. The design and operation of gas and leachate control and monitoring systems and their interaction with restoration and aftercare is emerging as a matter of considerable concern (Environment Agency, 1998). From a reclamation viewpoint, difficulties are also encountered due to settlement of the fill and, on many of the older workings subsequently proposed as landfills, a lack of suitable restoration materials.

RECLAMATION OPTIONS—AGRICULTURE

Agriculture is the most common afteruse for restored mineral workings in Britain, other than those which end up as lakes, and most landfill sites are also restored to an agricultural afteruse. For example, the 1994 statistics show that 58.5% of land reclaimed in England between 1988 and 1994 was to an agricultural afteruse (Department of the Environment, 1996b). The main driving force for this has been government policy, enforced by the Ministry of Agriculture, Fisheries, and Food (MAFF), which is to conserve the nation's stock of agricultural land, especially the "best and most versatile" category (Department of the Environment, 1996a). Most land taken for mineral extraction will have previously been in agricultural use and so there is also a certain logic to reclaim land to this widely acceptable and noncontroversial afteruse. Well-restored agricultural land is easily managed and may generate an income for the mineral or landfill operator. Also, land restored to an agricultural specification can be used for other purposes if necessary. On the other hand, the costs of agricultural reclamation are higher than for other reclamation options, and recent changes in government policy, reflecting current agricultural surpluses, mean that there is less pressure to restore land directly to agriculture. Thus, in cases where agricultural reclamation is difficult; e.g., on old workings and landfill sites where there is a lack of soils and other restoration materials, reclamation to an alternative afteruse may be preferable.

There is a considerable body of literature on the most appropriate techniques for agricultural reclamation (e.g., Ramsay, 1986; Corker, 1987; British Coal Opencast Executive, 1988; McRae, 1989; Rimmer, 1991; RPS Clouston and Wye College, 1996a; 1996b). The basic principles are (a) to ensure adequate site drainage by correct attention to the final topography as discussed above, (b) to reestablish a soil profile similar to the preworking condition or an acceptable alternative, and (c) to cause minimal damage to the soil resources during the operations.

Soil resource assessments are now a regular part of preworking investigations in Britain and the results are (or should be) taken into account in the design of the reclamation scheme. There should be separate stripping and subsequent handling of all the existing topsoil; i.e., the uppermost A horizon or cultivated layer usually about 250–300 mm thick, and subsoil or B horizon, which in Britain is normally taken as extending to about 1000 mm to 1500 mm from the surface. Material below this which is not part of the rooting zone of most vegetation is classed as residual overburden, but in some cases may be used as a soil substitute. These materials should be replaced in the correct sequence and thicknesses on the land being restored, with due consideration of how to deal with situations where the original soil profile was shallow, or perhaps had horizons with unfavorable textures or stoniness which could be replaced by other materials. Where a site contains contrasting soil types, then provisions have to be made for stripping and replacing them separately.

If the original soil resources are conserved there is usually little concern over any possible deterioration in chemical fertility, and any nutrient deficiencies can be easily rectified during the aftercare period. The reclamation of metalliferous mines and associated wastes has, of course, to take into account the likely presence of phytotoxic pollutants (Environmental Consultancy University of Sheffield and Richards, Moorehead and Laing Ltd, 1994). In most circumstances, however, the alteration in soil physical properties due to mechanical handling is of much more concern, notably loss of structure, increase in bulk density, and reduction in porosity and permeability. These changes, collectively referred to as compaction, are caused by the passage of heavy machinery such as earthscrapers or laden dump trucks over the soils, particularly during respreading operations.

The restored profile is normally constructed by respreading soils in a series of thin layers and compaction is caused by machinery running over the previously laid soil layers and over the soil which has just been respread (Figure 5.4). The restored soil profile thus has compacted layers at various depths. These impede drainage and the result is usually severe waterlogging during the first winter after soil replacement and thereafter. The compaction also restricts root growth, thus causing increased droughtiness during the summer. Various solutions for reducing compaction due to earthmoving machinery have been suggested. These include operating procedures which minimize trafficking, a general restriction on soil handling to dry conditions, and various soil loosening operations carried out during or after respreading.

Attempts to reduce trafficking over soils have included the so-called "single row" and "dump and turn" techniques (Bacon and Humphries, 1988). The first restricts the soil-handling traffic to a set route, along which the compaction is restricted to the vehicle wheelings. It is thought that most compaction occurs during the first passage and subsequent trafficking over the same soil causes little further compaction while the rest of the soil remains relatively free of compaction. The second requires machinery to run on the lowest available layer, turning onto the layer being spread only when actually depositing the soil and then immediately turning off. Neither is completely successful in reducing overall compaction.

It is known that dry soils are much more resistant to compaction than moist or wet soils. Thus soil handling is generally restricted to the dry period during the British summer and is prohibited during or immediately after heavy rainfall. More precise controls can be exercised by requiring the soils to be "dry and friable" before they are moved or, increasingly, to when they are drier than the lower plastic limit for that soil type.

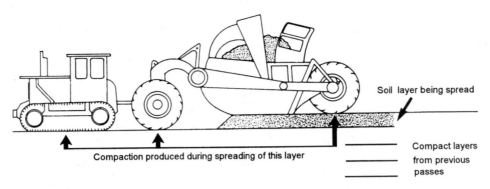

FIGURE 5.4. Compaction caused by the use of an earthscraper to replace soils. Running of any wheeled vehicles over soils during the respreading process will cause similar damage.

Even if the soils are moved when acceptably dry, some compaction will still be caused if trafficked. The compaction must be removed by a ripping or subsoiling operation in which tined implements are drawn through the soil to shatter, lift, and loosen compact layers (see Figures 5.5 and 5.6). This can be done sequentially during the respreading operations when the compact layers are still accessible to subsoiling equipment, most of which has a maximum effective operating depth of about 450 mm. Alternatively, a loosening operation can be attempted at the end of soil spreading. In this case the deeper compaction is not alleviated unless very large equipment is used, which causes additional compaction itself and/or an unacceptable amount of soil mixing.

There is increasing interest in Britain, especially on higher quality agricultural land, in a technique for completely avoiding compaction by entirely eliminating heavy wheeled traffic over the soils. This is the so-called "loose tipping" or "dump truck and back-acter" method. The operation is carried out in narrow bands, whose width of about 3 to 4 meters is dictated by the reach of the back-acting excavator (back-acter) being used. During stripping the excavator stands on the soil and loads into dump trucks drawn up alongside on a strip from which all the soils have already been removed. At the respreading area (Figure 5.7) the subsoil is dumped in small heaps and is leveled to the required thickness by a light, tracked bulldozer. The topsoil is then delivered to the edge of the strip and is dropped into position by a back-acter standing alongside, but not on, the replaced sub-soil, as shown in Figure 5.7. Alternatively, the topsoil is dumped partially on the subsoil and is graded out over it to the required thickness by bulldozer. Advocates of this method, such as Bransden (1991), emphasize the absence of compaction so that there is no need for any subsoiling operations, which are often not entirely successful. They believe that this outweighs the disadvantages of the slower, more complicated and more expensive procedure.

The main aim of the aftercare period following soil replacement in an agricultural rec-lamation is to promote the reestablishment of soil structure and hence overcome the poor soil physical conditions usually found on restored land. The aftercare legislation suggests a general statement of the aftercare strategy, followed by annual meetings at which the pre-vious years' cropping is reviewed, the proposed cropping for the following year is agreed, and any remedial works to be undertaken are discussed. Installation of under-drainage, virtually essential on all land restored to agriculture in Britain, is usually also undertaken as part of the aftercare. During the aftercare period it is expected that steps will be taken to

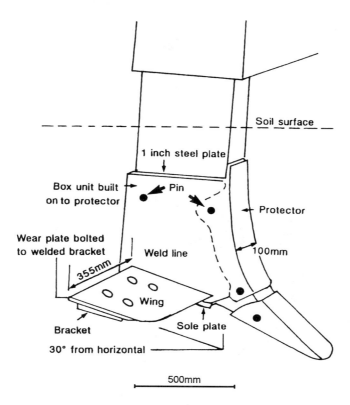

FIGURE 5.5. A subsoiling or ripping tine with "wings," used to relieve compact layers in respread soils.

start to recreate a typical British agricultural landscape on the reclaimed land, with a patch-work of fields, woodland, ponds, hedgerows, and tracks.

RECLAMATION OPTIONS—WOODLAND

Reclamation of mineral workings to forestry has been described by Moffat and McNeill (1994), summarizing a considerable body of work carried out over many years by the Forestry Commission and others. The techniques and concerns are similar to those already described for agricultural reclamation with particular attention paid to landform, site drainage, and the avoidance or alleviation of soil compaction. As with agricultural reclamation, there is increasing interest in the loose tipping of soils.

Reclamation to woodland in Britain is the norm where the land was previously under trees or where conditions after working are not conducive to agriculture; for example, where slopes are too steep or where vegetation has to be established directly on subsoil or overburden materials. In the latter case, the fertility can be built up by use of organic amendments such as sewage sludge (e.g., Bending, 1995). Recognition that Britain has a lower proportion of woodland than most other European counties, however, has led to encouragement of tree planting, particularly in lowland areas. The objective is no longer solely commercial forestry production but now includes landscape enhancement, nature conservation and amenity. Of particular interest is a lifting of the former "ban" on tree planting over landfills. In the past it was considered that trees and landfills were incompat-

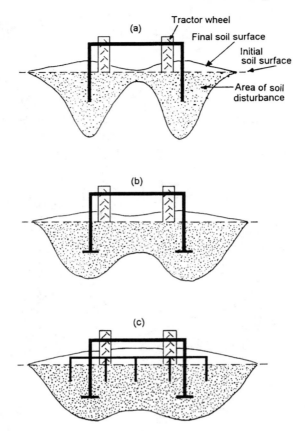

FIGURE 5.6. The action of a subsoiler or ripper, as shown by a section through the soil after the passage of a two-tined machine. (a) Conventional subsoiler. Note the cone-shaped zone of loosening and the lifting of the soil surface; (b) Winged subsoiler, which can disturb more of the soil than a conventional subsoiler with little increase in power required to pull it; (c) Winged subsoiler with leading shallower tines to preloosen the surface soil before the passage of the main tines, producing the maximum amount of loosening.

ible. Poor soil conditions and presence of landfill gases were not conducive to tree growth and there were fears that tree roots could damage the capping layers of containment landfills. Work by Dobson and Moffat (1993) has shown that, with proper landfill gas control on containment sites, adequate soil depths and proper soil handling, trees can grow successfully over landfills and that the risks to the capping layer are minimal. The result has been the acceptance in principle of woodland as a potential afteruse for British landfill sites (Environment Agency, 1998).

In all cases where trees are planted on or close to mineral workings or landfills, both for screening purposes and as part of the final vegetation, the correct species must be chosen for the purpose and conditions, and planting and maintenance should be properly carried out. There are, unfortunately, still many examples of the failure of trees on restored land in Britain, often due to poor planting techniques (in some cases the trees may actually be dead before planting) and lack of maintenance, with the young trees either overwhelmed by more vigorous vegetation or destroyed by rabbits or deer.

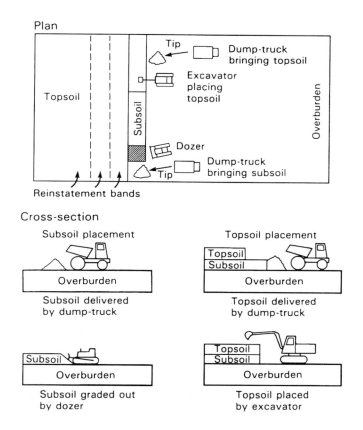

FIGURE 5.7. Soil replacement by loose-tipping in bands (dump truck and back-acter method). Subsoil is delivered by dump truck running only over overburden and graded out by bulldozer. Then topsoil is delivered by dump truck, also running only on overburden and lifted into place by back-acting excavator (back-acter) standing on the overburden.

RECLAMATION OPTIONS—NATURE CONSERVATION

Habitats with considerable nature conservation interest can arise by the abandonment of an open-pit site or associated waste heaps. To their credit, most mineral operators who realize that their worked-out site has a nature conservation value will do their best to protect or enhance it, and several are now the proud custodians of sites officially designated Sites of Special Scientific Interest because of the flora and fauna which has colonized them. In recent years there has been considerable interest in the deliberate creation of nature conservation sites to a preconceived plan. Most attention has been paid to wet pits, with the objective of providing suitable wetland habitats for aquatic birds and wildfowl (Andrews and Kinsman, 1990; Giles, 1992). Figure 5.8 shows the main features of a former sand and gravel site which has been restored specifically as a wildfowl reserve. Techniques include the shaping of banks and shorelines to provide suitable conditions for the species that it is intended to encourage. Careful attention is paid to the relative ground and water levels, with the creation of marshland and shallow water (less than 1.5 meters deep) seen as particularly beneficial. The mineral excavation and siting of waste tips, etc. which are to become part of the final landform has to be planned with the final landform in mind, since mistakes can be

1.	R. Great Ouse	17.	Canal overspill weir
2.	Pipe Bridge	18.	Main Lake outlet sluice
3.	Inlet sluice	19.	Main Lake
4.	Inlet sluice	20.	Marsh area
5.	Stanton Low Lake	21.	Near Hide
6.	Wet meadow	22.	Hedge to screen breeding sanctuary
7.	Mill Weir	23.	Ossier coppice
8.	Goose lawn and loafing spit	24.	Screened path to the hide
9.	Tern island	25.	Shingle loafing spit
10.	Outlet sluice	26.	The Front Ponds
11.	Security canal	27.	Reed bed
12.	Nestling meadow	28.	Car park
13.	Wader scrapes	29.	ARC Wildfowl Centre
14.	Far Hide	30.	Approach road
15.	The Hollow Island	31.	Connection to river
16.	Grazing pasture	32.	Experimental ponds
		33.	St Peters Lake

FIGURE 5.8. The main features of the Great Linford wildfowl reserve. The Main Lake (19) is used principally as a winter refuge and brood-rearing area, while St. Peters Lake (33) and Stanton Low Lake (5) are important nesting areas. Reproduced from *Wildlife after Gravel* with the kind permission of Game Conservancy Limited (Giles, 1992).

prohibitively expensive to correct subsequently. The choice of vegetation and its subsequent management are also seen as critical to ensure a self-sustaining ecosystem and the development of a food chain for the higher species. Wheeler and Shaw (1995) deal specifically with the restoration of peatlands, especially those affected by peat extraction.

Habitat creation techniques for grassland, woodland, heathland, and other vegetation assemblages are dealt with by, among others, Andrews and Kinsman (1990), Land Use Consultants (1992a; 1992b) and Land Use Consultants and Wardell Armstrong (1996). Again there is emphasis on creating the correct landform during the excavation stage, but also on the use of soils and other materials to provide suitable habitat conditions and the correct choice of species to sow or plant. For some vegetation types the objective is low fertility, the opposite of the objective for agricultural reclamation.

Particular attention has been paid to the treatment of the margins of worked-out pits; for example, a technique known as reclamation blasting, whereby attempts are made to replicate natural limestone cliffs and scree slopes by controlled blasting of limestone quarry faces (Gunn et al., 1992). Safety considerations need to be balanced with, in some cases, a need to maintain access to the former working faces; for example, for climbing or because they are of geological interest (Geoffrey Walton Practice, 1988; Land Use Consultants, 1992a; 1992b).

RECLAMATION OPTIONS—AMENITY AND RECREATION

The British public likes open spaces in which to walk, ride, or exercise their pets, and former mineral workings can cater very well to these needs, with the benefit to the mineral operator that a relatively low-specification and low-budget reclamation can suffice. In some situations, such as deep hard-rock quarries and wet pits, an amenity afteruse can be the only sensible choice, often with the additional objective of providing some form of nature conservation interest. Provision of facilities for more formal recreation such as golf and sports playing fields can be much more expensive and, at least in part, an agricultural-type specification is called for. Descriptions of the wide range of possible amenity uses for old mineral workings and other degraded land in Britain are discussed by Land Use Consultants (1992a; 1992b), with case studies, fact sheets, and specifications for various amenity afteruses.

Hundreds of wet pits, usually left by sand and gravel extraction in river valleys, are stocked with fish and used for angling (the most popular participatory sport in Britain) or fish-farming. These and nature conservation uses need to be segregated from areas for those who wish to indulge in water sports such as sailing, windsurfing, power-boating or water-skiing. Specially designed facilities need to be provided for these for which separate planning permission is required. Good examples are the National Water Sports Centre at Holme Pierrepont, near Nottingham and the Cotswold Water Park, a series of multipurpose water features; both based on former gravel workings. One of Britain's major theme parks, Thorpe Park on the southwest edge of London, has a similar origin.

RECLAMATION OPTIONS—BUILT DEVELOPMENTS

Obtaining permission to develop land for housing, commercial, or industrial development in Britain enhances its value very substantially. Many mineral operators would hope, perhaps secretly, that their worked-out sites could be granted permission for such uses.

Land use planning legislation, however, does not encourage what might be perceived as a "backdoor" method of establishing built developments. On the other hand, if land zoned for built development is known to contain minerals, their prior extraction is encouraged so as to avoid sterilization of the resource.

Assuming that appropriate planning permission has been obtained, the establishment of the actual buildings on the dry floors of worked-out pits is relatively simple, and the work usually consists only of leveling-out uneven topography and possibly treatment of potentially unstable side slopes. Good examples are the large housing development called the Chafford Hundreds in south Essex and out-of-town shopping centers at Lakeside (also in Essex) and Bluewater (in Kent). All of these are on the floor of former chalk pits.

There are many examples of successful smaller-scale developments on backfilled opencast coal workings or sites filled with inert fill; for example, many of the light industrial estates in west London. Built developments over or even close to sites filled with putrescible fill are much more problematical due to settlement and landfill gas emissions (Card, 1992). The engineering solutions are suitable only for prestige developments such as at Stockley Park, near Heathrow Airport (Ede, 1990).

TREATMENT OF TIPHEAPS

The main tipheaps in Britain come from deep-mined coal and from the open-pit extraction of china clay (Table 5.1). The techniques for stabilizing and revegetating them are well documented (e.g., Williamson et al., 1982; Geoffrey Walton Practice, 1991; Wardell Armstrong, 1993; Richards, Moorehead and Laing, 1995; 1996). Opencast coal and some other mineral extraction sites can have relatively long-term spoil heaps but these are almost always used eventually to backfill the associated void.

CONCLUSIONS

There is now little excuse for failing to carry out good restoration of mineral workings in Britain. Abundant advice is available and plenty of experience has been built up. However, as noted by McRae (1990), the chosen afteruse often lacks imagination and there is a need to realize that mineral extraction "wipes the slate clean" as far as original land use is concerned. It provides the opportunity for a creative and imaginative afteruse which can be of benefit to the community and of which the mineral extraction industry can be proud.

REFERENCES

Adams, J.G.U. *Determined to Dig.* Council for the Protection of Rural England, London, 1991, p. 28.

Andrews, J. and D. Kinsman. *Gravel Pit Restoration for Wildlife.* Royal Society for the Protection of Birds, Sandy, Bedfordshire, 1990, p. 184.

Bacon, A.R. and R.N. Humphries. Loose Soil Profiles Using Scrapers and Deep Ripping. *Mine Quarry Environ.*, 2, pp. 13-17, 1988.

Bending, N. Sewage Sludge and Land Reclamation: A Marriage Made in Heaven or Hell. *Environ. Manage. J.*, 3(3), pp. 27–30, 1995.

Bransden, B. Soil Protection as a Component of Gravel Raising. *Soil Use Manage.*, 7, pp. 139–145, 1991.

British Coal Opencast Executive. *Ten Years of Research—What Next? A Seminar on Land Restoration Investigation and Techniques.* British Coal, Mansfield, Nottinghamshire and University of Newcastle-Upon-Tyne, 1988, p. 130.

Card, G.B. Development on and Adjacent to Landfill. *J. Instit. Water Environ. Manage.*, 6, pp. 362–371, 1992

Coppin, N.J. and A.D. Bradshaw. *Quarry Reclamation.* Mining Journal Books, London, 1982, p. 112.

Corker, S.P. Bulk Soil Handling Techniques. *Mine Quarry Environ.*, 1, Part 1, pp. 15–17, 1987.

Department of the Environment. *Waste Management Paper 27. Landfill Gas*, 2nd ed. HMSO, London, 1991, p. 82.

Department of the Environment. *Waste Management Paper 26A. Landfill Completion.* HMSO, London, 1994, p. 289.

Department of the Environment. *Waste Management Paper 26B. Landfill Design, Construction and Operational Practice.* HMSO, London, 1995, p. 289.

Department of the Environment. *Minerals Planning Guidance Note MPG7. The Reclamation of Mineral Workings.* New Edition. HMSO, London, 1996a, p. 53.

Department of the Environment. *Survey of Land for Mineral Workings in England, 1994. Volume 1. Report on Survey Results.* HMSO, London, 1996b, p. 12.

Dobson, M.C. and A.J. Moffat. *The Potential for Woodland Establishment on Landfill Sites.* Report for the Department of the Environment. HMSO, London, 1993, p. 88.

Ede, B. The Stockley Park Project. *Landscape Design.* February 1990, pp. 42-47.

Environment Agency. *Waste Management Paper 26E. Landfill Restoration and Post Closure Management*, Stationery Office, London, 1998, in press.

Environmental Consultancy University of Sheffield and Richards, Moorehead and Laing Ltd. *The Reclamation and Management of Metalliferous Mining Sites.* Report for the Department of the Environment. HMSO, London, 1994, p. 178.

Geoffrey Walton Practice. *Handbook on the Hydrogeology and Stability of Excavated Slopes in Quarries.* Report for the Department of the Environment, HMSO, London, 1988, p. 42.

Geoffrey Walton Practice. *Handbook on the Design of Tips and Related Structures.* Report for the Department of the Environment. HMSO, London, 1991, p. 132.

Giles, N. *Wildlife after Gravel.* Game Conservancy Ltd., Fordingbridge, Hampshire, 1992, p. 135.

Gunn, J., D. Bailey, and P. Gagen. *Landform Replication as a Technique for Reclamation of Limestone Quarries.* Progress Report for the Department of the Environment. HMSO, London, 1992, p. 38.

Harris, J.A., P. Birch, and J.P. Palmer. *Land Restoration and Reclamation: Principles and Practice.* Addison Wesley Longman, Harlow, Essex, 1996, p. 230.

Land Use Consultants. *Amenity Reclamation of Mineral Workings: Main Report* for the Department of the Environment. HMSO, London, 1992a, p. 238.

Land Use Consultants. *The Use of Land for Amenity Purposes: A Summary of Requirements.* Report for the Department of the Environment, HMSO, London, 1992b, p. 127.

Land Use Consultants and Wardell Armstrong. *The Reclamation of Disturbed Land for Nature Conservation*, Report for the Department of the Environment. HMSO, London, 1996, p. 173.

McRae, S.G. The Restoration of Mineral Workings in Britain—A Review. *Soil Use Manage.*, 5, pp. 135–142, 1989.

McRae, S.G. Quarry Reclamation and the Environment: Creating an Imaginative Afteruse. *Quarry Manage.*, May, pp. 21–23, 1990.

Ministry of Agriculture, Fisheries and Food. *Code of Good Agricultural Practice for the Protection of Soil.* Ministry of Agriculture, Fisheries and Food, London, 1993, p. 55.

M.J. Carter Associates. *Low Level Restoration of Sand and Gravel Workings to Agriculture with Permanent Pumping.* Report for the Department of the Environment. HMSO, London, 1989, p. 89.

Moffat, A.J. and J.D. McNeill. *Reclaiming Disturbed Land for Forestry.* Forestry Commission Bulletin 110. HMSO, London, 1994, p. 103.

Ramsay, W.J.H. Bulk Soil Handling for Quarry Restoration. *Soil Use Manage.* 2, pp. 30-39, 1986.

Richards, Moorehead, and Laing. *Slate Waste Tips and Workings in Britain.* Report for the Department of the Environment. HMSO, London, 1995, p. 144.

Richards, Moorehead, and Laing. *Restoration and Revegetation of Colliery Spoil Tips and Lagoons.* Report for the Department of the Environment. HMSO, London, 1996, p. 161.

Rimmer, D.L. Soil Storage and Handling, in *Soils in the Urban Environment*, Bullock, P. and P.J. Gregory, Eds., Blackwell Scientific Publications, Oxford, 1991, pp. 76–86.

RMC Group. *A Practical Guide to Restoration*, 2nd ed., RMC Group, Feltham, Middlesex, 1987, p. 83.

Roy Waller Associates Ltd. *Environmental Effects of Surface Mineral Workings*. Report for the Department of the Environment. HMSO, London, 1992, p. 176.

RPS Clouston and Wye College. *The Reclamation of Mineral Workings to Agriculture*. Report for the Department of the Environment. HMSO, London, 1996a, p. 148.

RPS Clouston and Wye College. *Guidance on Good Practice for the Reclamation of Mineral Workings to Agriculture*. Report for the Department of the Environment. HMSO, London, 1996b, p. 87.

Wardell Armstrong. 1993. *Landscaping and Revegetation of China Clay Wastes*. Report for the Department of the Environment. HMSO, London, p. 198.

Wheeler, B.D. and S.C. Shaw. *Restoration of Damaged Peatlands*. Report for the Department of the Environment. HMSO, London, 1995, p. 211.

Whitbread, M. and C. Tunnell. *Review of the Effectiveness of Restoration Conditions for Mineral Workings and the Need for Bonds*. Report for the Department of the Environment. HMSO, London, 1993, p. 108.

Williamson, N.A., M.S. Johnson, and A.D. Bradshaw. *Mine Wastes Reclamation*. Mining Journal Books, London, 1982, p. 103.

Chapter 6

Lignite Surface Mining and Reclamation

L.R. Hossner

INTRODUCTION

Most of the economically important lignite deposits in the United States are located in the Gulf Coast and the Northern Great Plains Regions (Figure 6.1). The deposits are composed of relatively low-grade lignite (4,000–6,000 BTU) and are present in relatively thin layers (1–10 m thick). Multiple lignite layers that can be economically mined are commonly found in the geological column.

Geology

Near surface (0–60 m) Tertiary lignite recoverable by surface mining is located in the upland and alluvial floodplain systems in Texas, Louisiana, Arkansas, Alabama, and Mississippi. Lignite occurs principally in the Wilcox Group of Eocene age and to a lesser extent in the overlying sediments of the Claiborne and Jackson Groups, also of Eocene age, and underlying Midway Group of Paleocene age (Haley, 1960; Kaiser, 1974; Roland et al., 1976; Wielchowsky et al., 1977). Lignite occurs as a component facies of ancient fluvial, deltaic, and lagoonal rocks (Fisher and McGowen, 1967). The overburden consists of cross-bedded sands, silts, and clays from the Jackson, Claiborne, and Wilcox Groups. The uppermost strata of the Midway Group in portions of Mississippi and Alabama also contain lignite (Luppens, 1978). The clay mineralogy of the overburden sediments is dominated by expandable 2:1 minerals. Other layer silicate clays include kaolinite, mica, and chlorite, in that order (Senkayi and Dixon, 1988).

Most of the lignite reserves in the Northern Great Plains are located in the Tongue River and Sentinel Butte Formations in the Fort Union Group of Paleocene Age, and part of the Tertiary System (Sandoval et al., 1973). These formations consist of alternating layers of lignite, soft shales, and some sandstone. The Tongue River Formation is about 304 m thick in southwestern North Dakota, and has lignite seams 1.5 to 12 m in thickness. The Northern Great Plain deposits and overburden sediments were laid down along margins of extensive Cretaceous freshwater depositional basins and swamps (Royse, 1970; 1972). The overburden sediments throughout this region consist of mainly interbedded silts and clays that occur in beds ranging from a few millimeters to tens of meters thick. The overburden can be moderately saline, and highly sodic (Sandoval et al., 1973). Layer silicate mineralogy of the overburden sediments in North Dakota are mostly smectite and illite. Smaller amounts of vermiculite, kaolinite and chlorite, in that order, are also present (Klages and Hopper, 1982).

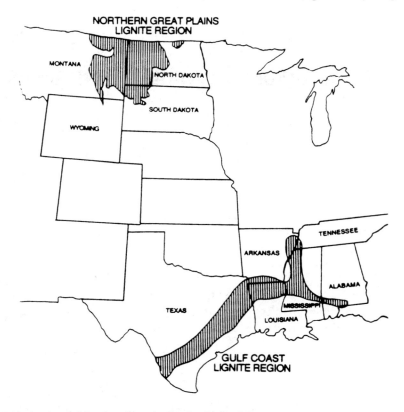

FIGURE 6.1. Distribution of lignite deposits in the United States.

Climate

The climate of the Gulf Coast Region is conducive to rapid reclamation. Average annual rainfall in most mining areas will exceed 90 cm. In southwestern Texas where there is some mining activity, the total annual rainfall may be as low as 50 cm. Mild winters, high summer temperatures, and relatively high humidity promote abundant plant growth (White, 1978; Whiteley and Hossner, 1980; O'Shay and Hossner, 1985). Temperatures are generally hot in the summer and mild in the winter. A major consideration is rainfall intensity and distribution. Intense rainfall followed by extended periods of drought are common in the region. The average frost-free growing season is at least 210 to 270 days across the entire region with 270 to 300 days available in southern Texas.

The Northern Great Plains region is characterized by a semiarid climate with annual rainfall ranging between 25 and 66 cm. In Western North Dakota, rainfall is the dominant factor controlling crop yields. Temperatures range between 38°C in the summer and −23°C in the winter.

Vegetation

Natural vegetation in the important Gulf Coast Lignite Region range from mixed ash-pine forest to Ash-Savannah. Current land uses include native pastures, improved pasture, wildlife habitat and forestry in the uplands, and improved pasture, forestry and row crops in the alluvial areas (Gould, 1975; Lyle, 1985). Vegetation in the Northern Great Plains is largely pastureland and rangeland with gently sloping topography and grass and shrub cover.

Mining Techniques

Area mining is the predominant method used in the lignite mining areas of the Gulf Coast and the Northern Great Plains where the coal lies in relatively thin seams under flat or gently rolling land surfaces (Figure 6.2). Electrically powered draglines constitute the primary earthmoving equipment. Area mining begins with surface preparation. After removal and storage of topsoil, where required, an initial cut is made down to the coal, and the spoil is cast into a stockpile to one side away from the planned mine. This cut, referred to as the box cut, is extended across the property to the limits of the planned mine and the coal is removed. Additional cuts are then made parallel to the first, with the spoil from each succeeding cut being placed in the previous mined-out pit. Each cut may be several hundred or thousands of meters in length. The final cut may be more than a kilometer from the initial cut. Reclamation requirements include refilling the final cut, smoothing the surface in a timely manner to conform to the surrounding topography, replacement of the original topsoil or a substitute topsoil, and revegetation. Bulldozers are commonly used for regrading of spoil piles and topsoils.

RECLAMATION

A variety of methods is used to replace topsoil or to place selected overburden strata on the surface for revegetation. Topsoil can be removed and stored or the topsoil can be moved directly by trucks or scrapers to a spoil area that has been smoothed, contoured, and prepared for topsoil replacement. Conveyor belts are commonly used to transport topsoil around or across the pit for placement in the area to be reclaimed. In Texas, draglines are the most common equipment used to remove overburden from the lignite seam. Where unconsolidated overburden is present, bucket wheel cross-pit spreaders (with a 205-m cross pit conveyor belt), remove and transport topsoil and selected overburden directly across the pit to be deposited on top of unsuitable overburden that had previously been removed by draglines. This unique method of handling overburden material eliminates multiple handling and compaction common to other types of equipment. The bucket wheel cross-pit spreader can also be used to form a bench up to 18 m below the original surface that is suitable for dragline operations to remove overburden and extract lignite from deeper deposits.

Overburden Characteristics

Physical and chemical properties of overburden can vary widely with depth both within and between mine areas (Doll et al., 1984). Particle size distribution of overburden strata in the Gulf Coast Region ranges from sand to clay. Clay mineralogy of the overburden is primarily smectitic with lesser amounts of kaolinite, vermiculite, mica, and chlorite (Dixon et al., 1980). The overburden consists of an oxidized zone of varying thickness (approximately 5 to 20 m depending on texture) which overlies unoxidized overburden (Dixon et al., 1982). The physical and chemical properties of the oxidized and reduced zones are quite different (Figure 6.3). Nutrient levels, with the exception of nitrogen, and perhaps phosphorus, are adequate. The pH of the unoxidized zone is higher than the oxidized zone

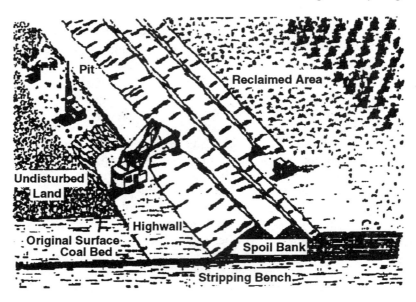

FIGURE 6.2. Area strip mining operation with concurrent reclamation.

(pH 4.5 to 6.0) and may range from 6.0 to 7.5 if iron sulfides are not present and allowed to oxidize. Iron sulfides (Figure 6.4) are commonly found as pyrite and marcasite in the reduced overburden They are largely associated with the lignite seams and appear to be concentrated at the margins of the lignite (Figure 6.5). The greatest limitation to reclamation of randomly mixed overburden in the Gulf Coast Region is the presence of FeS_2 on the leveled surface and the subsequent formation of acid spoils due to the oxidation of pyrite and marcasite

$$FeS_2 + 7/_2H_2O + 15/_4O_2 = Fe(OH)_3 + 2H_2SO_4$$

Carbonates occur infrequently in materials that overlie Wilcox lignite of East Texas. Calcite has been identified in a few layers. Siderite ($FeCO_3$) is important in local concentrations in the unoxidized overburden and appears to be widely disseminated in the overburden as grain-size particles (Frisbee and Hossner, 1995). Complete mixing of the overburden column results in a minesoil of loamy texture, smectitic mineralogy, intermediate pH (5.0 to 7.5), low salt content (conductivity < 1 mmho cm^{-1}), and an abundance of unweathered minerals.

The Fort Union Group overburden in the Northern Great Plains consists of alternating layers of lignite, soft shales, silts, and some sands (Sandoval et al., 1973). Smectite is the dominant clay with appreciable amounts of illite and vermiculite and much lower amounts of kaolinite and chlorite. Overburden materials tend to be nonsaline to moderately saline (EC below 8) but are often sodic (SAR above 20). The Na content is variable, with SAR values varying from 2 to 70. In those areas where the Fort Union stratified materials are overlain by glacial drift and/or aeolian materials, these materials are generally low in soluble salts and Na (Sandoval et al., 1973; Schroder, 1976). The most important properties to consider in the overburden material are sodium content, salinity, clay content, and mineralogy (Doll et al., 1984).

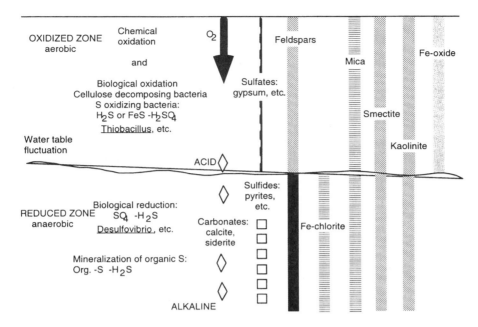

FIGURE 6.3. Schematic diagram of lignite overburden showing zones of oxidation and reduction and mineral components that may occur in each (Source: Dixon et al., 1982).

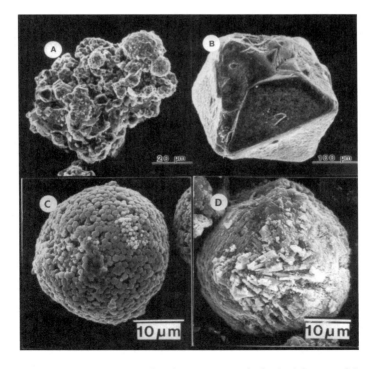

FIGURE 6.4. Scanning electron micrographs showing morphological forms of iron sulfides (FeS_2) separated from lignite overburden: (A) polyframboidal pyrite; (B) Low surface area pyrite octahedron; (C) Framboidal pyrite; (D) Marcasite.

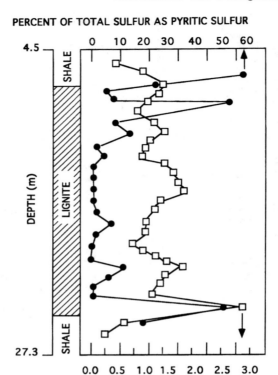

PERCENT OF TOTAL SULFUR AS PYRITIC SULFUR

FIGURE 6.5. Distribution of total sulfur and pyritic sulfur in relationship to a lignite seam in Texas (Source: Arora et al., 1980).

Properties of Native Soils

Native soils are derived from geologic materials, from overlying glacial till, or from alluvial or aeolian deposits. The Bernaldo (fine-loamy, siliceous, thermic Glossic paleudalfs) and Freestone (fine-loamy, siliceous, thermic Gossaquic Paleudalfs) series are representative of the better soils of the upland position (DeMent and Cooney, 1992). These soils are very deep, well drained and moderately well drained soils on uplands. Both have a fine sandy loam surface ranging from 20 to 50 cm thick and a loam or clay loam argillic horizon. They are developing in cross-bedded sands, silts, and clays from the Wilcox group of Eocene age. Soils representative of the alluvial valley position include Roetex c (fine, mixed, thermic, Udertic Haplustoll) and Kaufman c (very fine, montmorillonitic, thermic, Typic Pelludert) (Hossner et al., 1992). The Roetex c soil is formed on fine textured alluvium along streams carrying sediments from Permian red beds. The soil ranges from mildly alkaline through moderately alkaline. The Kaufman c soils are extensive and are located on level to gently sloping floodplains of streams draining the Blackland Prairies. They are moderately acid at the surface and calcareous at depths of 60 cm or greater.

Representative native soils in the mining region of North Dakota include Cabba (loamy, calcareous, frigid, shallow, mixed, typic Ustorthents) and Sen (fine-silty, mixed Typic Haploborolls) (Hofmann et al., 1983) series. Most prime soils occur on nearly level or concave portions of the landscape. Omodt et al. (1975) state that the most productive soils in western North Dakota are located on concave slopes and concave landscape positions; are medium or fine in texture; and contain less than 10% free lime. The topsoil contains 2% or more organic matter and is no more than slightly saline or somewhat sodic. State and federal laws require separate handling of prime and nonprime soils (Schroeder, 1992).

Options for Revegetation of Minelands

Mining for lignite in Texas and North Dakota, the two largest regions of lignite mining in the United States, began in the early 1970s. State regulations were in effect at this time to monitor the reclamation process. State and federal regulations, including the 1977 Federal Surface Mining and Control and Reclamation Act (SMCRA), were largely based on the concept that soils should be reconstructed to as near a premine condition as possible. Regulations were developed with a major input from pedologists who dealt largely with soil classification and had less appreciation of soil-plant relationships. The National Research Council (1981) proposed that "An appropriate goal for reclamation is to ensure that society does not lose important land-use opportunities that are available prior to soil disturbance or that can be generated in the reclamation process. This does not necessarily imply restoring precisely the characteristics of the premine soil and landscape but rather involves the establishment of geological and hydrologically stable landscapes capable of supporting a natural mosaic of ecosystems. Such landscapes provide the widest range of options for future land use." Efforts to restore the soils to their original characteristics resulted in a number of problems: restoration to original topography was expensive and probably not appropriate in most cases, and surface soils were not necessarily the best material for sustaining vegetation. Replacing surface soils resulted in excessive compaction, while subsurface layers of the soils may contain excessive salt, sodium, or clay which could result in an inferior growing medium.

Prime Farmlands

Coal mining companies and regulatory agencies are responsible for ensuring that nonprime and prime farmlands are restored to their premining productivity. To ensure the proper reclamation of these lands, mining companies must demonstrate in permit applications that they have the technological capability to properly mine and reclaim.

Current lignite mining activities in the United States are largely restricted to the Northern Great Plains and the Gulf Coast Region of Louisiana and Texas. Prime farmland soils are distributed in the upland areas of the Northern Great Plains and in the upland and alluvial valleys of the Gulf Coast Region. Surface mining and reclamation operations in prime farmland soils must be conducted in a manner which ensures that the agricultural utility and level of productivity in the affected areas are reestablished. In addition, the important characteristics of soil structure and porosity supporting the essential hydrologic functions must be preserved during and after mining.

Properties of Reconstructed Soils

Numerous studies have been conducted that deal with the reclamation of mine soils. Kohnke (1950) discussed the reclamation aspects of spoil from surface mining and described the changes in soil physical conditions as they affect plant growth and subsequent land use. Mine soils can be constructed with variable chemical and physical properties by amending the original soil with suitable material from below the surface horizon (Dancer and Jansen, 1981; Dunker et al., 1982; Hargis and Redente, 1984). Topsoil materials have generally produced better plant growth but materials from the B and C horizons have provided adequate revegetation substrate in some instances (Dancer and Jansen, 1981; McSweeney et al., 1987). Reduction in yield is commonly due to compaction by equipment with a resulting high soil bulk density and reduced rooting volume for the crop (McSweeny et al., 1987; McSweeney and Jansen, 1984).

Physical and chemical properties of reconstructed soils can significantly influence the future utility and productivity of mine soils. High clay content and clay pans present in alluvial soils in Texas were ameliorated by deep mixing. Improved soil physical conditions resulted in higher crop yields and tree survival compared to native soils (Table 6.1) (Hossner et al., 1992). Clay content in prime farm soils in North Dakota is considerably lower than in Texas soils and ranges from 7 to 26%; silt ranges from 5 to 53%; bulk density from 0.9 to 1.6 Mg m^{-3}; and slope from 0.5 to 9.8% (Schroeder, 1994). Postmine soil mapping of mine soils formed by selective handling of overburden materials indicates that superior soil properties and productivity can be a result of the reclamation process. DeMent and Cooney (1992) reported that 65.9% of the reclaimed land area of the Winfield surface lignite mine was declared prime farmland by the U.S. Soil Conservation Service compared to the 38.3% present within the permit area prior to mining (Table 6.2). These data suggest that substitution of selected overburden for topsoil is a valid option, particularly in areas where native topsoils have low to moderate productivity.

The primary cause of reduced productivity in prime farmland reclamation is excessive soil compaction (Allen, 1992). Compaction caused by heavy equipment during the respreading of topsoil and subsoil after mining may have a range of effects on soil properties and potential productivity. Bulk density values of mine soils are generally higher than prior to mining as a result of decreased porosity. The reduced porosity usually comes at the expense of macropores (Schroeder, 1993). Thus, a compacted soil may have poor aeration, low nutrient and water availability, low water infiltration, slow water permeability, and mechanical impedance to root growth (Raney et al., 1955). The amount of water storage can also be determined by pore characteristics.

Coal companies in North Dakota are required to separate primeland topsoil from nonprimeland topsoil before mining (Schroeder, 1992). Following mining, the prime topsoil must be replaced in a prime location and the nonprime topsoil in a nonprime location. Spring wheat grain yields have generally correlated to total water use (Halvorson and Doll, 1991). Topographic position can dramatically affect productivity of reclaimed prime farmlands in a semiarid climate (Doll et al., 1984). Schroeder (1992) reported that small grain yields on reconstructed prime soils were 30 to 80% higher on the lower topographic positions (such as the footslope or toeslope) compared to upslope positions (such as the shoulder). Differences in yield were related to available water at planting and point out the necessity to maximize water availability by adjusting topographic effects during reclamation. Schroeder (1995) showed that downslope topographic positions of reclaimed primelands with uniform depth of topsoil replacement had nearly 300% more available soil water at planting and significantly greater spring wheat yields than upslope positions. The results suggest that the uniform depths of replaced topsoil and subsoil have not affected soil water and wheat yields differently on reclaimed mine lands than on undisturbed landscapes. Wollenhaupt and Richardson (1982) presented evidence that even microtopographic differences can be an important factor in determining crop yields on reclaimed land.

Table 6.1. Coastal Bermudagrass Yields and Tree Survival on an Undisturbed Kaufman Clay Soil and on a Deep Mixed Alluvium at a Trinity River Site.[a]

Years After Planting	Crop	Kaufman Clay	Mixed Alluvium[b]
	Grass Yield (kg ha^{-1})		
3	Coastal bermudagrass (*Cynodon dactylon*)	13,010 a[c]	25,760 b
4	Coastal bermudagrass (*Cynodon dactylon*)	11,730 a	23,520 b
5	Coastal bermudagrass (*Cynodon dactylon*)	13,150 a	15,730 b
	Tree Survival (%)		
2	Water oak (*Quercus nigra*)	40 a	54 a
3	Pecan (*Carya illinoensis*)	38 a	100 b
4	Autumn olive (*Elaeagnus unbelleta*)	36 a	52 a
4	Green ash (*Fraxinus pennsylvanica*)	72 a	92 b

[a] Source: Hossner et al., 1992.
[b] Random mix of alluvium to a depth of 6.2 m.
[c] Means for grass yield or tree survival for a given crop and year followed by the same letter are not significantly different at the 10% confidence level.

Table 6.2. Distribution of Prime Farmland and Nonprime Farmland Soils at the Winfield Mine in Texas Before and Following Mining and Reclamation.[a]

	Prime Farmland Soils[b]	Nonprime Farmland Soils
	%	
Premine	38.8	61.2
Postmine	65.9	34.1

[a] Adapted from DeMent and Cooney, 1992.
[b] Designated by U.S. Soil Conservation criteria and including minesoils with <5% slope.

Species Selection and Performance

Federal and state laws require that cover and productivity on reclaimed surface mined lands be equal to or greater than the premined condition. A major portion of the surface mined land in the Northern Great Plains will be revegetated with perennial grasses for permanent grazing use (Hofmann et al., 1983). Federal rules and regulations (*Federal Register* 47(56):12602) for U.S. mine land reclamation standards of revegetation success require that areas developed for use as grazing or pasture land shall have the ground cover and production of living plants on the revegetated area equal to that of a reference area or

Table 6.3. Comparison of Yields of Various Crops Grown on Unmined Prime Farmland Compared to Yield on Mined/Reclaimed Lands.

Crop	Yield Unmined	Yield Mined/Reclaimed	Location	Reference
	——— kg ha^{-1} ———			
Corn	[a]	2,890	Texas	Hons et al., 1978
Grain Sorghum	[a]	8,830	Texas	Hons et al., 1978
Soybeans	[a]	750	Texas	Hons et al., 1978
Coastal Bermudagrass[b]	11,280	19,340	Texas	Hossner et al., 1992
Cotton lint[b]	460	567	Texas	Hossner et al., 1992
Coastal Bermudagrass	12,540	17,920	Texas	DeMent and Cooney, 1992
Wheat	2,350	2,620	Texas	DeMent and Cooney, 1992
Barley	672	989	North Dakota	Doll et al., 1984
Barley	2,174	1,037	North Dakota	Doll et al., 1984
Wheat	2,230	1,420	North Dakota	Doll et al., 1984
Corn Silage	12,470	14,260	North Dakota	Doll et al., 1984
Corn Silage	24,864	18,592	North Dakota	Doll et al., 1984

[a] Not normally grown.
[b] 3 yr average.

other approved success standard. Vegetation on reclaimed sites include smooth bromegrass (*Bromus inermis* Leyss.), crested wheatgrass [*Agropyron desertorum* (Fisch. ex Link) Schult.], intermediate wheatgrass [*Agropyron intermedium* (Host) Beauv.], and alfalfa (*Medicago sativa* L.). Native species include blue grama [*Bouteloua gracilis* (H.B.K) Lag. ex Griffiths], and sedges (*Carex* sp.), green needlegrass (*Stipa viridula* Trin.), needle-and-thread (*Stipa comata* Trin. & Rupr.) and western wheatgrass (*Agropyron smithii* Rydb (Hofmann et al., 1981; Powell, 1988). Small grains are the major field crop grown in that region.

The primary agricultural crops on prime farmlands in the Gulf Coast Region are cotton (*Gossypium hirsutum*), corn (*Zea mays* L.), grain sorghum (*Sorghum bicolor*), trees, and improved pasture (commonly *Cynodon dactylon*). Upland areas are commonly not used for row crop production even when the soils can be classified as prime farm soils using Soil Conservation Service criteria. One of the positive aspects of mining is the increase in prime farm soils after reclamation is completed. Theoretically, these soils could be used for row crop production after mining when they were completely unsuited for this use prior to mining. Hons et al. (1978), Askenasy et al. (1980), and Hossner et al. (1992) have demonstrated increased productivity of field crops, forages, and trees on mined land that has been properly reclaimed and managed.

Yield of several crops on prime farmlands compared to the productivity of the same crop on lands that have been reclaimed or reclaimed and reclassified as prime farm soils is presented in Table 6.3. Yields of crops in side-by-side comparisons are similar on reclaimed lands compared to crop yield on native prime soils. In those instances where reclamation has created prime farm soils (Hons et al., 1978) in nonprime soil areas, the potential to produce crops that are not naturally grown because of adverse soil conditions is greatly enhanced.

REFERENCES

Allen, M. A Review of Procedures OSM Uses to Evaluate and Improve State Regulatory Programs Regarding Prime Farmland Reclamation of Mined Soils, in *Prime Farmland Reclamation. The Surface Mining Control and Reclamation Act: 15 years of Progress.* Dunker, R.E. et al., Eds. Department of Agronomy, University of Illinois, Urbana, IL, 1992, pp. 169–172.

Arora, H.S., C.E. Pugh, L.R. Hossner, and J.B. Dixon. Forms of Sulfur in East Texas Lignitic Coal. *J. Environ. Qual.*, 9, pp. 383–386, 1980.

Askenasy, P.A., L.R. Hossner, and E.L. Whiteley. Row Crop Production on Leveled Lignite Spoil Banks, in *Reclamation of Surface-mined Lignite Spoil in Texas.* Hossner, L.R., Ed. The Texas Agricultural Experiment Station and The Center for Energy and Mineral Resources, College Station, TX, 1980, pp. 49–52.

Dancer, W.S. and I.J. Jansen. Greenhouse Evaluation of Solum and Substratum Materials in the Southern Illinois Coal Field. 1. Forage Crops. *J. Environ. Qual.*, 10, pp. 396–400, 1981.

DeMent, J.A. and S. Cooney. Development of a Prime Farmland Minesoil in Selectively Cast Overburden in Northeast Texas, in *Prime Farmland Reclamation. The Surface Mining Control and Reclamation Act: 15 Years of Progress.* Dunker, R.E. et al., Eds. Department of Agronomy, University of Illinois, Urbana, IL, 1992, pp. 177–181.

Dixon, J.B., H.S. Arora, F.M. Hons, P.E. Askenasy, and L.R. Hossner. Chemical, Physical, and Mineralogical Properties of Soils, Mine Spoil, and Overburden Associated with Lignite Mining, in *Reclamation of Surface-Mined Lignite Spoil in Texas.* Hossner, L.R., Ed. The Texas Agricultural Experiment Station and The Center for Energy and Mineral Resources, RM-10. College Station, TX, 1980, pp. 12–21.

Dixon, J.B., L.R. Hossner, A.L. Senkayi, and K. Egashira. Mineralogical Properties of Lignite Overburden as They Relate to Mine Spoil Reclamation, in *Acid Sulfate Weathering.* Special Pub. No. 10. Kittrick, J.A., D.S. Fanning, and L.R. Hossner, Eds. Soil Sci. Soc. Am. Madison, WI, 1982, pp. 169–191.

Doll, E.C., S.D. Merrill, and G.A. Halvorson. Soil Replacement for Reclamation of Strip-Mined Lands in North Dakota. North Dakota Agr. Exp. Stn. Bull. 514. Fargo, 1984.

Dunker, R.E., I.J. Jansen, and M.D. Thorne. Corn Response to Irrigation on Surface-Mined Land in Western Illinois. *Agron. J.*, 74, pp. 411–414, 1982.

Fisher, W.L. and J.H. McGowen. Depositional Systems in the Wilcox Group of Texas and Their Relationship to Occurrence of Oil and Gas. *Gulf Coast Assoc. Geol. Soc. Trans.*, 17, pp. 105–125, 1967.

Frisbee, N.M. and L.R. Hossner. Siderite Weathering Under Carbon Dioxide, Air, and Oxygen. *J. Environ. Qual.*, 24, pp. 856–860, 1995.

Gould, F.W. Texas Plants—A Checklist and Ecological Summary. Texas Agri. Exp. Sta. MP-585. College Station, TX, 1975.

Haley, B.R. Coal Resources of Arkansas, 1954. *U.S. Geol. Survey Bull.*, 1072-P, pp. 795–831, 1960.

Halvorson, G.A. and E.C. Doll. Topographic Effects on Spring Wheat Yields and Water Use. *Soil Sci. Soc. Am. J.*, 55, pp. 1680–1685, 1991.

Hargis, N.E. and E.F. Redente. Soil Handling for Surface Mine Reclamation. *J. Soil Water Cons.*, 39, pp. 300–305, 1984.

Hofmann, L., R.E. Ries, and J.E. Gilley. Relationship of Runoff and Soil Losses to Ground Cover of Native and Reclaimed Grazing Land. *Agron. J.*, 75, pp. 599–602, 1983.

Hofmann, L., R.E. Ries, and R.J. Lorenz. Livestock and Vegetation Performance on Reclaimed and Nonmined Rangeland in North Dakota. *J. Soil Water Conserv.*, 36, pp. 41–44, 1981.

Hons, F.M., P.E. Askenasy, L.R. Hossner, and E.L. Whiteley. Physical and Chemical Properties of Lignite Spoil Material as it Influences Successful Revegetation, in *Gulf Coast Lignite Conference: Geology, Utilization, and Environmental Aspects.* Kaiser, W.R., Ed. Bureau of Economic Geology Report of Investigations No. 90. Austin, 1978, pp. 209–217.

Hossner, L.R., S.G. Tipton, and D.G. Purdy. Reclamation of Alluvial Valley Soils in Texas Using Mixed Overburden, in *Prime Farmland Reclamation. The Surface Mining Control and Reclamation*

Act: 15 Years of Progress. Dunker, R.E., et al., Eds. Department of Agronomy, University of Illinois, Urbana, IL, 1992, pp. 25–30.

Kaiser, W.R. Texas Lignite: Near-Surface and Deep Basin Resources. Texas Bur. Econ. Geology. Rept. Inv. 79. Austin, TX, 1974.

Klages, M.G. and R.W. Hopper. Clay Minerals in Northern Plains Coal Overburden as Measured by X-Ray Diffraction. *Soil Sci. Soc. Am. J.* 46, pp. 415–419, 1982.

Kohnke, H. The Reclamation of Coal Mine Spoils, *Advances in Agronomy*, Norman, A.G., Ed. Academic Press. 2, pp. 317–349, 1950.

Luppens, J.A. Exploration for Gulf Coast United States Lignite Deposits: Their Distribution Quality and Reserves, in *Proceedings of the Second International Coal Expl. Symp.*, Argall, G.O., Ed. Denver, CO, 1978, pp 195–210.

Lyle, E.S. Vegetation and Land Use. Lignite Surface Mine Reclamation of the Southern Gulf Coast Region. Southern Cooperative Series Bull. 294. Arkansas Agr. Exp. Sta. Fayetteville, AK, 1985, pp. 9–23.

McSweeney, K. and I.J. Jansen. Soil Structure and Associated Rooting Behavior in Minesoils. *Soil Sci. Soc. Am. J.*, 48, pp. 607–612, 1984.

McSweeney, K., I.J. Jansen, C.W. Boast, and R.E. Dunker. Row Crop Productivity of Eight Constructed Minesoils. *Reclamation Revegetation Res.*, 6, pp. 137–144, 1987.

National Research Council. *Surface Mining: Soil, Coal, and Society.* National Academy Press, Washington, DC, 1981.

Omodt, H., F.W. Schroeder, and D.D. Patterson. The Properties of Important Agricultural Soils as Criteria for Mined Land Reclamation. North Dakota Agric. Exp. Sta. Bull. Fargo, 1975, p. 492.

O'Shay, T. and L.R. Hossner. Lignite Distribution in the Gulf Region and Associated Precipitation and Temperature. Lignite Surface Mine Reclamation of the Southern Gulf Coast Region. Southern Cooperative Series Bull. 294. Arkansas Agric. Exp. Sta. Fayetteville, AK, 1985, pp. 1–7.

Powell, J.L. Revegetation Options, in *Reclamation of Surface-Mined Lands*, Hossner, L.R., Ed. CRC Press. Boca Raton, FL, 1988, pp. 49–91.

Raney, W.R., T.W. Edminster, and W.H. Allaway. Current Status of Research on Soil Compaction. *Soil Sci. Soc. Am. Proc.*, 19, pp. 423–428, 1955.

Roland, H.L., Jr., G.M. Jenkins, and D.E. Pope. Lignite: Evaluation of Near-Surface Deposits in Northwest Louisiana. Louisiana Geol. Survey. Mineral Resources Bull. 2. Baton Rouge, LA, 1976.

Royse, C.F. A Sedimentologic Analysis of the Tongue River-Sentinel Butte Interval (Paleocene) of the William Basin, Western North Dakota. *Sediment. Geol.*, 4, p. 19, 1970.

Royse, C.F. The Tongue River and Sentinel Butte Formations (Paleocene) of Western North Dakota: A Review, in *Depositional Environments of the Lignite-Bearing Strata in Western North Dakota.* Ting, F.T.C., Ed. Misc. Ser. No. 50. North Dakota Geological Survey. Fargo, ND, 1972.

Sandoval, F.M., J.J. Bond, J.F. Power, and W.O. Willis. Lignite Mine Spoils in the Northern Great Plains-Characteristics and Potential for Reclamation, in *Some Environmental Aspects of Strip Mining in North Dakota*, Mohan K. Wali, Ed. North Dakota Geol. Survey Ed. Series 5. Fargo, ND, 1973, pp. 1–24.

Schroeder, F.W. Chemical and Physical Characterization of Coal Overburden. *North Dakota Agr. Exp. Sta. Farm Res.*, 34, pp. 5–11, 1976.

Schroeder, S.A. Reclaimed Topography Effects on Small Grain Yields in North Dakota, in *Prime Farmland Reclamation. The Surface Mining Control and Reclamation Act: 15 Years of Progress*, Dunker, R.E., et al., Eds., Department of Agronomy, University of Illinois, Urbana, IL, 1992, pp. 31–34.

Schroeder, S.A. Tillage Influences on Physical Properties and Forage Yields. Land Reclamation Research Center. 1993 Mine-Land Reclamation Research Review. March 15, 1993. Mandan, ND, 1993, pp. 17–20.

Schroeder, S.A. Reliability of SCS Curve Number Method on Semi-Arid, Reclaimed Minelands. *Intern. J. Surface Mining, Mining, Reclamation Environ.*, 18, pp. 41–45, 1994.

Schroeder, F.W. Topographic Influences on Soil Water and Spring Wheat Yields on Reclaimed Mineland. *J. Environ. Qual.*, 24, pp. 467–471, 1995.

Senkayi, A.I. and J.B. Dixon. Mineralogical Considerations in Reclamation of Surface-Mined Lands, in *Reclamation of Surface-Mined Lands*. Hossner, L.R., Ed. CRC Press, Boca Raton, FL, 1988, pp. 105–124.

White, R.L. Land Reclamation in Texas—An Opportunity, in *Proc. Gulf Coast Lignite Conf. Geology, Utilization, and Environmental Aspects*, Kaiser, W.R., Ed. Univ. Texas Bur. Econ. Geol., Rep. Invest. No. 90. Austin, TX, 1978, pp. 199–208.

Whiteley, E.L. and L.R. Hossner. Lignite Resources and Environmental Considerations in Texas, in *Reclamation of Surface-Mined Lignite Spoil in Texas*, Hossner, L.R., Ed. Texas Agr. Exp. Sta. and the Center for Energy and Mineral Resources Publication RM-10. College Station, TX, 1980, pp. 5–11.

Wielchowsky, C.C., G.F. Collins, L.K. Gerahian, and E.J. Calhoun. Frontier Areas and Exploration Techniques: Frontier Lignite Exploration in the South-Central United States, in *Geology of Alternate Energy Resources in the South-Central United States*, Campbell, M.D., Ed. Houston Geological Society. Houston, TX, 1977, pp. 125–159.

Wollenhaupt, N.C. and J.L. Richardson. The Role of Topography in Revegetation of Disturbed Lands, in *Mining and Reclamation of Coal Mined Lands in the Northern Great Plains, Proc.* Billings, MT. Montana Agric. Exp. Stn. Res. Rep. 194. Bozeman, MT, 1982, pp. C-2-1 to C-2-11.

Chapter 7

Reclamation of Abandoned Bentonite Mined Lands

G.E. Schuman

INTRODUCTION

Large-scale mining for bentonite in the Northern Great Plains of the United States began in the 1930s; however, only limited reclamation was achieved prior to the passage of state reclamation laws in the early 1970s. The Northern Great Plains states of Montana, South Dakota, and Wyoming produce 90% of the nation's supply of bentonite (Ampian, 1980). In 1973, more land was disturbed by bentonite mining than coal mining in Montana and more abandoned mine spoils had accumulated over the years from bentonite mining than from coal (National Academy of Sciences, 1974). This example is indicative of the Northern Great Plains in general.

Revegetation, natural and man-assisted, of nontopsoiled bentonite mine spoils has resulted in no or very poor success (Dollhopf and Bauman, 1981; Sieg et al., 1983). This historical poor record of revegetation stressed the need for spoil amendment to ameliorate the high clay content, high salinity, and sodicity of these spoils if they were to be successfully reclaimed. Revegetation of abandoned bentonite spoils has typically met with little success unless organic amendments were utilized (Hemmer et al., 1977; Bjugstad et al., 1981; Dollhopf and Bauman, 1981; Schuman and Sedbrook, 1984). Borrowing of topsoil from unmined adjacent lands is not considered an option because of the limited topsoil development. Topsoil borrowing also results in double the amount of land area requiring revegetation. Research evaluating the success of organic amendments (woodchips, wood residues, manure, and straw) reported good plant establishment with woodchips and wood residues but reported more limited success with other more readily decomposable materials (Dollhopf and Bauman, 1981; Schuman and Sedbrook, 1984). Dollhopf and Bauman (1981) also evaluated inorganic amendments (gypsum, calcium chloride, and sulfuric acid); however, these amendments resulted in poor initial seedling establishment. These preliminary studies demonstrated that spoil amendments that resulted in immediate improvements in water infiltration into the spoil also promoted vegetation establishment.

The research reported in this chapter was initiated in 1981 to define the levels of wood residue, gypsum, and nitrogen fertilizer amendments necessary to achieve good reclamation of the bentonite spoils and to study the effects of the amendments on salt leaching, sodicity, and long-term sustainability of the revegetated plant community and 'newly developing soil.'

Table 7.1. Physical and Chemical Characteristics of Pretreatment Bentonite Spoil Samples, 1981.[a]

Parameter	Mean and Standard Error[b]	
Particle-size separates, %		
Sand	10.8±	0.8
Silt	29.6±	0.8
Clay	56.6±	1.1
NO_3-N, mg kg^{-1}	7.7±	0.4
NH_4-N, mg kg^{-1}	2.6±	0.1
Kjeldahl-N, mg kg^{-1}	751.1±	5.8
P, mg kg^{-1}	8.1±	0.3
C, g kg^{-1}	10.0±	1.0
pH	6.8±	0.1
EC_{se}, dS m^{-1}	13.4±	1.1
Soluble cations, mg kg^{-1}		
Ca	187.9±	9.2
Mg	73.6±	4.2
Na	3613.7±	101.3
K	32.0±	0.8
Sodium adsorption ration (SAR)	63.1±	1.2

[a] From Smith et al., 1985.
[b] Particle-size separates obtained from five observations, all other parameters are a mean of 144 samples.

MATERIALS AND METHODS

Study Site Description

The study site was located on abandoned bentonite mine spoils in northeastern Wyoming, USA. The area is characterized by broad, flat valleys covered with sagebrush (*Artemisia tridentata* ssp. *wyomingensis*) and grass, separated by low Ponderosa pine (*Pinus ponderosa*) covered hills. Soils overlying the bentonite deposits are formed from very fine clay and shale particles and have low inherent fertility. The bentonite deposits in this area were formed in shallow marine lakes, resulting in saline-sodic soils. The area was mined in the mid-1950s, and no regrading or revegetation was attempted. The spoils were typical of the abandoned spoils associated with the Mowry shale formation (Table 7.1).

The long-term average annual precipitation of the area is 370 mm, with approximately 60% occurring during April through July. The average frost-free period is 165 days.

Experimental Design

An area of abandoned spoil approximately 2 ha in size was regraded to an approximate level topography. The experimental design was a split plot type with split block within the main treatment (wood residue). Treatments included: four wood residue application rates (0, 45, 90, and 135 Mg ha^{-1}), four nitrogen fertilizer application rates (0, 2.5, 5.0, and 7.5 kg N Mg^{-1} wood residue), and two grass seed mixtures (native and introduced species, Table 7.2). Four plots (29.3 by 39.0 m) were established in each of the three replicate blocks.

Table 7.2. Plant Species Mixtures Seeded on Abandoned Bentonite Spoils, Upton, WY, October 1981.[a]

Drill

Scientific Name	Common Name	Seeding Rate PLS $m^{-2,b}$	Growth Form[c]
Native grass mixture			
Agropyron smithii Rydb.	Western wheatgrass	130	CSRG
Agropyron dasystachym (Hook.)Scribn.	Thickspike wheatgrass	130	CSRG
Agropyron trachycaulum (Link)Malte	Slender wheatgrass	130	CSBG
Agropyron riparium Scribn. & Smith	Streambank wheatgrass	130	CSRG
Stipa viridula Trin.	Green needlegrass	130	CSBG
Atriplex nuttalli S. Wats.	Nuttall saltbush	32	S
Introduced grass mixture			
Agropyron desertorum (Fisch.)Schult.	Crested wheatgrass	130	CSBG
Agropyron elongatum Host.	Tall wheatgrass	130	CSBG
Agropyron intermedium (Host.)Beauv.	Intermediate wheatgrass	130	CSRG
Agropyron trichophorum (Link)Richt.	Pubescent wheatgrass	130	CSRG
Bromus inermis Leyss.	Smooth brome	130	CSRG
Atriplex nuttalli S. Wats.	Nuttall saltbush	32	S

[a] From Smith et al., 1986.
[b] Pure live seed.
[c] CSBG = cool-season bunchgrass; CSRG = cool-season rhizomatous grass; S = shrub.

Wood residue consisting of Ponderosa pine bark (45%), woodchips (20%), and sawdust (35%) was obtained from a local sawmill. Each wood residue rate was applied to one randomly selected main plot. Each plot was then subdivided into eight 9.8 by 14.6 m subplots, each representing a specific combination of N-fertilization and seed mixture at that wood residue level. In general, N-fertilizer (ammonium nitrate) was applied on the basis of wood residue application rate to maintain a uniform C:N ratio across wood residue treatments (Table 7.3). However, where no wood residue was applied (control), N was added at rates equivalent to the per hectare rates applied on the 45 Mg ha^{-1} wood residue treatment (0, 112, 224, and 336 kg N ha^{-1}). This enabled statistical comparisons between the control and 45 Mg ha^{-1} treatments and the 45, 90, and 135 Mg ha^{-1} treatments. Phosphorus (treble

Table 7.3. Wood-Residue and N-Fertilizer Rates, with C:N Ratios, Applied to Abandoned Bentonite Spoils.[a]

Wood-Residue Level Mg ha^{-1}	N-Fertilizer Rate		C:N Ratio
	kg Mg^{-1} Wood Residue	kg N ha^{-1}	
0	—	0	—
	—	112	—
	—	224	—
	—	336	—
45	0	0	469:1
	2.5	112	137:1
	5.0	224	81:1
	7.5	336	57:1
90	0	0	469:1
	2.5	224	137:1
	5.0	448	81:1
	7.5	672	57:1
135	0	0	469:1
	2.5	336	137:1
	5.0	672	81:1
	7.5	1008	57:1

[a] From Schuman and Belden, 1991.

superphosphate) was uniformly applied to all plots at 90 kg P ha^{-1}. All fertilizer was broadcast onto the spoil surface and tilled into the surface 30 cm during incorporation of the wood residue. Residue incorporation and seedbed preparation were accomplished by chiseling (30-cm centers and at least 30 cm deep) and disking the plot area. Each subplot was drill seeded with either the native or introduced grass mixture in October 1981.

In May 1983, litter bags were constructed and placed at 5-cm depths in the plots to assess the rate of wood residue decomposition. Details of the litter bag technique and study design are discussed by Schuman and Belden (1991). The objective of this phase of the research was to determine the rate of decomposition of the wood residue as affected by rate of application and N-fertilizer application. Our interest was to determine if the wood residue longevity would be long enough to ensure sustainability of the revegetated landscape and to use wood residue decomposition as a measure of microbial function in the reclaimed bentonite spoils.

In response to observed increases in the exchangeable sodium percentage (ESP) of the revegetated spoil, despite significant leaching of salts, a third objective was incorporated into this research. In April 1987, surface-applied gypsum (56 Mg ha^{-1}) was amended to about half of each of the native grass seeded plots. This level of gypsum was adequate to reduce the ESP of the spoil to 15% from a pretreatment level of 48% (Schuman and Meining, 1993).

Spoil Sampling

Baseline spoil samples were taken in September 1981 prior to the application and incorporation of the amendments. Samples were obtained to 45 cm depth and separated into 0–15, 15–30, and 30–45 cm increments. Spoil samples were also collected each spring from 1982 to 1990 to assess the effects of treatments on salt leaching, nitrogen status, and sodicity. The spoil samples were analyzed for Kjeldahl-N (Schuman et al., 1973), pH, electrical conductivity (U.S. Salinity Lab Staff, 1954), soluble salts (Rhoades, 1982), SAR (sodium-adsorption-ratio)/ESP (U.S. Salinity Lab Staff, 1954), and gravimetric soil moisture content.

Vegetation Sampling

Aboveground biomass was estimated during the period of peak standing crop (July–August) by harvesting all of the vegetation within four quadrats (0.25 m^2) randomly located within each nitrogen by wood residue by species mixture subplot (1982–1986) or one quadrat (0.50 m^2) located randomly within each nitrogen by wood residue by gypsum subplot (1988–1990). Plant material was clipped to ground level within each quadrat, separated by species and/or classes (i.e., annual grasses, annual forbs, etc.), and oven-dried at 60°C for 48 hr.

Data Analysis

Analysis of variance was conducted on all data and tested at the 0.05 level of probability, except the sodicity/ESP phase of the study which was tested at the 0.10 level of probability. Mean separations were evaluated using Fisher's protected least significant difference.

RESULTS AND DISCUSSION

General Vegetation and Spoil Responses

Seedling densities observed in 1982 were significantly enhanced by the wood residue amendment. Seedling density was 13, 41, 60, and 70 plants m^{-2} for the 0, 45, 90, and 135 Mg ha^{-1} wood residue treatments, respectively. Although these differences were significant for all levels of wood residue treatment, the biggest increases observed were between the control and 45 Mg ha^{-1} treatment, with adequate densities and less change observed between the three wood residue rates. Seedling density showed no consistent response to fertilizer N rates, except it was significantly greater with N fertilization than without N. Seedling density was not different between the native and introduced grass mixture.

Some of the grass species did not establish or they became established to a limited extent only under specific wood residue-nitrogen fertilizer treatment combinations. The successfully established grass species (*Agropyron smithii, A. dasystachyum, A. riparium, A. elongatum, A. intermedium,* and *Bromus inermus*) were all rhizomatous except one; suggesting that sod-forming grasses are generally better suited for revegetation of high clay spoils (Smith et al., 1986). Rhizomes have been noted to exhibit resistance to breakage and the capacity for regrowth and increased production from new rhizomes if breakage occurs (White and Lewis, 1969). Weaver and Albertson (1956) noted that clay soils in the region were dominated by sod-forming grasses. From these observations and others noted in their research, Smith et al. (1986) recommended that species used to revegetate abandoned bentonite spoils should possess at least some of the following characteristics: sod-forming morphology, drought and salt tolerant, adaptation to clay texture, and adaptation to a shallow, poorly drained root zone.

Table 7.4. Perennial Grass Production (kg ha^{-1}, Averaged Across Species Mixtures and N-Fertilizer Treatments) in Response to Wood Residue Rate, Upton, WY, 1983 to 1986.

Year	Wood Residue Rates (Mg ha^{-1})[a]			
	0	45	90	135
	kg ha^{-1}			
1983	59a[b]	669b	1748c	2550d
1984	80a	361a	1220b	1956c
1985	10a	55a	148a	448a
1986	15a	116a	324a	886b

[a] From Schuman, 1995.
[b] Means among wood residue rates within a year followed by the same letter are not significantly different, P ≤ 0.05.

Perennial grass production increased as wood residue rate increased, with maximum production occurring at the 135 Mg ha^{-1} wood residue rate, for the years 1983 to 1986 (Table 7.4). The large reduction in grass production in 1985 was due to a severe drought experienced that year. However in late 1985 and 1986, normal precipitation resulted in much improved forage production in 1986. Visual observations of the research plots in 1986 indicated that some plant mortality occurred during the drought; therefore, predrought production levels probably would not be expected or at least not achieved for several years. Grass production responded to N fertilization in 1983 and 1984, with peak production occurring at the 2.5 and 5.0 kg N Mg^{-1} of wood residue, respectively. The lack of response to N in 1985 and 1986 may have resulted from the lack of adequate moisture in 1985 and the observed plant mortality. Grass production was not significantly different between the native and introduced species mixture treatments.

The spoil responded in three ways to the wood residue amendment: increased water infiltration and storage, decreased salinity due to leaching and increased sodicity. Schuman and Sedbrook (1984) in an earlier study showed water storage was significantly enhanced by wood residue amendment (Table 7.5). The increased water movement into spoil resulted in significant leaching of soluble salts from the surface 15 cm of the amended spoil (Figure 7.1). The drought in 1985 that resulted in significant reductions in grass production also caused significant upward migration of soluble salts. Although leaching of soluble salts was enhanced by the addition of the wood residues, SAR was increased in the spoil profile (Figure 7.2). Since sodium comprised 91% of the soluble cations in the system, as leaching occurred it became more dominant in comparison to the relatively low proportions of calcium and magnesium, thereby increasing the SAR. The significant increases observed in SAR can have long-term effects on plant nutrition, spoil physical properties and sustainability of the revegetated plant community. Therefore, chemical amendments such as gypsum, calcium chloride, or phosphogypsum would be required to ensure amelioration of the sodic conditions and ensure long-term reclamation success.

Wood Residue Decomposition

Restoration of decomposition processes in revegetated mine spoils is necessary to ensure the long-term success of the reclamation. Evaluation of wood residue decomposition

Table 7.5. The Effect of Wood Residue Amendments on the Soil-Water Content (1980 and 1982 Average) of Bentonite Spoils, Upton, WY.[a]

	Soil Depth, cm		
	0–20	20–40	40–60
Wood Residue Treatment		Water Content	
—— Mg ha⁻¹ ——		—— g kg⁻¹ ——	
0	115a[b]	138a	139a
112	212b	166b	143a
224	232b	180b	155a

[a] From Schuman and Sedbrook, 1984.
[b] Means among wood residue levels within a spoil depth followed by the same letter are not significantly different, P ≤ 0.05.

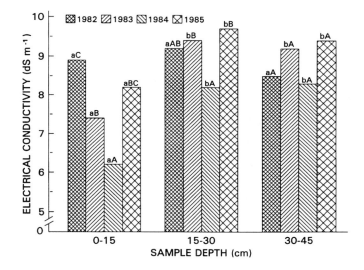

FIGURE 7.1. Mean EC averaged across wood residue and N fertilizer treatments, for 1982 to 1985, at three spoil depths. Means with the same letter among years (lowercase) or within years (uppercase), are not significantly different, P ≤ 0.05 (Belden, 1987).

indicated that microbial functions have begun to develop. Wood residue decomposition exhibited significant increases over years and also with increased N-fertilizer application (Figures 7.3 and 7.4). The slow rate of wood residue decomposition indicates that the wood residue longevity should be adequate for initial soil development, enable plant community development and long-term, sustainable reclamation of these spoils. Whitford et al. (1989) suggest that more resistant sources of organic mulches are superior to readily decomposed material because they provide slow release of organic particles that serve as energy sources for the microflora. Therefore, the wood residue and the plant litter and root turnover from the established vegetation together provide relatively recalcitrant and readily available sources of organic carbon to help ensure long-term success through sustained nutrient cycles.

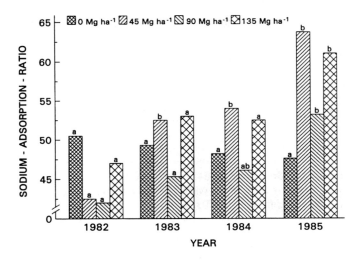

FIGURE 7.2. Mean SAR averaged across N fertilizer treatments and sample depths, at four wood residue levels for 1982 to 1985. Means among years with the same lowercase letter are not significantly different, P ≤ 0.05 (Belden, 1987).

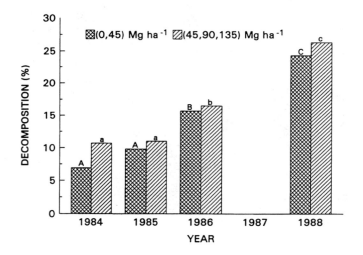

FIGURE 7.3. Decomposition of wood residue in amended bentonite spoil for a five-year period. Data analyzed separately for 0- and 45; and 45-, 90-, and 135 Mg ha^{-1} wood residue treatments. Means within the 0- and 45-; and 45-, 90-, and 135 Mg ha^{-1} treatments with the same upper- and lowercase letter, respectively, are not significantly different, P ≤ 0.05 (Schuman and Belden, 1991).

Gypsum Amendment Effects

The inorganic amendment phase of this study was designed to evaluate the effectiveness of gypsum in ameliorating the sodicity of revegetated mined land and in a more general sense to determine if gypsum would be an effective amendment under a natural rainfall environment. Questions of gypsum dissolution in a semiarid environment (370 mm precipitation) needed to be addressed. The surface-applied gypsum increased the EC

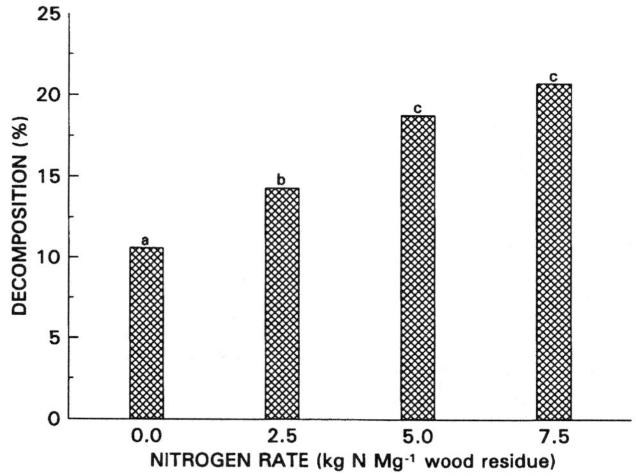

FIGURE 7.4. Decomposition of wood residue amended to bentonite spoils as a function of N fertilizer application rate (averaged across all wood residue rates and years). Means with the same letter are not significantly different, $P \leq 0.05$ (Schuman and Belden, 1991).

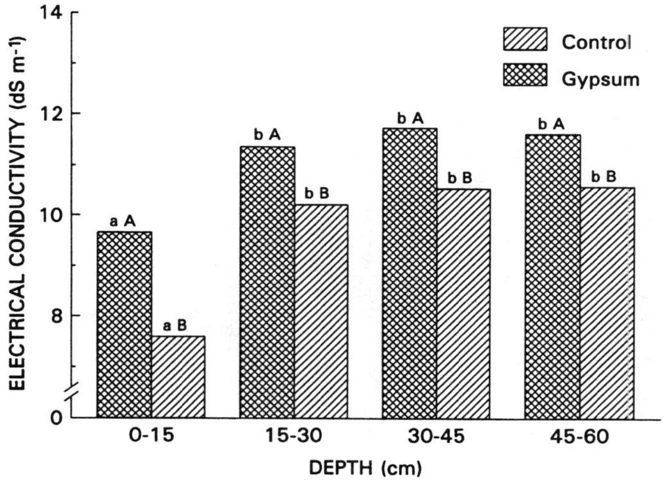

FIGURE 7.5. The effect of gypsum amendment on the EC of wood residue-amended bentonite spoil at four depths, averaged across years 1988 to 1990. Means with the same letter within a treatment (lowercase) or within a depth (uppercase) are not significantly different, $P \leq 0.10$ (Schuman and Meining, 1993).

at all soil depths (Figure 7.5). Such increases in EC might cause reductions in seed germination and seedling survival if it had been applied with the other amendments because of the inherently high levels of salinity in the spoil (Table 7.1). However, incorporation of the gypsum into 30 cm of spoil would also have reduced the impact it would have had on the spoil EC. Gypsum amendment increased spoil water content at all depths (Figure 7.6). This response to gypsum and the lack of a significant response to wood residue indicates that the improved spoil physical characteristics due to gypsum was greater than

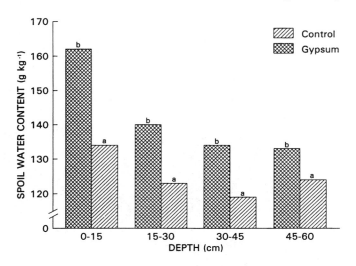

FIGURE 7.6. Response of spoil-water content to gypsum amendment of revegetated saline-sodic bentonite mine spoil, averaged across years 1988 to 1990. Means within a spoil depth with the same lowercase letter are not significantly different, $P \leq 0.05$ (Schuman and Meining, 1993).

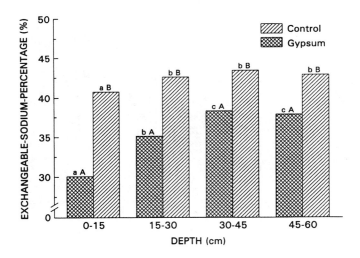

FIGURE 7.7. The effect of gypsum amendment on the ESP of wood residue-amended bentonite spoil at four depths, averaged across years—1988 to 1990. Means with the same letter within a treatment (lowercase) or within a depth (uppercase) are not significantly different, $P \leq 0.05$ (Schuman and Meining, 1993).

that contributed by the wood residue amendment during the period 1988 to 1990. Although gypsum amendment increased the spoil EC, the enhanced soil moisture resulted in a significant increase in grass production (Schuman et al., 1994). ESP was significantly reduced by gypsum amendment in the entire 60 cm profile (Figure 7.7). However, the greatest changes occurred in the top 30 cm, with no differences being evident between the two lower depths.

CONCLUSION

This chapter reviews and summarizes the findings of an intensive and long-term research project that led to the development of a reclamation technology for abandoned bentonite mine spoils. Wood residue amendment resulted in the immediate improvement of water infiltration and leaching of soluble salts in sodic, clay spoils. This improvement in water infiltration enabled successful plant establishment and good forage production on these amended spoils. The research also demonstrated the need for, and effectiveness of, gypsum amendment for ameliorating sodicity in these spoils. The wood residue decomposition data indicates that some level of microbial activity has been reestablished in these spoils, helping to promote the long-term success of the reclamation.

This technology was readily adaptable to large-scale reclamation of abandoned bentonite mine spoils and over 6000 hectares have been revegetated in the region. The average cost of reclaiming abandoned bentonite mine lands using this technology is about $12,000 ha (US$), with amendments accounting for about 21% of the cost (Richmond, 1991). Reclamation of these abandoned lands has eliminated a significant environmental problem and improved the aesthetics of a major tourist region. Much of the results of this research is applicable to active bentonite mining and other mining that involves high clay content and sodic spoils.

ACKNOWLEDGMENT

The author is grateful to the conference organizers and sponsors for the invitation and support to participate in the International Conference on the Remediation and Management of Degraded Lands.

REFERENCES

Ampian, S.G. Clays, in *Mineral Facts and Problems*. USDI Bureau of Mines Bull. 671. U.S. Government Printing Office, Washington, DC, 1980.

Belden, S.E. Edaphic Responses to Wood Residue and Nitrogen Amendment of Abandoned Bentonite Spoil. M.S. thesis, University of Wyoming, Laramie, WY, 1987.

Bjugstad, A.J., T. Yananota, and D.W. Uresk. Shrub Establishment on Coal and Bentonite Clay Mine Spoils, in *Shrub Establishment on Disturbed Arid and Semi-Arid Lands*. Wyoming Game and Fish Commission, Laramie, WY, 1981, pp. 104–122.

Dollhopf, D.J. and B.J. Bauman. *Bentonite Mine Land Reclamation in the Northern Great Plains*. Montana Agric. Exp. Stn. Res. Rpt. 179. Montana State University, Bozeman, MT. 1981.

Hemmer, D., S. Johnson, and R. Beck. *Bentonite Mining Related Reclamation Problems in the Northwestern States*. Old West Regional Commission Rpt., Montana Department of State Lands, Helena, MT, 1977.

National Academy of Sciences. *Rehabilitation Potential of Western Coal Lands*. Ballinger Publishing Co., Cambridge, MA, 1974.

Rhoades, J.D. Soluble Salts, in *Methods of Soil Analysis*, Part 2, 2nd ed., Page, A.L. et al., Eds., Agron. Monogr. 9. Amer. Soc. Agron. and Soil Sci. Soc. Amer., Madison, WI, 1982, pp. 167–180.

Richmond, T.C. Reclamation of Abandoned Bentonite Mines in Wyoming with Wood-Waste and Calcium Amendments to Mitigate Sodic Soils, in *Proc. Natl. Meet. Amer. Soc. Surface Mining and Reclamation, Durango, CO. 14–17 May 1991*. Oaks, W.R. and J. Bowden, Eds., Amer. Soc. Surface Mining and Reclamation, Princeton, WV, 1991, pp. 455–478.

Schuman, G.E. Revegetating Bentonite Mine Spoils with Sawmill By-Products and Gypsum, in *Agriculture Utilization of Urban and Industrial By-Products*, Karlen, D.L. et al., Eds., Amer. Soc. Agron. Spec. Publ. 58, Amer. Society of Agron., Madison, WI, 1995, pp. 261–274.

Schuman, G.E. and S.E. Belden. Decomposition of Wood-Residue Amendments in Revegetated Bentonite Mine Spoils. *Soil Sci. Soc. Am. J.*, 55, pp. 76–80, 1991.

Schuman, G.E., E.J. DePuit, and K.M. Roadifer. Plant Responses to Gypsum Amendment of Sodic Bentonite Mine Spoil. *J. Range Manage.* 47, pp. 206–209, 1994.

Schuman, G.E. and J.L. Meining. Short-Term Effects of Surface-Applied Gypsum on Revegetated Sodic Bentonite Spoils. *Soil Sci. Soc. Am. J.*, 57, pp. 1083–1088, 1993.

Schuman, G.E. and T.A. Sedbrook. Sawmill Wood Residue for Reclaiming Bentonite Spoils. *Forest Products J.*, 34, pp. 65–68, 1984.

Schuman, G.E., M.A. Stanley, and D. Knudsen. Automated Total Nitrogen Analysis of Soil and Plant Samples. *Soil Sci. Soc. Am. Proc.*, 37, pp. 480–481, 1973.

Sieg, C.H., D.W. Uresk, and R.M. Hansen. Plant-Soil Relationships on Bentonite Mine Spoils and Sagebrush-Grassland in the Northern High Plains. *J. Range Manage.*, 36, pp. 289–294, 1983.

Smith, J.A., E.J. DePuit, and G.E. Schuman. Wood Residue and Fertilizer Amendment on Bentonite Mine Spoils: II. Plant Species Responses. *J. Environ. Qual.*, 15, pp. 427–435, 1986.

Smith, J.A., G.E. Schuman, E.J. DePuit, and T.A. Sedbrook. Wood Residue and Fertilizer Amendment of Bentonite Mine Spoils: I. Spoil and General Vegetation Responses. *J. Environ. Qual.*, 14, pp. 575–580, 1985.

U.S. Salinity Laboratory Staff. *Diagnosis and Improvement of Saline and Alkali Soils.* USDA Agric. Handbook 60. U.S. Government Printing Office, Washington, DC, 1954.

Weaver, J.E. and F.W. Albertson. *Grasslands of the Great Plains.* Johnson Publishing Co., Lincoln, NE, 1956.

White, E.M. and J.K. Lewis. Ecological Effect of a Clay Soil's Structure on Some Native Grass Roots. *J. Range Manage.*, 22, pp. 401–404, 1969.

Whitford, W.G., E.F. Aldon, D.W. Freckman, Y. Steinberger, and L.W. Parker. Effects of Organic Amendments on Soil Biota on a Degraded Rangeland. *J. Range Manage.*, 42, pp. 56–60, 1989.

Chapter 8

Paper Mill Sludge as a Revegetation Tool in an Abandoned Sandpit: Project Outline and Preliminary Results

A. Fierro, D.A. Angers, and C.J. Beauchamp

INTRODUCTION

Mine soils have no pedogenic profile development and are characterized mainly by the kind of material extracted and its handling procedures (Macyk, 1987). Sandpits are among the most extensive mine soils in the Canadian province of Quebec. Complete ecosystem restoration is not required under local legislation, but a viable plant cover must be established and maintained when the abandoned pit is not assigned to other uses (MENVIQ, 1984).

Successful revegetation of abandoned sandpits is generally hindered by serious deficiencies in retention and supply of water and nutrients. The mining process disrupts the cycling of carbon (C), nitrogen (N) (Reeder and Sabey, 1987), and most essential nutrients by altering their forms and amounts in the soil-plant-microbial ecosystem to the point of impeding the reestablishment of a self-sustained plant cover. Decomposition and nutrient mineralization are essential processes in the cycling of nutrients (Mary et al., 1996); therefore, an increase in the organic matter content of soil through a single organic amendment can reinitiate nutrient cycling and may also improve water retention capacity.

Paper mill sludges are a readily available source of organic matter. Deinking sludge, which is very similar to primary sludge, is composed of wood fiber but also contains fillers, ink, and chemicals used to disassociate this material from the pulp fibers (NCASI, 1991). Both primary and deinking sludges are mainly disposed of by incineration and landfilling. Disposal alternatives which are more environmentally sound include land application (NCASI, 1992), use as pot culture media (Chong and Cline, 1993), and mine soil reclamation (Feagley et al., 1994).

The nature of the organic amendment and its behavior once incorporated into the disturbed soil influence establishment and subsequent growth of the plant cover (LRN, 1993). Ideally, decomposition and nutrient mineralization should match plant needs while avoiding significant losses of N (leaching, volatilization, nitrification) and other nutrients. In situ decomposition patterns and nutrient dynamics of paper sludge will determine the length of the stabilization period, the nutrient needs of the vegetation cover, and the risks of groundwater contamination by nitrate leaching during that period. Here we present a summary of the preliminary results on sludge decomposition and N release patterns, water-holding capacity, and plant cover establishment and growth, as well as a more detailed report on the evolution of nitrates in soil solution following the amendment.

MATERIALS AND METHODS

Experiment Outline

The study was conducted in an abandoned sandpit at St-Lambert-de-Lévis (Quebec, Canada) which is about 20 km south of Quebec City. Topsoil and adjacent layers of sand were removed up to approximately 2 m depth. The exposed mine soil left was a C horizon with approximately 94% medium-textured sand, a pH of 5.1, an electrical conductivity of 0.1 dS m^{-1}, and total C and N of 0.1% and 0.02%, respectively. The one-time intervention consisted of incorporating deinking paper sludge (0 and 105 dry t ha^{-1}) into the first 0.21 m of soil. In addition, three rates of N (3, 6, and 9 kg N t^{-1} dry sludge) and two rates of P (0.5 and 1.0 kg P t^{-1} dry sludge) were applied. Finally, tall wheatgrass [*Agropyron elongatum* (Host) Beauv.] was seeded at a density of 680 pure live seeds m^{-2}. Deinking sludge came from the Daishowa, Inc. paper mill (Quebec, Canada) (pH = 9, electrical conductivity = 0.12 dS m^{-1}, dry bulk density = 0.12 g mL^{-1}, total C = 38%, total N = 0.3%). Additional N and P were applied as urea and superphosphate, respectively. Potassium chloride was applied to all plots at 85 kg K ha^{-1}. Experimental plots were established following a split-split-plot design with four blocks; sludge treatments were assigned to main plots, N treatments to subplots, and P treatments to sub-subplots.

Nylon litter bags containing air dry sludge were inserted in soil on the day of sludge incorporation. Three bags per plot were retrieved periodically over a 27-month period and their contents were oven-dried (50°C/72 h) and weighed. Subsamples were used to determine water content, ash content, total N, total C, and carbon fractions. Analysis of soil water retention was conducted on samples collected at different dates over a two-year period; only the highest and the lowest N rates (3 and 9 kg t^{-1} sludge) were evaluated. Analytical methods are described elsewhere (Fierro et al., 1997). Total dry standing biomass of grass and ground cover were determined by the end of the first and the second growing seasons (year 1 and year 2, respectively), using two quadrats (1 m^2) selected randomly in each plot.

Nitrates in Soil Solution

Soil solution was sampled with porous-cup tension lysimeters (Mod. 1920. Soil Moisture Equipment Corp., Santa Barbara, California) at 0.25 m depth, just below the zone of amendment incorporation. One lysimeter was installed at the center of each plot of the two central blocks for a total of 24. Solution samples (volume-weighted) were collected weekly, at which time lysimeters were drained and evacuated to –50 kPa with a hand vacuum pump. From month 12 onward, water from a rivulet permanently flowing along the eastern border of the sandpit was also sampled weekly both upstream and downstream. Monthly composite samples of the soil solution and rivulet water were conserved frozen. Analyses for NO$_3$-N were performed on filtered (45 µm) subsamples by automated ionic chromatography (DX-500, Dionnex Co., Sunnyvale, California). No statistics were performed on this variable since there were only two replications and at several times some lysimeters yielded no water because of low soil water content or because their porous cup became progressively clogged.

Table 8.1. Mean NO_3-N Concentrations (mg L^{-1}) of Soil Solution Sampled at 25 cm Depth in the Amended Sandpit.

Month[a]	Soil without Sludge N Application Rate (kg t^{-1})			Soil with Sludge N Application Rate (kg t^{-1})		
	3	6	9	3	6	9
1	130.4[b]	110.6	195.7	14.9	12.6	85.5
2	249.5	183.6	443.0	17.9	27.0	192.0
3	202.8	98.4	204.1	4.7	129.1	130.2
11	3.9	19.8	20.9	2.3	6.6	53.1
12	4.5	3.3	9.0	11.7	6.3	0.3
13	1.6	2.2	7.3	0.5	0.0	1.1
14	1.9	1.9	56.3	0.3	0.1	0.2
15	0.2	—[c]	1.3	1.0	0.1	0.4
23	0.7	2.7	1.6	—[c]	0.1	0.4
24	0.2	1.6	1.2	—[c]	0.3	0.2
26	0.1	1.4	5.5	—[c]	0.0	0.4

[a] Months after sludge and/or fertilizer incorporation.
[b] Concentrations in monthly composite samples (means of P treatments and two replications).
[c] No sample collected (dry soil or clogged lysimeter).

RESULTS AND DISCUSSION

Vegetation and Soil Responses

Aboveground plant biomass (all standing biomass) was greater in the presence of sludge than in its absence; this effect was observed after one and two years (Fierro et al., in preparation). In some cases, vegetation growth was four times greater in the presence of sludge than in its absence. Aboveground biomass continued to increase during the second year of this study in the presence of sludge, while in its absence there was no further growth. Similar results were reported from a study where the organic amendment for a surface mine soil was sawdust and continuous growth was measured over the five years following the amendments (Roberts et al., 1988). In the presence of sludge, the highest N rate increased growth, especially two years after fertilizer applications when the difference was about twofold compared to lowest N rate. This is in agreement with the N content of decomposing sludge (discussed below) which was still higher at that time for this treatment, indicating that remineralization might have supported a better growth. Under N-limiting conditions, immobilization lasts longer (Mary et al., 1996). So the remineralization at the lowest N rate was limited. The establishment of the vegetation cover (i.e., growth during the first year) was improved by the highest P rate in all treatments, indicating the importance of this nutrient in the establishment of most plants. No effect of P was observed in subsequent growth. In a preliminary greenhouse study, P status of plants growing in sand-sludge mixtures was improved as P addition rate increased, indicating that P was limiting growth (Fierro et al., 1997).

Ground cover was considered excellent in the presence of sludge for both years, ranging between 57 and 87% among fertilizer treatments. In the absence of sludge, ground cover was rather poor, ranging between 20 and 50%. Ground cover increased over time in the presence of sludge, whereas it decreased after one year in the absence of sludge. Similar

trends were reported on abandoned pyritic mine lands where ground cover was still increasing after nine years but only in organic-amended plots (Pichtel et al., 1994). Fertilizer
treatments did not affect ground cover.

Decomposition of sludge in litter bags occurred rapidly through the first 16 months
when a phase of slow decomposition began (Fierro et al., in preparation). Lignified matter
was dominant at that time, accounting for half of the mass remaining which partially ex-
plains the slower decomposition rates observed in this phase (Cheshire and Chapman,
1996). By the end of this study (month 27), total mass remaining ranged from 42 to 47%
among treatments. Decomposition pattern of sludge appears to be faster than that of temperate hardwood leaf litter (Melillo et al., 1982; Prescott, 1995) but slower than that of
cereal straw (Andrén and Paustian, 1987; Cheshire and Chapman, 1996). Total mass of
sludge remaining in soil and its C fractions (lignin, cellulose, hemicelluloses, and solubles)
were not affected by N or P additions at any time. Over the entire course of this study, the
sludge decomposed at approximately the same rate for all N or P treatments. Jack pine
(*Pinus banksiana*) needles also decomposed at the same rate in control and in treatments
fertilized with N and/or P (Prescott, 1995). Nitrogen content of sludge decomposing in
soil increased with N addition rates at all sampling dates. As discussed above, approximately the same amount of decomposed material remained in the soil whatever the addition of N, but this remaining material was richer in N for the highest N addition rate. In all
cases, sludge N content showed an important decrease within the first month (estimates of
initial content included fertilizer N), followed by a small accumulation phase (immobilization) that was apparent only with the lowest N rate. Then a gradual release phase (mineralization) was continuously observed. These three phases of the N dynamics are not always
evident during decomposition process, depending mainly on the composition of the material (Berg and Staaf, 1981).

Water-holding capacity was higher in soil with sludge than in unamended soil at all
matric potentials ranging from 0 to –1500 kPa (Fierro et al., in preparation). This effect
was apparent immediately after incorporation of sludge as well as one and two years later.
This difference in water retention varied from onefold at the low potentials to threefold at
the high potentials. Plant available water, that is the amount of water held between –33
and –1500 kPa, was about doubled in the presence of sludge. These results agree with
those of Smith et al. (1986) who showed that large wood-residue amendments (90 and 135
t ha^{-1}) increased soil water content of bentonite mine spoils. Water desorption curves of
sandpit soil were not affected by N and P treatments. Along with nutrient retention and
supply, water-holding capacity of the sandpit soil is an important limiting factor for revegetation. The days following seeding of the grass were very rainy and seedling emergence
was uniform in the absence and in the presence of sludge. Therefore, differences in plant
response were not due to poor germination. An increased water-holding capacity accounted
for a better survival rate since about twice as much water was continuously available to
plants in the presence of sludge. In addition, this certainly ensured a better flow of nutrients to the roots.

Nitrates in Soil Solution

Nitrate concentrations measured in soil solution collections below the rooting zone
are good indicators of potential leaching losses (Knoepp and Swank, 1995). In this study,
soil solution was collected just below the zone of sludge and fertilizer incorporation,
which also corresponded to the rooting zone of the seeded grass. Values in Table 8.1

represent NO_3-N concentrations in soil solution collected and accumulated over one calendar month following the incorporation of sludge and fertilizers (i.e., month 1 = August 1994, month 26 = October 1996). In all cases, nitrate concentration increased markedly in months 2 and 3. From that time onward, concentrations dropped consistently with the exception of two moderate peaks of a few milligrams per liter in the presence of sludge (months 12 and 15) and one peak in the absence of sludge (month 14). In general, the highest N application rate resulted in higher nitrate concentrations than the other two rates. In the absence of sufficient plant uptake, denitrification, or microbial immobilization, significant amounts of NO_3-N can be leached from the soil profile (Montagnini et al., 1986). The initial high nitrate concentrations observed corresponded to the period of lowest uptake and immobilization. The germinating grass in the sandpit could not account for any significant N uptake at that time, and the period of microbial colonization (Moorhead and Reynolds, 1991) following the massive sludge amendment delayed the onset of net N immobilization. Indeed, initial microbial populations were probably very limited in this system because of its very low content of indigenous organic matter and its disturbed status. In addition, a violent rainstorm happened the day after sludge and fertilizers incorporation, which certainly accounted for important N losses through leaching.

At all sampling dates, nitrate concentration was higher in the absence of sludge than in its presence. This difference between sludge treatments was about tenfold for the first three months and decreased with time. Two factors can be invoked to explain this early effect in mitigating nitrate leaching: (1) the high water-holding capacity of sludge, especially as the internal drainage in this soil is high; and (2) microbial net immobilization of N in the presence of sludge during the rapid decomposition phase. Paper mill sludge may thus be an effective tool for nitrate leaching control that could be used in soils presenting high nitrate leaching potential.

Water from the rivulet had NO_3-N concentrations varying from 0.3 to 9.1 mg L^{-1} (Table 8.2). Concentrations in water sampled downstream were usually slightly higher than in samples from upstream. The difference in nitrate concentrations between upstream and downstream (input) of the rivulet is attributed to the N application treatments in the sandpit. The rivulet is fed by a shallow water table flowing across the sandpit. The highest inputs of nitrates (about 2 to 4 mg L^{-1}) occurred in the most rainy months (14, 15, 23, and 24) which corresponded to September and October 1995, as well as July and August 1996. However, the load of nitrates to groundwater table at the sandpit was below hazardous limits (10 mg L^{-1}) at least from month 12 to month 26 after the amendments. Thus, during the period of monitoring, amounts of N applied to the sandpit did not represent a risk of groundwater contamination by nitrates. However, during the first three months following the amendments, the load of nitrates to the groundwater might have been considerable.

In conclusion, sludge application to an abandoned sandpit sustained a viable plant cover for at least 27 months. Plant growth and ground cover were still increasing at the end of this study in the presence of sludge, while vegetation was in decline after only one year in the absence of sludge even at the highest fertilization rates. Decomposition processes may dominate the nutrient status of this system since it is not buffered by the presence of significant quantities of soil organic matter, plant litter, or mineral nutrients. The longer the effects of sludge on maintenance of plant cover, the higher the chances for the system to reach a sustainable equilibrium since organic matter generated by the vegetation will eventually replace the decomposing sludge in its role on retention and supply of water and nutrients.

Table 8.2. Variations in NO$_3$-N Input and Concentrations in the Rivulet Flowing Along the Sandpit (mg L^{-1}).

	Months after Amendment Application						
	12	13	14	15	23	24	26
Upstream	0.9	0.4	0.3	0.6	2.4	1.9	9.1
Downstream	0.9	1.0	2.7	2.5	5.4	5.8	9.0
Input[a]	0	0.6	2.4	1.9	3.0	3.9	−0.1

[a] input from the sandpit = (Downstream concentration − Upstream concentration)

ACKNOWLEDGMENTS

Support was provided by Les Composts du Québec, Inc. Thanks are also addressed to Ministère de l'Éducation du Québec and to Consejo Nacional de Ciencia y Tecnología (Mexico) for a joint scholarship to A. Fierro.

REFERENCES

Andrén, O. and K. Paustian. Barley Straw Decomposition in the Field: A Comparison of Models. *Ecology*, 68, pp. 1190–1200, 1987.

Berg, B. and H. Staaf. Leaching, Accumulation and Release of Nitrogen in Decomposing Forest Litter, in Terrestrial Nitrogen Cycles, Clark, F.E. and T. Rosswall, Eds., *Ecol. Bull.*, 33, pp. 163–178, 1981.

Cheshire, M.V. and S.J. Champan. Influence of the N and P Status of Plant Material and of Added N and P on the Mineralization of ^{14}C-Labelled Ryegrass in Soil. *Biol. Fert. Soils*, 21, pp. 166–170, 1996.

Chong, C. and R.A. Cline. Response of Four Ornamental Shrubs to Container Substrate Amended with Two Sources of Raw Paper Mill Sludge. *HortScience*, 28, pp. 807–809, 1993.

Feagley, S.E., M.S. Valdez, and W.H. Hudnall. Bleached Primary Papermill Sludge Effect on Bermudagrass Grown on a Mine Soil. *Soil Sci.*, 157, pp. 389–397, 1994.

Fierro, A., J. Norrie, A. Gosselin, and C.J. Beauchamp. Deinking Sludge Influences Biomass, Nitrogen and Phosphorus Status of Several Grass and Legume Species. *Can. J. Soil Sci.*, 77, pp. 693–702, 1997.

Knoepp, J.D. and W.T. Swank. Comparison of Available Soil Nitrogen Assays in Control and Burned Forested Sites. *Soil Sci. Soc. Am. J.*, 59, pp. 1750–1754, 1995.

Land Resources Network Ltd. (LRN). Organic Materials as Soil Amendments in Reclamation: A Review of the Literature. Report #RRTAC 93. Alberta Conservation and Reclamation Council, 1993.

Macyk, T.M. An Agricultural Capability System for Reconstructed Soils. Alberta Land Conservation and Reclamation Council Report #RRTAC 87-13, 1987, p. 46.

Mary, B., S. Recous, D. Darwis, and D. Robins. Interactions Between Decomposition of Plant Residues and Nitrogen Cycling in Soil. *Plant Soil*, 181, pp. 71–82, 1996.

Melillo, J.M., J.D. Aber, and J.F. Muratore. Nitrogen and Lignin Control of Hardwood Leaf Litter Decomposition Dynamics. *Ecology*, 63, pp. 621–626, 1982.

Ministère de l'environnement du Québec (MENVIQ). La réhabilitation de carrières et sablières: un coup de main è l'environnement. Envirodoq 840091, 1984.

Montagnini, F., B. Haines, L. Boring, and W. Swank. Nitrification Potentials in Early Successional Black Locust and in Mixed Harwood Forest Stands in the Southern Appalachians, USA. *Biogeochemistry*, 2, pp. 197–210, 1986.

Moorhead, D.L. and J.F. Reynolds. A General Model of Litter Decomposition in the Northern Chihuahuan Desert. *Ecol. Model.*, 56, pp. 197–219, 1991.

NCASI (National Council for Air and Stream Improvement). Characterization of Wastes and Emissions from Mills Using Recycled Paper. Technical Bulletin No. 613, 1991.

NCASI. Solid Waste Management and Disposal Practices in the U.S. Paper Industry. Technical Bulletin No. 641, 1992.

Pichtel, J.R., W.A. Dick, and P. Sutton. 1994. Comparison of Amendments and Management Practices for Long-Term Reclamation of Abandoned Mine Lands. *J. Environ. Qual.*, 23, pp. 766–772, 1994.

Prescott, C.E. Does Nitrogen Availability Control Rates of Litter Decomposition in Forests? *Plant Soil*, 168–169, pp. 83–88, 1995.

Reeder, J.D. and B. Sabey. Nitrogen, in *Reclaiming Mine Soils and Overburden in the Western United States. Analytic Parameters and Procedures*, Williams, R.D. and G.E. Schuman, Eds., Soil Conservation Society of America, 1987, pp. 155–184.

Roberts, J.A., W.L. Daniels, J.C. Bell, and D.C. Martens. Tall Fescue Production and Nutrient Status on Southwest Virginia Mine Soils. *J. Environ. Qual.*, 17, pp. 55–62, 1988.

Smith, J.A., E.J. DePuit, and G.E. Schuman. Wood Residue and Fertilizer Amendment on Bentonite Mine Spoils: II. Plant Species Responses. *J. Environ. Qual.*, 15, pp. 427–435, 1986.

Chapter 9

Potential of Fodder Tree Species for the Rehabilitation of a Surface Mine in Mpumulanga, South Africa

N.F.G. Rethman and J.P. Lindeque

INTRODUCTION

The coal strip mines situated on the plateau of the Mpumulanga Province in South Africa disturb several hundred hectares of land each year. The objective of environmental management programs is to restore land capability to "as good, if not better" than the original condition. This implies no loss in production potential and no increase in erodibility.

Apart from considerable inputs into the landscaping of topography to ensure minimum slopes and good surface hydrology, and the prestripping and return of top soil to ensure a good rooting depth, particular emphasis has been placed on the establishment of a stable and productive vegetation (Anon., 1981). Such vegetation has invariably taken the form of a planted pasture, which will protect the area against erosion, improve soil conditions, and contribute to production systems based on rehabilitated lands. The ideal, in terms of ecological and economic stability (sustainability), is to incorporate both grasses and legumes in such pastures.

There has, however, been considerable difficulty with the establishment and/or persistence of herbaceous legumes in this region (Barnes et al., 1986). In the light of promising results reported by Rethman and Lindeque (1996) it was decided to evaluate the potential of leguminous fodder trees and shrubs for the rehabilitation of such disturbed areas.

MATERIALS AND METHODS

The investigation was conducted on the Kromdraai Colliery in the Witbank district of Mpumulanga Province (25°48'S, 29°05'E). The site is located at 1,510 m above sea level and is typified by an annual precipitation of 700 mm, which is concentrated in the summer months from October to March. While the wet summers are characterized by mild temperatures (mean maximum in the hottest month is 27°C), the long dry winter period often experiences killing frosts from May to August (mean minimum in the coldest month is 3°C, with the absolute minimums as low as –5°C). The trial was laid out on a deep (> 2 m), well-drained, acidic (pH[H_2O] <5), and sandy (<12% clay) soil. The site received supplementary irrigation, using gypsiferous water, in varying amounts from no irrigation to full irrigation. This water was produced by lime treating acid mine drainage. Prior to planting the site received eight tonnes of dolomitic lime, one tonne of single superphosphate (10.3% P), and a N:P:K planting mixture containing 34 kgN, 51 kg P, and 34 kg K ha^{-1}.

Seedlings of five temperate leguminous shrubs (*Chamaecytisus palmensis, Cytisus maderiensis, Lupinus arborea, Medicago arborescens,* and *Teline stenopetala*) and four subtropical species (*Albizzia julibrissin, Leucaena leucocephala, L. pulverulenta,* and *Sesbania sesban*) were planted in the late summer/autumn of 1994 and were evaluated for survival, productivity, quality, and root development in the subsequent two growing seasons (1994/95 and 1995/96).

RESULTS AND DISCUSSION

Survival

Two and one-half years after planting, there were marked differences in survival/persistence under local conditions. Of the temperate species, *C. palmensis* and *T. stenopetala* (85% and 75% survival, respectively) had the best persistence. *M. arborescens* (55%) and *C. maderiensis* (50%) were intermediate, while the lupin (*L. arborea*) was devastated by disease. Among the subtropical species, both *Albizzia* and *L. pulverulenta* proved to be cold tolerant, with survival rates of 100% and 90%, respectively. In contrast, *L. leucocephala* suffered severe top kill, but had a good coppice regrowth in spring, while *S. sesban* behaved as an annual, with a total winter kill and no coppice regrowth.

Productivity

From the results presented in Table 9.1 it would appear that, in the short term, the temperate species *Chamaecytisus, Cytisus,* and *Teline* exhibit good production potential. These observations must, however, be tempered by information on persistence. This is particularly well illustrated by the lupin species evaluated, which yielded well in the first season but did not persist into the second season. The dwarf *Medicago* compares poorly in terms of dry matter production.

Although the subtropical species were slow starters, by the second season all four species were looking very promising. Sesban has the obvious disadvantage of behaving as an annual under local conditions but the productivity would warrant further investigation of propagation or reseeding techniques.

Quality

Apart from parameters such as the proportion of edible material (leaves and fine twigs <5 mm) in the summer months, and leaf retention in the winter, nutritive parameters such as in vitro digestibility, crude protein, acid detergent fiber, neutral detergent fiber and lignin were also assessed. While there were small differences between temperate and subtropical species with respect to the percent edible material in summer (47% and 51%, respectively), in the winter months temperate species were characterized by good leaf retention while the more tropical species were completely deciduous under local conditions. This is very important in an area where protein-rich forage is in such short supply in winter.

From the data presented in Table 9.2 it would appear that although there was no difference in the crude protein content of temperate and tropical species, the former had a markedly better digestibility and lower fiber fraction. Together with the good leaf retention in winter, this definitely counts in favor of this group.

Table 9.1. Dry Matter Production of Leguminous Fodder Trees and Shrubs in the 1994/95 and 1995/96 Growing Seasons (kg per tree per year).

Species	Growing Season	
	1994/95	1995/96
Chamaecytisus palmensis	1.15	2.85
Cytisus maderiensis	4.25	3.13
Lupinus arborea	2.01	0.00
Medicago arborescens	0.07	0.35
Teline stenopetala	1.07	2.12
Albizzia julibrissin	0.00	1.53
Leucaena leucocephala	0.35	2.94
Leucaena pulverulenta	0.07	1.25
Sesbania sesban	0.22	3.70

Table 9.2. Average Nutritive Value of Leaves of Temperate and Subtropical Leguminous Fodder Trees and Shrubs, Sampled at the End of the Growing Season.

Parameter	Temperate Species	Subtropical Species
% Crude protein	20	20
In vitro digestibility	64	50
Acid detergent fiber	23	28
Neutral detergent fiber	37	39
Lignin	10	14

Root Development

An investigation of the vertical and horizontal spread of roots indicated that the roots of all species were characterized by very shallow (<25 cm) root systems. This might be ascribed to chemical characteristics of the deeper horizons which had a much lower pH and nutrient status (Table 9.3). The direct effect was a strong lateral root growth. This strong, shallow lateral root development would inevitably result in strong competition with any herbaceous species established in combination with such trees or shrubs. It would be preferable for leguminous trees/shrubs, growing in combination with grasses, to have a deeper, more complementary and less competitive root system. Over the longer term it is hoped that irrigation with gypsiferous water will have a beneficial effect on root distribution as the pH and calcium content of the deeper soil horizons are improved.

The results of soil analyses in Table 9.4 indicate that after two seasons of irrigation with gypsiferous water, which was characterized by high levels of Ca and SO_4 (Table 9.5) the pH, calcium, and sulfate content of all sampled horizons (to a depth of 60 cm) were markedly higher. Whether there are other factors which might be affecting vertical root development has yet to be determined, but the generally more favorable pH should also improve the availability of other nutrients.

Table 9.3. Soil Analyses at Three Depths on Nonirrigated Sites.

Depth (cm)	pH(H$_2$O)	Ca	SO$_4$	Mg	P	K (mg kg^{-1})
0–20	5.9	207	244	9	6	19
20–40	4.9	61	152	4	2	15
40–60	4.7	45	85	3	1	10

Table 9.4. Average Soil pH (Water) and Calcium and Sulfate Status of 0–20, 20–40, and 40–60 cm Soil Horizons from Composite Soil Samples on Nonirrigated and Irrigated Sites.

Parameter	Horizon (cm)	Nonirrigated	Full Irrigation
pH (water)	0–20	5.9	6.8
	20–40	4.9	6.1
	40–60	4.7	5.1
Ca (mg kg^{-1})	0–20	207	627
	20–40	61	343
	40–60	45	197
SO$_4$ (mg kg^{-1})	0–20	244	266
	20–40	152	277
	40–60	85	212

Table 9.5. Chemical Composition of Gypsiferous, Lime Treated Acid Mine Water (Average over Two Seasons).

pH (H$_2$O)	6.7
EC (mS m^{-1})	275
NH$_4$ (mg L^{-1})	30
NO$_3$	312
P	10
K	79
Ca	434
Mg	30
SO$_4$	1326
Mn	2.35
Fe	0.41
Na (mmol L^{-1})	0.3
Cl (mmol L^{-1})	2.4

CONCLUSIONS

From the results obtained in this exploratory trial it is evident that there are large differences between the species evaluated, with respect to virtually all parameters considered. Among the temperate species, *Chamaecytisus*, *Cytisus*, and *Teline* warrant further investigation as evergreen, quality forage in an area often characterized by dry, poor quality roughage for a large part of the year. Although the tropical group is deciduous and generally has a poorer quality, their role as summer browse and in terms of leaf mulch should not be underestimated.

REFERENCES

Anonymous. *Guidelines for the Rehabilitation of Land Disturbed by Surface Coal Mining in South Africa.* Chamber of Mines of South Africa, Johannesburg, South Africa, 1981, p. 140.

Barnes, D.L., N.F.G. Rethman, C.C. de Witt, and G.D. Kotze. Exploratory Trials on Re-Inforcement of Veld with Legumes in the South-Eastern Transvaal Highveld. *J. Grassland Soc. Southern Africa,* 3(3), pp. 90–95, 1986.

Rethman, N.F.G. and J.P. Lindeque. A Comparison of Tropical and Mediterranean Fodder Trees and Shrubs for Possible Use in Production Systems of Sourveld Areas of the Northern Provinces of South Africa. *Proceedings of the Fifth Regional Conference of the Southern African Association for Farming Systems Research-Extension.* Arusha, Tanzania, 1996.

Chapter 10

Impact of Rare-Earth Mining on the Environment and the Effects of Ecological Recovery Measures on Soils

L.F. Xu and M.L. Liu

INTRODUCTION

There are abundant rare-earth resources of medium and heavy types in Guangdong and Jiangxi, and especially, rare-earth deposits having ion adsorption properties is one of the distinct characteristics of the mineral resources in South China. In recent years, with the extensive use of rare-earths, domestic and international demand has increased, and with the economic development of mountain areas to eliminate poverty, rare-earth mining is developing rapidly. However, this has led to a series of ecoenvironmental problems, such as forest destruction, soil and water loss, and environmental pollution. At present, there are a few reports on the environmental impact of rare-earth mining and possible ecological remediation measures and their effects on corresponding soil properties (Bradshaw 1990; Guo et al., 1990). This chapter, taking the rare-earth mining in Pingyuan County of Guangdong Province as an example and from the aspect of sustainable resources utilization, investigates and analyzes the impact of rare-earth mining on the environment, evaluates the existing effect of ecological remediation measures on soils, and finally, suggests some alternative vegetation recovery measures.

MATERIALS AND METHODS

On the basis of collecting relevant material about the impact of rare-earth mining on the environment, this chapter focused on the investigation of rare-earth mining in Huanshe and Renju towns where mining is widespread.

The study reports the main effects of the existing ecological remediation measures (Table 10.1), analyzes the effects of ecological remediation measures on soil properties, including the changes in soil mechanical composition, soil organic matter, available N, P, and K contents and pH value.

The following methods were used for soil analysis: the hydrometer method for determining soil mechanical composition; the $K_2 Cr_2 O_7$ volumetric procedure for soil organic matter; the alkali-hydrolyzadle diffusion method for available N, 0.5M $NaHCO_3$ immersed colorimetric method for available P, 1N NH_4 OAc immersed for available K, and flame photometry for the K determination; the water extract method for soil pH value, with water:soil 2.5:1 (Nanjing Agricultural College, 1980).

Table 10.1. Designing of Main Vegetation Recovery Quadrates.

No.	Vegetation Treatments
1	Former natural vegetation (Control 1)
2	Mining mud piles (tailings) (Control 2)
3	Planting of grasses (*Stylidium uliginosum* Swartsz., *Mielinis minutiflora* Beauv.)
4	Planting of bamboo
5	Mixed planting of grasses, bamboo, and pine
6	Planting of fir
7	Mixed planting of grasses, bamboo, pine, and schima

RESULTS AND DISCUSSION

Rare-Earth Mining Situation of Pingyuan Country

Rare-earth resources in Guangdong Province are abundant and are characterized by many varieties, large reserves, and extensive distribution. Hence Guangdong has become one of the main rare-earth production regions in China. In Pingyuan, the main rare-earth is the ion-adsorption type. The mining began in 1986, and the main mining sites distributed in Huanshe, Renju, and Bachi Towns. In 1989, the county's ability to produce oxidized rare-earth reached a maximum, with output of 1600 t yr^{-1}. From 1990 to 1992, many mining sites stopped production and went bankrupt because of falling prices for rare-earths. In the latter half of 1993, the price rose again and more than 50 sites resumed production (Figure 10.1). It was estimated that there were 200 illegally-run mining sites in 1995.

Effects of Rare-Earth Mining on the Environment

Rare-earth mining, especially the mining of the ion type, resulted in a series of ecoenvironmental problems such as serious vegetation destruction, as well as soil and water loss.

Destruction of Vegetation

To mine rare-earth of the ion-type, it is necessary to cut vegetation and strip the topsoil, then unearth the mineral and extract the oxidized rare-earth. Therefore, no natural vegetation remains and the green mine sites become "yellow slopes." According to our investigation, the average biomass is 1.045 kg m^{-2}. If the mining sites occupy 211.5 ha, the damaged vegetation amounts to 2211.8 t. In the region, there are no endangered species of plants.

Occupying Hills and Farmland

Rare-earth mining naturally occupies some valuable agricultural lands. The land occupation by rare-earth mining in Pingyuan county in recent years is shown in Figure 10.2. In addition, the tailings sometimes destroyed farmland following torrential rain. For example, on August 2, 1990, the flood destroyed some mining banks and about 100 ha of farmland.

FIGURE 10.1. Change in rare-earth output in recent years in Pingyuan county.

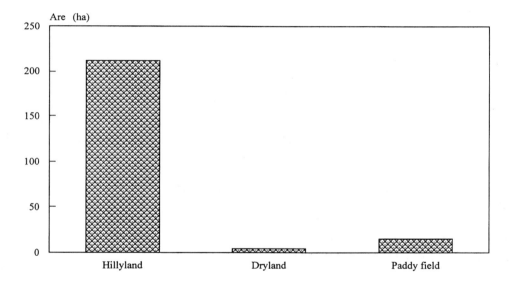

FIGURE 10.2. Land occupation by rare-earth mining.

Serious Loss of Soil and Water

Soil and water loss is a main environmental problem of rare-earth mining. To mine each t of oxidized rare-earth, 1,300–1,600 m^3 of waste rock is unearthed. If there is no bank to protect the large amount of tailings, and torrential rain falls, mud with stones will scour the farmland, silt up rivers, pools, and reservoirs, and pollute water resources. In the light of our observation at two typical mine sites, soil erosion reaches 4.18×10^4 t km^{-2} on the loose tailing piles with no remediation treatment. If the land occupation area is assumed to be 211.5 ha, soil erosion amounts to 8.8×10^4 t yr^{-1}.

Effect on Water Quality of Rivers and Reservoirs

Mining rare-earth of the ion-type requires the use of a large amount of ammonium sulfate and oxalic acid, one t of mixed oxidized rare-earth needing 6–7 t of ammonium and 1.2–1.5 t of oxalic acid. Because the waste mine sludge absorbed a large amount of extracting solution, acid wastewater constantly seeped from tailing sites; pH of water in drainage ditches was 4.6 to 5.3, polluting lower reaches and reservoirs. According to the test in the area, the river water pH was 5.3 to 5.8, chemical-consumed oxygen was high, the highest was 44 mg L^{-1}, the sulfate radical reached 33.06 mg L^{-1}, beyond the standard of surface water quality. Moreover, the water in the river was muddy regardless whether raining or not, and the suspended amount was also high.

Analysis of Effect of Ecological Remediation Measures on Soil Properties

The ecological principle of the mining remediation measures is based on the theory of ecological succession. Vegetation recovery in the degraded ecological system is the most important work of recovery ecology, because the recovery and rebuilding of all natural ecological systems is always controlled by ecological succession (Lan and Shu, 1996; Peng, 1996).

Effect of Different Vegetation Recovery on Soil Erosion

Vegetation recovery is a key to controlling soil erosion. This investigation showed that the different vegetation recovery measures played an important role in controlling soil erosion of rare-earth tailings. Using the investigation quadrates and the U.S. soil loss equation (A=R · K · LS · C · P) in which the parameters were adjusted and revised, the authors measured and calculated soil erosion under different vegetation recovery measures (Figure 10.3). Figure 10.3 indicates that soil erosion of the tailing piles was very serious, with erosion reaching 41800 t km^{-2}, 50 times greater than that under the original vegetation cover. Among the recovery measures, the mixed species planting and planting of fir (close planting) were the best treatments in reducing soil erosion, although it did not fall to premining levels.

Effect of Different Vegetation Recovery Measures on Soil Physical and Chemical Properties

Different vegetation covers influence soil physical and chemical properties. In this study, various vegetation recovery measures had been in place for 4–5 years, and the analytical results of topsoil samples showed that the soil properties of different treatments had changed reasonably.

Soil clay (<0.01 mm) content varied with treatments, with the highest content (30.8%) in the control 1 (under former natural vegetation) and the lowest (4.2%) in the tailings. The clay contents of other treatments also increased, related to the tailings, particularly on the grassed treatment, which had a clay content six times as much as that of the tailings. Thereby, vegetation recovery measures significantly increased soil clay content (Figure 10.4).

Vegetation recovery measures had a positive effect on the recovery of soil organic matter (Figure 10.5). Compared with the tailings, grasses increased organic matter content the

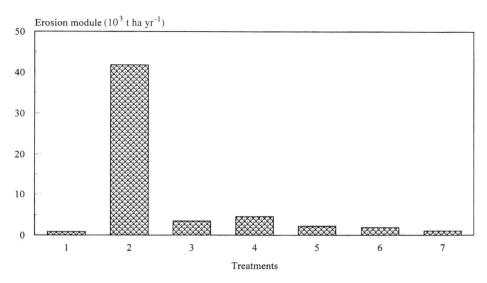

FIGURE 10.3. Effect of different vegetation recovery measures on soil erosion modules.

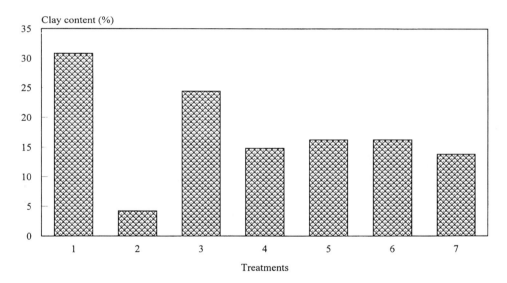

FIGURE 10.4. Effect of vegetation recovery on soil clay content.

least (55.7%). The largest increase was found in the mixed planting of treatment 7, increasing more than 12 times over the tailings, with other treatments intermediate in effect. It is therefore concluded that vegetation recovery measures effectively promote the recovery of soil organic matter content.

Changes in rapidly available N, P, and K in soil are shown in Figure 10.6. The contents of rapidly available N in the soils by planting grasses and mixed planting of grass, bamboo, pine, and schima in treatment 7 were higher than that of other treatments, but less than half as much as that of control 1; the order was treatment 5>4>6; even the highest content of rapidly available N of treatment 5 was 22.6% of that of control 1. Figure 10.6 also showed

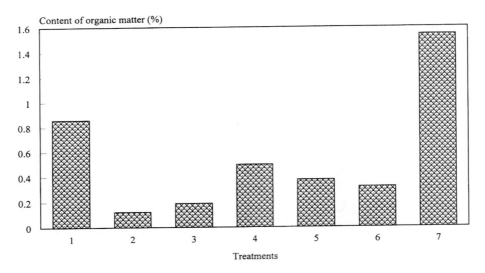

FIGURE **10.5.** Effect of vegetation recovery on soil organic matter.

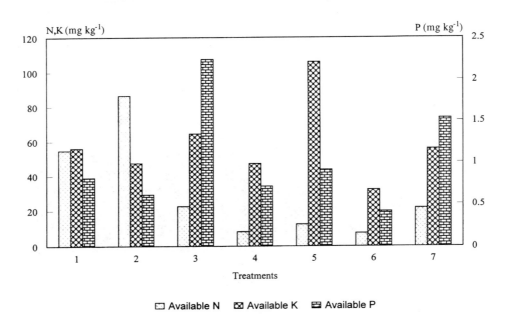

☐ Available N   ⊠ Available K   ▦ Available P

FIGURE **10.6.** Impact of vegetation recovery on rapidly available N, P, and K in soil.

that the available N content of control 2 was higher than that of control 1, because in the course of extracting minerals, $(NH_4)_2\ SO_4$ extracting solution is used to raise the NH_4^+ content of the tailings. It also indicates that after planting vegetation, the content of available N dropped sharply, and never returned to the level of the original natural vegetation. The changes in available P and K were relatively small. For treatments of planting grass and mixed planting of grass, bamboo, pine, and mixed planting of grass, bamboo, pine, and schima, the contents of P and K were similar or even higher than that of the natural veg-

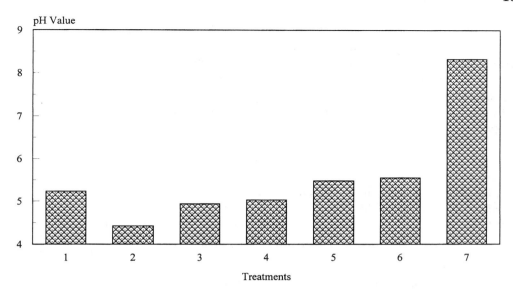

FIGURE 10.7. Impact of vegetation recovery on pH value of soil.

etation. But for treatments of planting fir and planting bamboo, the contents were lower, sometimes even lower than that of control 2 (tailings). That was because during the vegetation recovery process, since the tailings have a high content of NH_4^+, in other words a high content of available N, organic fertilizer as well as P and K fertilizer were applied to maintain soil nutrient status. Therefore, the P and K contents were ever higher than that of the soil under the original vegetation. For the treatment of planting fir and planting bamboo, since the vegetation grew well, the trees are the predominant plant type. Canopy closure had occurred, therefore inhibiting the under-layers where plants grow. The absorption of soil nutrients increased; therefore, the P and K contents were relatively low.

During the production of rare-earths, a large quantity of oxalic acid ($H_2C_2O_4$) was required to react with the sulfuric acid rare-earth solution, in order to extract oxalic acid rare-earth. The solution has a high acidity, which causes soil pH to fall. The impacts of recovery measures on pH values of the soil varied. All recovery measures caused the pH value of the soil to rise to varying degrees. Among all the measures, planting fir, mixed planting of grass, bamboo, and pine, and mixed planting of grass, bamboo, pine, and schima increased the pH value most, even higher than that of natural vegetation soil. Therefore, the vegetation recovery measures play a significant role in controlling the acid pollution of soil.

CONCLUSION

From the above-mentioned analysis we can conclude that the mining of rare-earth has caused relatively serious ecoenvironmental problems, i.e., destruction of vegetation, soil erosion, occupation of cultivated land, and environmental pollution to a certain extent, among which soil erosion of tailings is the most serious. In order to effectively control the ecological destruction caused by rare-earth mining, certain engineering measures, such as building flood control dams, are required but the most important is to provide vegetative cover.

The plant types presently used for vegetation recovery can survive and grow well after a series of management measures such as applying fertile soil and organic fertilizers. The various vegetation recovery measures have played a role in raising soil clay content; improving soil composition; increasing organic matter and available N, P, and K content; improving soil nutrients; and increasing soil pH.

Further investigation of the effects of vegetation replacement on mined lands is needed to optimize remediation of these degraded lands.

REFERENCES

Bradshaw, A.D. Management and Restoration of Wasteland in Western Europe. *Acta Ecologica Sinica*, 10, pp. 28–35, 1990.

Guo, H.C., D.R. Wu, and H.X. Zhu. Land Restoration in China. *Acta Ecologica Sinica*, 10, pp. 24–27, 1990.

Lan, C.Y. and W.S. Shu. Amelioration of Soil Media During Vegetative Rehabilitation in the Mining Wasteland. *Chinese J. Ecol.*, 15, pp. 55–59, 1996.

Nanjing Agricultural College. *Soil Agrochemical Analysis.* Agriculture Press, Beijing, China, 1980, pp. 381.

Peng, S.L. Restoration Ecology and Vegetation Reconstruction. *Ecologic Sci.*, 15, pp. 26–31, 1996.

Chapter 11

Rehabilitation of Copper Mine Tailings at Zhong Tiao Shan and Tong Ling, China

L.B. Zhou, R. van de Graaff, H.W. Dai, Y.J. Wu, and L.N. Wall

INTRODUCTION

The two pilot areas selected for researching suitable strategies and methodologies for rehabilitating mine tailings were the Mao Jia Wan tailings pond in Zhong Tiao Shan, a mountain range about 1,000 km south of Beijing, and the Wu Gong Li tailings pond in the floodplain of the Yangtsg River on the outskirts of Tong Ling, a major copper-producing center and city 180 km southwest of Nanjing. In earlier rehabilitation works in the Zhong Tiao Shan area, some older tailings ponds had been covered with 1 to 1.5 m of locally quarried reddish brown loess material, with a dominantly silty clay loam texture (34% clay), prior to release of the remade land to farmers. A few samples of this material proved to have a cation exchange capacity of about 40 meq 100 g^{-1} and between 0.2 and 4.0% finely disseminated calcium carbonate and occasional gypsum crystals. It is understood that cropping on these rehabilitated tailings ponds was successful from the start. However, covering tailings with such a thick layer of imported soil material entails major costs. Moreover, the loess is not necessarily as plentiful in other locations where it could be used for rehabilitation.

At Tong Ling, none of three tailings ponds visited by the Chinese-Australian team had undergone any rehabilitation treatment. The dominant natural soil in the area was a Terra Rossa developed on limestone. Natural soil materials for amending tailings were, however, almost unobtainable.

Because of the high cost of laboratory analyses, the number of samples analyzed was kept as low as possible, and analytical results are generally regarded as indicative only. Soil or tailings samples represented composited samples from at least ten individual samples.

The tailings at both locations were predominantly very fine to fine sandy in texture (particle size from 0.2 to 0.02 mm), but where slimes have accumulated, the textures were silty (0.02 to 0.002 mm). At Tong Ling they contained between 3.5 and 8.0% of carbonate, while at Zhong Tiao Shan they had about 30% carbonate. At Zhong Tiao Shan, the bulk mineralogy of the tailings as well as that of the clay fraction (less than 1.5%) was determined. In the bulk sample quartz, dolomite and feldspar were codominant, with muscovite subdominant, and calcite forming a minor constituent. Possible traces of pyrite and talc were identified. The clay fraction was dominated by chlorite, muscovite, and vermiculite with minor and trace amounts of calcite, dolomite, feldspar, and quartz. Physical and mineralogical analyses were carried out by the State Chemistry Laboratory (SCL) in Werribee, Victoria, and the CSIRO Division of Soils Laboratory in Adelaide, South Australia.

At Zhong Tiao Shan, plant-available potassium (SCL Method) in the tailings was around 360 mg kg^{-1}, and considered to be adequate for crops. Plant-available phosphorus (Olsen et al., 1954) was always below detection (<0.1 mg kg^{-1}). Total Kjeldahl nitrogen ranged from 10 to 68 mg kg^{-1}, except for old tailings at the surface with a sparse weed growth which had 127 mg kg^{-1}. Phosphorus and nitrogen were considered to be deficient. pH values (1:5 H$_2$O) ranged between 7.5 and 7.7.

Total and EDTA-extractable heavy metals were determined in all tailings samples, but only total concentrations will be discussed in this chapter. At both pilot locations the tailings contained between 200–600 mg kg^{-1} Cu, 50–150 mg kg^{-1} Co, 5–35 mg kg^{-1} As, as well as minor concentrations of Mn, Pb, Se, and Zn, and very low levels of Cd.

The tailings were thus judged to make a poor soil for plant growth due to: (1) low water-holding capacity, (2) low cation exchange capacity, (3) virtual absence of N and P, and (4) low interparticle cohesion, leading to dust generation and possible sandblasting of seedlings. Due to the high pH and the probable high concentration of Ca in the soil solution, it was believed that Cu uptake by the crop would be minimal (Römkens, 1994) and that Co would be the only other heavy metal of potential concern.

It follows that any rehabilitation experiments should attempt to overcome the deficiencies of the tailings; that is, amending the tailings by increasing the clay content and the availability of N and P, and, in the longer term, by building up the organic matter level and promoting soil microbial life.

At Zhong Tiao Shan the restricted availability of loess from a local quarry led to an experiment with four tailings treatments using three reduced levels of added loess and one control. A single fertilizer treatment was given to all plots, and, superimposed on this scheme, four crops and three green manures were grown. The fertilizer rate was based on local district application rates and doubling this to overcome the lack of nutrient capital in the tailings. The loess also was very low in plant-available N and P, and posed no heavy metal problems.

At Tong Ling, where soil could not be obtained, we used only the double fertilizer application rate and set up trial plots for individual species of grasses, white clover, lucerne, and a number of tree and shrub species.

OBJECTIVES

Mao Jia Wan Tailings Dam, Zhong Tiao Shan

- To develop cheap but effective methods for converting the unproductive tailings materials into productive agricultural soil by adding suitable amendments and using minimal loess for satisfactory results;
- To find crop species that are suited to the new soils created by modifying and remediating the tailings;
- To analyze all types of produce normally used for human consumption in terms of heavy metal concentrations that may affect human health.

Wu Gong Li Tailings Dam, Tong Ling

- To develop suitable revegetation methods of tailings to control dust generation in a setting where no imported soil can be used to modify or cover the tailings;
- To determine which grass and tree species can be grown on tailings having shallow groundwater, which are also suitable or desirable for street and park plantings.

MAO JIA WAN TAILINGS DAM PILOT PROJECT

Methods

An experimental area of approximately 100 x 100 m was selected on the tailings, covering a level area on the crest of the dam of the Mao Jia Wan tailings pond, and fenced. The area was smoothed by bulldozer and subdivided into 144 square plots of 5 x 5 m, separated by 1- or 2-m wide paths for access. In a third of these—48 plots—imported loess soil was brought in and placed on top of the tailings at four levels, with a control receiving none, and the others receiving 15 cm, 30 cm, and 40 cm thick layers. These tailings treatments are called T1, T2, T3, and T4, respectively. These treatments were intended to give tailings to loess ratios of 1:0, 1:1, 1:3, and 0:1, as the latter was too deep to mix with tailings. T4 resembled functionally the traditional method but used far less loess. Each of the treatment areas had 12 plots to be planted to four crops in three replicates.

Due to logistical constraints, all the plots with the one level of loess placement were adjacent to each other. After placement of the loess, the soil was cultivated to the maximum depth achievable with ox-drawn plows and a small ride-on tractor, or with shovels by hand. The intention was to mix the surface loess layer with as much of the underlying tailings as was possible. The maximum depth of cultivation with the available equipment was about 30 cm. In treatment T2, additional tailings were brought in on top of the loess to achieve a 1:3 mixing ratio. Based on the average calcium carbonate content of the tailings (30%), the loess (1.2%), and the mixed soils of T2 and T3, it was estimated that following mixing rations were achieved:

T1: Tailings only
T2: Loess:Tailings = 2:1
T3: Loess:Tailings = 2.8:1
T4: Loess only

Three green manure treatments were superimposed on additional areas of 48 plots each of the T1 (zero loess) and T4 (40 cm loess) treatments. These were a first season (summer 1994) planting of, respectively, sudax (a sorghum-Sudan grass hybrid) and lupins (*Lupinus* sp.), perennial ryegrass (*Lolium perenne*) and white clover (*Trifolium repens*), and mung beans, sometimes referred to as Chinese lentils, (*Vigna radiata*), to be followed by cropping for winter wheat in 1994–1995. Apart from the sudax and lupins, which came from Australia, all other green manure crops were locally obtained varieties.

Each area of the first mentioned T1, T2, T3, and T4 was planted to two leguminous crops—soybeans and peanuts—and two graminaceous crops—maize and sorghum. Each crop therefore had three replicates.

On the upwind side of the experiment a windbreak was planted of a mixture of sudax and sorghum. The windbreak was 15 m deep and extended the full width of the experimental field.

The only fertilizers available locally were a complete NPK fertilizer, ammonium phosphate, and ferric sulfate. All plots received the following in kg ha^{-1}:

Nitrogen N	Phosphorus P	Potassium K	Sulfur S	Follow-up Nitrogen, N
180	239	36	177	245

These application rates, representing twice the usual application rate used by farmers in the district, were intended to make up for the lack of nutrients in the tailings-loess materials. The use of ferric sulfate was strongly recommended by the local Agricultural Extension Service for new soils. The Chinese-Australian team considered that ferric sulfate would probably not be harmful to plants growing on alkaline calcareous soils, and gave the Extension Service an opportunity for involvement in the experiment.

Details on row spacing, seed spacing, germination, and crop development are reported in unpublished project reports available from CARIM in Beijing.

Results and Discussion

Crop Yields on the Mao Jia Wan Tailings Dam, Zhong Tiao Shan

Although average rainfall in the Zhong Tiao Shan area is 630 mm per year, with most rain falling in summer, the summer of 1994 was reported to be the driest for 30 years. Unfortunately, a rain gauge was not installed on the experimental site until the following year. Early rains were good and all crops, including soybeans and maize, developed to full size, subjectively measured as knee-high and head-high, respectively, at flowering time. Follow-up rains did not arrive and the soybeans failed to set seed, while maize cobs appeared to be rather small. The same phenomena were observed on farmland in the surrounding district.

Peanuts and sorghum, however, produced respectable yields of the same magnitude as occurred in the district, as shown in Table 11.1. The data are based on sampling the entire plots, calculating an average per mu (1/15th ha) and aggregating the results for the whole treatment.

Sinclair and Serraj (1995) suggested that soybeans belong to one of two groups of legumes and this group is unable to transport nitrogen effectively from the roots into the above-ground plant under drought conditions. Peanuts belong to the other group which utilizes a different mechanism for N transport. They classed soybean as a legume particularly prone to yield loss under drought conditions.

We believe that part of the variability between plot yields within the same tailings treatment and for the same crop is due to uneven fertilizer distribution. Weighed amounts of fertilizer were broadcast by hand by inexperienced laborers and at the time, the uneven application was easily observed.

At planting time there appeared to be some moisture already in the tailings and the loess. Possibly there was enough water to start the swelling of the seeds. On the 23rd of May 1994, several days after planting finished, there was good rainfall and in treatments T1, T2, and T3, seedlings emerged, but none could be seen in T4. T4 seedlings emerged on June 5th. It is hypothesized that in a dry summer, water was more readily available in the sandier substrates of T1 to T3 than in the silty clay loam of T4.

It can be seen that peanuts could yield as well as the district average on the tailings if supplied with adequate fertilizer. Peanuts may not require much water. Sorghum is known to be a drought-resistant grain crop. It performed above expectations in T2. There is an apparent trend for yields to decline with an increase in loess content of the root zone, with T1 having the highest yields, with peanuts being the only exception yielding best in T3. A hypothesis that water became less readily available with an increase in clay content and too easily lost by drainage in the pure tailings as the summer progressed would seem to be an obvious explanation. Soil water monitoring by neutron probe did not start until the 1996 growing season.

Table 11.1. 1994 Summer Season Plot and Average Yields in kg ha^{-1} at Zhong Tiao Shan (Sun-Dried Weights).

Crop	Long Term District Ave.[a]	T1 Treatment	T2 Treatment	T3 Treatment	T4 Treatment
Peanuts (in shell)	2250–3000	1596; 2628; 2836; *ave. 2353*	3504; 3128; 2920; *ave. 3184*	2636; 2992; 2520; *ave. 2716*	1812; 1648; 1448; *ave. 1636*
Sorghum (grains)	7500	5620; 2624; 5520; *ave. 4588*	8312; 7640; 8296; *ave. 8082*	6380; 7028; 6700; *ave. 6702*	4016; 4740; 4856; *ave. 4537*
Maize (kernels)	5100	2996; 1484; 2748; *ave. 2409*	3356; 4496; 4956; *ave. 4269*	3268; 4216; 4476; *ave. 3986*	2848; 4392; 2492; *ave. 3244*
Soybean (beans)	2400	108; 88; 68; *ave. 88*	136; 232; 92; *ave. 153*	160; 120; 108; *ave. 129*	240; 272; 196; *ave. 236*

[a] Information obtained from Yuanqu County Department of Agriculture

Table 11.2 plots the trends of yields through the summer of 1994, the winter wheat harvest of June 8, 1995, followed by the summer crops of 1995 and the winter wheat harvested in early June 1996. Winter wheat was planted on the entire experimental area and also received complete fertilizer at double the district practice. The table also suggests that the green manure had an effect in raising yields. T1 with sudax/lupins yielded almost twice as much wheat as did T1 cropped in the previous summer, even though the lupin component was almost nonexistent. The same sudax/lupins green manure on T4 also gave an improved yield. Perennial ryegrass and white clover established very poorly; hardly any clover plants became established. This type of green manure therefore could not demonstrate any possible beneficial effects. The mung beans green manure showed a very small, possibly nonsignificant beneficial effect.

On the T4 sudax green manure area, the yield of 1995–96 winter wheat was more than 50% higher than on the T4 cropping area. This suggests that the effect of the sudax green manure lasts at least two years.

Heavy Metal Concentrations in Produce for Human Consumption

The initial analyses of tailings and loess for EDTA-extractable and total heavy metal concentrations indicated to the team that copper and cobalt were the only metals of concern. The analytical data are all in unpublished project reports held by CARIM in Beijing, AusAid in Canberra, and one of the junior authors, R.H.M. van de Graaff in Melbourne. EDTA-extraction of heavy metals is routine in Australian environmental auditing, but there is no useful general correlation between EDTA-heavy metal levels and crop uptake in Australia or China. Similarly, there is no such correlation between total concentrations and uptake. The latter, however, is more useful in case of direct ingestion of soil.

Uptake of heavy metals by plants is governed by solubility in the soil water and hence by speciation. Since the root zone of the tailings and the amended tailings may be characterized as a generally oxidizing and alkaline environment, the predicted species are (Hutchinson and Ellison, 1992):

Table 11.2. Crop Yields in Relation to Yuanqu County Yields, 1994-1996.

	T1 (kg ha⁻¹)	T2 (kg ha⁻¹)	T3 (kg ha⁻¹)	T4 (kg ha⁻¹)	T1-GM(s/l) (kg ha⁻¹)	T1-GM(cl) (kg ha⁻¹)	T1-GM(pr/c) (kg ha⁻¹)	T4-GM(s/l) (kg ha⁻¹)	T4-GM(cl) (kg ha⁻¹)	T4-GM(pr/c) (kg ha⁻¹)	Normal Year Average normal farmland (kg ha⁻¹)	1994-95 Average district yields (kg ha⁻¹)
Summer crops 1994												
Peanuts in shell	2353	3184	2716	1636							2250–3000	n.a.
Sorghum grain	4588	8082	6702	4537	All plots covered by a green manure crop						7500	n.a.
Maize kernels	2409	4269	3986	3244							5100	n.a.
Soybeans	88	153	129	236							2250–2400	n.a.

Notes:Bold numbers where yield exceeded the normal year average
The 1994 summer season was the driest on record for 30 years

	T1	T2	T3	T4	T1-GM(s/l)	T1-GM(cl)	T1-GM(pr/c)	T4-GM(s/l)	T4-GM(cl)	T4-GM(pr/c)	normal farmland	district yields
Winter wheat crop 1994-95												
Wheat	675	1275	1215	1455	1170	870	675	1845	1485	1290	3750–4500	1500–1875

Notes:GM(s/l) = green manure of sudax and lupins
GM(cl) = green manure of Chinese lentils (mung beans)
GM(pr/c) = green manure of perennial ryegrass and white clover
No rain fell at all after the beginning of May '95 till the wheat was harvested on 8 June '95, severely depressing that season's yield

	T1	T2	T3	T4	T1-GM(s/l)	T1-GM(cl)	T1-GM(pr/c)	T4-GM(s/l)	T4-GM(cl)	T4-GM(pr/c)	normal farmland	district yields
Summer crops 1995												
Peanuts *) in shell	1698	0	0	0	np	np	np	np	np	np	Most farmers	
Sorghum grain	np	np	np	np	np	np	np	np	np	np	did not plant	
Maize*) kernels	0	0	0	0	990	866	0	0	0	0	summer crops	
Soybeans	np	np	np	np	np	np	np	np	np	np	due to lack of rain	

Notes:The 1995 summer was again a very dry summer, all seeds germinated one month too late, and yields were poor.
Crops in several plots were pulled out to ready the land for winter wheat sown in October 1995.
Plot T1(pr/c) missed out on fertilizer and was a total failure.
*) Peanuts were sown into a standing winter wheat crop; 90 day maize variety in view of short season after wheat
np = not planted in this season
               n.a. = data not obtained from Yuanqu Department of Agriculture

	T1	T2	T3	T4	T1-GM(s/l)	T1-GM(cl)	T1-GM(pr/c)	T4-GM(s/l)	T4-GM(cl)	T4-GM(pr/c)	normal farmland	district yields
Winter wheat crop 1995-96												
	np	1437	1665	1257	np	np	390	1964	1021	1369		2250

For cobalt: solid phase (s) and suspended solid (ss)—$Co(OH)_2(ss)$, $Co_3O_4(s)$, and $Co_2O_3(s)$
ionic forms—Co^{+2}, $Co_2(OH)^{+3}$
For copper: solid phase—$Cu_2O(s)$, $Cu(OH)_2(s)$, $CuCO_3(s)$
ionic forms—Cu^{+2}, $Cu(OH)^+$, $Cu_2(OH)_2^{+2}$

Since the tailings and the amended tailings, as well as the loess, have free calcium carbonate, it is likely that both metals commonly occur as carbonates in the solid phase.

Römkens (1994) researching the behavior of copper in Dutch soils heavily dressed with pig manure found that there was an interaction between dissolved organic carbon (DOC) and Ca^{+2} ions in the soil moisture, resulting in a precipitation of DOC when Ca^{+2} concentrations increased. At zero calcium concentration, the DOC concentration was about 120 mg L^{-1} and this decreased linearly to about 20 mg L^{-1} at a calcium concentration of about 90 mg L^{-1}. Copper in the soil solution appears to be largely complexed to DOC, so that with the increase of calcium in the soil solution there was a proportional decrease in dissolved copper.

Applying this finding to the tailings rehabilitation, we predicted that it was most unlikely for crops to take up excessive copper from their root zones, particularly as over time, organic matter contents will rise.

Baker (1990) describes the existence of copper in soils as being in six 'pools': soluble ions; soluble inorganic and organic complexes; exchangeable Cu; stable organic complexes, adsorbed Cu on hydrous oxides of Mn, Fe and Al; adsorbed Cu on the clay-humus colloidal complex; and crystal lattice-bound Cu. The copper in the tailings is likely to exist chiefly as the crystalline carbonate, malachite, and, in the presence of an abundance of dolomite and calcite it will be stable. However, as the tailings weather, iron oxyhydroxide coatings are forming on the particles as evidenced by color changes from light grey to yellow browns and rusty colors. These iron compounds should capture and immobilize Cu ions from the water phase.

Because of its siderophile nature, Co in the tailings may be present in iron- and manganese-bearing minerals substituting for Fe and Mn. This Co is not readily available to plants. The solid phase forms of cobalt are all very insoluble (Smith, 1990) and Co is expected to be immobile in alkaline environments. Co released by weathering may be sorbed by other soil minerals, especially Fe and Mn oxides. Adsorption of Co to MnO_2 is well documented by Gilkes and McKenzie (1988), and increases with pH (McKenzie, 1967). Thus, excessive uptake of cobalt by the crops grown on the amended tailings is not expected.

Table 11.3 summarizes heavy metal contents in the soil and amended tailings in which crops were grown, as well as in the seeds bought locally for the experimental crops, and then in the edible produce, the harvested grains, beans, or peanuts. Note that uptake in the produce of Cu and Co appears to be unrelated to the levels of these metals in the T1 to T4 soil/tailings media. This is expected on the basis of established geochemical principles.

Table 11.3 also shows that Cu concentrations in the produce are very similar to those in the bought seeds, which presumably were grown on 'normal' soils in the district. The leguminous crops appear to have a tendency to take up much more Cu than the graminaceous crops. For Co, the picture differs in that the produce contains four to eight times as much of this metal than did the original seed. Again, the leguminous crops accumulate more Co than the grains.

In terms of the quoted tolerable daily intakes by humans, which are official Government standards in The Netherlands, (Vermeire et al., 1991), Table 11.3 suggests that the daily consumption of about 0.6 kg of maize, or about 0.3 kg of sorghum, or 0.16 kg of

Table 11.3. Soil-Food Chain Relationships Between Heavy Metal Concentrations in Seeds and Produce at Zhong Tiao Shan.

Material Type and Location	Sample No.	Depth (cm)	Total As (mg kg^{-1})	Total Cd (mg kg^{-1})	Total Co (mg kg^{-1})	Total Cr (mg kg^{-1})	Total Cu (mg kg^{-1})	Total Hg (mg kg^{-1})	Total Ni (mg kg^{-1})	Total Pb (mg kg^{-1})	Total Zn (mg kg^{-1})
Soil materials											
Soil - T1 treatment											
spring sample '94	12329/94	0–10	19	<0.6	n.d.	29	330	<0.2	33	<30	20
ditto autumn sampling '94	21093/94				91		320				19
ditto autumn sampling '95	14624/95				80		270				40
Soil - T2 treatment											
spring sample '94	12330/94	0–10	14	<0.6	n.d.	36	290	<0.2	39	<30	37
ditto autumn sampling '94	21094/94				80		240				41
ditto autumn sampling '95	14625/95				130		320				30
Soil - T3 treatment											
spring sample '94	12331/94	0–10	18	<0.6	n.d.	35	230	<0.2	35	<30	39
ditto autumn sampling '94	21095/94				64		190				41
ditto autumn sampling '95	14626/95				130		350				20
Soil - T4 treatment											
spring sample '94	12332/94	0–10	15	<0.6	n.d.	40	32	<0.2	30	<30	55
ditto autumn sampling '94	21096/94				14		29				54
ditto autumn sampling '95	14627/95				20		60				40
Seeds purchased for exp't											
Seeds - maize	12325/94	n.a.	<0.2	0.01	<0.03	<0.6	2.1	n.d.	n.d.	<0.2	15
Seeds - sorghum	12326/94	n.a.	<0.2	0.01	0.14	0.6	5.3	n.d.	n.d.	<0.2	20
Seeds - peanut	12327/94	n.a.	0.2	0.02	0.12	<0.6	14	n.d.	n.d.	<0.2	41
Seeds - soybean	12328/94	n.a.	<0.2	0.01	0.41	0.7	17	n.d.	n.d.	<0.2	47
Seeds - winter wheat '94–'95	14623/95	n.a.	0.1	0.02	0.08	1	8	<0.02	1	<0.3	30
Produce harvested '94											
Produce - maize - T1	21077/94	n.a.	0.2	<0.02	0.13	1	4		2	<0.3	19
Produce - maize - T2	21078/94	n.a.	<0.2	<0.02	0.12	1	4		2	<0.3	19
Produce - maize - T3	21079/94	n.a.	0.2	<0.02	0.12	1	4		2	0.3	19
Produce - maize - T4	21080/94	n.a.	0.2	<0.02	0.18	1	3		<2	<0.3	14
Produce - sorghum - T1	21081/94	n.a.	0.7	<0.02	0.25	1	5		0.7	0.3	16
Produce - sorghum - T2	21082/94	n.a.	0.2	<0.02	0.32	1	5		0.2	0.3	18
Produce - sorghum - T3	21083/94	n.a.	0.2	<0.02	0.21	1	5		0.2	<0.3	18
Produce - sorghum - T4	21084/94	n.a.	<0.2	<0.02	0.44	1	5		<0.2	0.3	16
Produce - peanuts - T1	21085/94	n.a.	0.2	0.02	0.74	1	16		5	<0.3	31
Produce - peanuts - T2	21086/94	n.a.	<0.2	0.02	0.61	1	15		4	<0.3	33

	Sample no.										
Produce - peanuts - T3	21087/94	n.a.	<0.2	0.02	0.73	2	15		5	<0.3	33
Produce - peanuts - T4	21088/94	n.a.	0.5	0.03	0.41	1	14		4	<0.3	28
Produce - soybeans - T1	21089/94	n.a.	<0.2	<0.02	3.3	1	19		14	<0.3	43
Produce - soybeans - T2	21090/94	n.a.	<0.2	<0.02	2.8	1	20		12	<0.3	42
Produce - soybeans - T3	21091/94	n.a.	0.3	<0.02	2.6	2	21		12	<0.3	38
Produce - soybeans - T4	21092/94	n.a.	<0.2	<0.02	2.7	2	19		11	<0.3	35
Produce harvested '95											
winter wheat '95 per treatment											
T1 plot - no green manure	14621/95	n.a.	0.1	0.01	0.95	1	10	<0.02	1	<0.3	25
T1 rye grass clover plot	14622/95	n.a.	0.1	0.01	1.00	1	11	<0.02	1	<0.3	28
T1 sudax lupin plot	14616/95	n.a.	0.1	0.01	0.71	1	9	<0.02	1	<0.3	27
T1 Chinese lentil (mung bean) plot	14617/95	n.a.	0.2	0.01	0.89	1	9	<0.02	1	<0.3	28
T2 plot	14620/95	n.a.	<0.1	0.02	0.57	1	10	<0.02	1	<0.3	32
T3 plot	14619/95	n.a.	0.1	0.02	0.73	1	9	<0.02	1	<0.3	31
T4 plot - no green manure	14618/95	n.a.	0.1	0.07	1.10	1	10	<0.02	2	<0.3	28
T4 rye grass clover plot	14615/95	n.a.	<0.1	0.05	0.61	1	9	<0.02	2	<0.3	29
T4 sudax lupin plot	14613/95	n.a.	0.1	0.05	0.68	1	9	<0.02	2	<0.3	22
T4 Chinese lentil (mung bean) plot	14614/95	n.a.	<0.1	0.05	0.76	1	9	<0.02	2	<0.3	25
Produce bought in the market											
Houma peanuts	11388/95	n.a.	0.1	0.03	<0.04	1	13	<0.02	<2	<0.3	n.d.
Houma soybeans	11389/95	n.a.	<0.1	0.02	0.17	1	17	<0.02	2	<0.3	n.d.
Houma red beans	11390/95	n.a.	<0.1	<0.01	0.04	1	10	<0.02	<2	<0.3	n.d.
Houma Chinese lentil (mung beans)	11391/95	n.a.	<0.1	<0.01	0.04	1	11	<0.02	<2	<0.3	n.d.
Mao Jia Wan peanuts	11392/95	n.a.	<0.1	0.02	<0.04	1	12	<0.02	<2	<0.3	n.d.
Standard for general foods (Standard A12, Australia) (mg person⁻¹)[a]			1.0	0.05	?	?	10	0.03	?	1.5	150
Tolerable daily intake (mg adult⁻¹)[a]			?	<0.057–0.71	?	<0.2	<2	<.043	<0.25	<0.3	<15
Tolerable daily intake											
TDI (mg kg⁻¹ body wt.)			0.0021	0.001	0.0014	0.005	0.14	0.00061	0.05	0.0036	1
TDI (mg adult⁻¹ of 70 kg)[a]			0.147	0.07	0.098	0.35	9.8	0.0427	3.5	0.252	70

[b] from WHO-FAO

[a] TDI value per adult calculated from X μg/kg × 70. (Vermeire et al., 1991).

peanuts would approximate the TDI for Co, but only a much smaller proportion of the TDI for Cu. Vermeire et al. arrived at these human TDI criteria by taking the subacute toxicity concentrations for experimental rats and dividing this by 1000. Their procedure may be overly conservative. However, even with this degree of conservatism it would be concluded that the produce grown on the remediated tailings at Zhong Tiao Shan is generally acceptable for human consumption, especially since it is unlikely that a farming family would consume all its food from crops grown entirely on their own farm.

WU GONG LI TAILINGS DAM PILOT PROJECT

Methods

Rehabilitation of this tailings pond was constrained because tailings slurries were still being discharged in it when the work started, and most of the 12-hectare area was too wet and soft to support pedestrian traffic. In addition, it was not possible to obtain large quantities of natural soil materials for mixing with the tailings. Small trial plots were established in the driest corner of the pond and extended the following year to a wet site, but without standing water. Several grass and legume species were trialed in pure stands and in a mixture. A number of tree species were also planted to observe establishment and growth.

Analyses of the tailings had also indicated extremely low levels of available N and P and it was decided to apply nutrients to the plots at the same rate as in Zhong Tiao Shan. The trees were given split fertilizer dressings with half in the planting hole and half on the soil surface around the tree.

Herbaceous species used were: couch grass (*Cynodon dactylon*), tall fescue (*Festuca arundinaria*), perennial ryegrass (*Lolium perenne*), white clover (*Trifolium repens*), lucerne (*Medicago sativa*), and crown vetch (*Coronilla varia*). In 1996, additional species were added: brome grass (*Bromus inermis*), cocksfoot (*Dactylis glomerata*), zoysia (*Zoysia japonica*), yellow lotus clover (*Lotus corniculatus*), and serradella (*Ornithopus perpusifolia*), as these were reputedly suitable species for sandy, infertile soils.

Tree species included in the trial are: poplar (*Populus* spp.), honey locust or Chinese scholar (*Robinia pseudoacacia*), willow (*Salix* spp.), camphor tree (*Camphora japonica*), holly (*Ilex* spp.), *Amorpha fruticosa*.

Results and Discussion

A striking result was the spontaneous establishment of weeds around the planted trees, which was attributed to the sudden availability of nutrients. Weeds and sown grasses also established themselves downstream from the fertilized trial plots, with nutrients washed off the plots. The tailings apparently are so low in fertility that weed seeds will not grow unless fertilizer is added.

Of the sown grasses, the couch grass, tall fescue, and cocksfoot were successful and have persisted to the present day. The legumes failed almost totally. Three lucerne plants survive today.

In the spring of 1996, two hectares of the drier part of the tailings were sown to a mixture of grasses and white clover. The mixture included the more successful species of the plot trials, but the clover failed again, with only six surviving plants having been counted. It is thought that the lack of suitable rhizobia is the cause. The grasses gave a good cover, but this has not been quantified.

The most suitable trees proved to be poplar, honey locust, and willow.

In the 1997 spring, the remaining ten hectares were sown to a grass and clover mixture; this time the clover seeds was treated either with a rhizobial culture purchased in Beijing or with an imported culture from Australia. It is understood that clovers are becoming established, but estimates of plant numbers have not yet been made.

Hare droppings now abound in the grassed area and insects and birds are now colonizing what used to be a bare expanse of grey tailings. Senior Tong Ling copper mine officials have grasped the ecological significance of this green revolution.

CONCLUSIONS

Great progress has been made in four years in developing an understanding of the many related geochemical, mineralogical, and agricultural problems associated with the remediation of mine tailings in China, and with the fostering of a holistic approach to developing solutions through the partnership of Chinese and Australian organizations.

It has been demonstrated that infertile, dust-generating tailings with residual heavy metals can be made to produce crops suitable for human consumption or be revegetated in a straightforward and logical manner after an analysis of the limitations of the tailings and the site. The utility of chemical principles governing the behavior of heavy metals in the soil for predicting rehabilitation outcomes has also been proved.

That some cropping and revegetation experiments have produced low yields or disappointing results has led to a keener appreciation of the factors controlling plant growth and provided valuable lessons for the future.

REFERENCES

Baker, D.E. Copper. Ch. 8, in *Heavy Metals in Soils*, Alloway, B.J., Ed., Blackie, Glasgow & London, 1990.

CARIM. 1993–1997 Unpublished Project Reports by van de Graaff, R.H.M. and L.B. Zhou. China-Australia Research Institute for Mine Waste Management, Beijing, P.R. China.

Gilkes, R.J. and R.M. McKenzie. Geochemistry of manganese in soil. Ch. 2, in *Manganese in Soils and Plants*, Graham, R.D., R.J. Hannam, and N.C. Uren, Eds., Kluwer, Dordrecht, The Netherlands, 1988.

Hutchison, I.P. and R.D. Ellison. *Mine Waste Management*. Lewis Publishers, Boca Raton, FL, 1992.

McKenzie, R.M. The Sorption of Cobalt by Manganese Minerals in Soils. *Australian J. Soil Sci.*, 5, (2), pp. 235–246, 1967.

Olsen, S.R., C.V. Cole, F.S. Watanabe, and L.A. Dean. *Estimation of Available Phosphorus in Soils by Extraction with Sodium Bicarbonate*. U.S. Department of Agriculture, Circular No. 939, 1954.

Römkens, P.F. Interaction Between Calcium and Dissolved Organic Carbon: Implications for Metal Mobilisation, in *Abstracts 3rd International Symposium on Environmental Geochemistry*, Faculty of Geology, Geophysics and Environmental Protection, University of Kraków, Kraków, Poland, 1994.

Sinclair, T.R. and R. Serraj. Legume Nitrogen Fixation and Drought. *Nature*, 378, p. 344, 1995.

Vermeire, T.G., M.E. van Apeldoorn, J.C. de Fouw, and P.J.C.M. Janssen. *Voorstel voor de humaantoxicologische onderbouwing van C-(toetsings)waarden*. [Proposal for the justification of critical C-values on human toxicological criteria] Report No. 725201005, National Institute of Public Health and Environmental Protection, Bilthoven, The Netherlands, 1991.

Chapter 12

Direct Revegetation of Salt-Affected Gold Ore Refining Residue: Technology Evaluation

G.E. Ho, M.K.S.A. Samaraweera, and R.W. Bell

INTRODUCTION

Iron, bauxite, gold, nickel, diamond, and mineral sands mined and processed in Western Australia represent a significant proportion of the world's output and generate 70% of the export income of the state (Department of Resources Development, 1996). In 1994, Western Australia produced 193.6 tonnes of gold, which is 74% of Australian production and 8% of the world total. Much of the gold mined in Western Australia is in the arid hinterland of the state. A significant amount is, however, produced in the southwestern part of the state, near Boddington, in the eastern jarrah forest, 125 km southeast of Perth. Evaluation of the technology for direct revegetation of the ore refining residues from the Boddington gold mining operation is the subject of the present chapter.

Location and Site-Specific Conditions of Case Study

Gold was discovered in 1980 directly under the boundary that demarcates bauxite lease areas held by Alcoa of Australia and Worsley Alumina which have developed the Boddington Gold Mine (BGM) and Hedges Gold (HG) mines, respectively. The gold deposits are found in the surface 15 m to 60 m depth in the profile in association with highly weathered soft clay and saprolite materials. Mining for gold ore is by the open-cut method. Gold extraction produces a predominantly clay residual ore material. The refining residue is deposited into purpose-built residue storage areas (RSA) as a slurry comprising approximately 40% solids in saline process water. Residue storage areas will eventually cover 1000 ha requiring revegetation. As properties of the residue produced by the two companies were similar, BGM and HG jointly engaged Murdoch University for a three-year study to prepare a revegetation prescription to generate a sustainable ecosystem in the RSA.

A mixture of rainwater and water from the Hotham River are used in gold extraction. The water of the Hotham River is saline and the dissolved salts content can vary from 2,000 to 6,000 mg L^{-1} (electrical conductivity, 320 to 960 mS m^{-1}, with an average value of 880 mS m^{-1}), with the lowest concentrations in winter. As the process water is continuously recycled and NaOH/NaCN are added during processing, it is estimated that the final salinity of the residue soil solution will be 10,000 mg dissolved salts L^{-1} (electrical conductivity, 1,600 mS m^{-1}).

Boddington receives an average annual rainfall of 810 mm, equivalent to only half the mean annual evaporation of the area. During the period January 1989 to December 1994, the residue storage areas experienced a large variation in the yearly rainfall from 1,079 mm in 1991 to 658 mm in 1994. The average annual rainfall during this period was 865 mm. Generally, 75–82% of the rain falls in the winter and the early spring (May to October). The rainfall pattern in 1994 was unusual, as 95% of the year's below-average total was received during the period May to October.

The natural vegetation cover of the area surrounding the mine is open eucalyptus (*E. marginata*) forest. The ecosystems that exist within the forest are self-sustaining, characterized by a high level of biodiversity and efficient nutrient cycling and water use. Relatively little water or nutrients are lost from the ecosystem. Decommissioned RSA, on the other hand, will receive more water during the rainy season than the same area would have received if it remained as forest, because of the clearing of the surrounding forest and the increased run-on. The residue is not permeable enough to allow infiltration and deep percolation of the excess incoming water, and hence the water ponds on the surface. This will be further aggravated by the buildup of upward pressure in the perched water in the RSA due to the consolidation of the lower layers of residue in the RSA. Studies by consulting engineers have indicated that the water in the upper two-thirds of the depth of the residue will remain under upward pressure for a long period.

Residue storage areas are constructed by building dikes across naturally existing valleys. Residue is piped into the RSA as a slurry comprising 40% finely ground ore, mainly non-swelling clay, and 60% water of variable salinity. The residue consolidates upon standing, increasing in density to about 70% solids. Decant water is returned to the plant and reused as process water. Surface consolidation to 55 to 60% solids typically occurs by the time freestanding water ceases to cover the residue. Vertical cracking occurs as the residue dries by evaporation to 60% solids at the surface. Cracks develop to a depth of 1 to 2 m and salt crusts form on the surface. Cracks comprise up to 15–20% of the surface area of the RSA, and occupy up to 10% of the residue of the surface to a depth of 2 m. Fifteen years after decommissioning, the residue is predicted to have a solids content of 76–84%. Because the depth of the residue increases from zero at the edge to 35 m at the deepest point, it is predicted that a natural surface gradient of 0.5 to 2% will occur when the RSA is full and the residue eventually ceases consolidating.

The environmental management programs of the two companies are somewhat similar and include the rehabilitation of the mine pit, overburden, premises of the processing plant and related debris, and the residue storage areas. Issues such as protection of the surrounding catchment from the spread of jarrah dieback (*Phytophthora cinnamomi*), stability of slopes and roads, avoidance of groundwater contamination due to spills and leakage from RSA and reservoirs, and minimizing dust formation are common to both operations and resolved after consultation with the community and relevant government agencies.

Managing RSA requires special attention and consideration. The residue in addition to being alkaline, sodic, and saline contains metallo-cyanide complexes and traces of environmentally sensitive elements such as copper and arsenic. Regular monitoring of the water quality of the nearby 34 Mile Brook, Hotham River, and a series of bore holes located around the RSA is being done in order to assess the impact of the RSA on surface and groundwater quality.

The two companies have a commitment to minimize the environmental impacts by revegetating the RSA after decommissioning. The major issues which have to be addressed during decommissioning are the hydrology of the surrounding forest, the probable influence of RSA on the quality of the groundwater, time taken for the RSA to complete its consolidation, and the success of revegetation of the RSA.

Guidelines of the Department of Minerals and Energy of Western Australia require the RSA to be covered with 500 mm of inert waste, followed by the establishment of a self-regenerating vegetation compatible with the surrounding vegetation cover. The long-term stability of any perennial vegetation established in an inert material overlying the unamended residue is questionable, as the residue poses severe constraints to plant root development. In addition, the limited availability of inert materials and their high cost made it necessary for the owners of the mines to explore other revegetation options.

In this chapter we present results of our field trials, which show that it is feasible to directly revegetate on saline gold-refining residue. Results of laboratory and glasshouse studies are added where relevant.

Constraints to Plant Growth

Salinity and waterlogging are major constraints to plant establishment and growth when directly revegetating the residue. During the winter, the RSA are waterlogged by perched water with moderate salinity. Unless appropriate ameliorations are imposed, it is likely that the combined effects of salt and waterlogging would have a highly detrimental influence on plant establishment, growth, and survival.

While the residue is moderately saline in summer and autumn, it forms a strong salt crust at the surface upon evaporation and drying. Evaporation results in the upward capillary movement of saline water from the lower layers, and leaves the salts behind on the surface. Therefore, during the summer the electrical conductivity at the surface, measured using a soil to water extract of 1:5 [$EC_{1.5}$], may be as high as 6,500 mS m^{-1}. The residue has a pH of 9.3. Generally, an $EC_{1.5}$ of 140–350 mS m^{-1} and pH above 8.5 are considered to severely limit growth of most plants. Much of the salts in the water taken from the river, and in the chemicals added during processing (NaOH and NaCN) result in sodium being the predominant cation, and consequently the residue is highly sodic.

The dried residue has a fine texture, no structure, and is lacking macropores needed for water and air movement. However, voids created by the cracks represent channels through which drainage water and air can move through the surface 1–2 m of the residue, provided they can be stabilized to avoid slumping and in-filling. In addition, the salt crust formed on the surface of the residue during the summer may be subject to wind erosion. If nothing is done to improve infiltration, the winter rains will dissolve the salt that is on the surface but not remove it by leaching. A possible sequence of events in unamended residue during winter is: slow dissolution of salt crust, filling of cracks with rainwater contaminated with salt, and surface ponding and waterlogging of residue due to stagnating water. Furthermore, the consolidated residue is brick-like in character, making it difficult for plant roots to penetrate.

Because of the processing it has undergone, the residue does not contain organic matter, plant nutrients, or any biological activity that enhances plant growth and subsequent functioning of the ecosystem.

Approach to Revegetation

Our model for revegetation is based on the following approach: (1) identification of physical, chemical, and biological properties of the residue; (2) development of methods of ameliorating the residue properties so that plant growth is possible. This involves the use of chemical amendments and leaching of salts; (3) sequential introduction of plants; initially salt and waterlogging tolerant species are selected as pioneers to be followed by more sensitive tree species once residue properties have been sufficiently improved; (4) reshaping of the residue surface to facilitate draining of excess water and drying of residue; (5) sequential revegetation of the RSA commencing on the perimeter as drying of the residue begins at the outside edges; rehabilitation is aimed at accelerating the drying process and ameliorating the periphery; and (6) avoiding the use of heavy equipment and backfilling of the residue unless it is absolutely essential, as costs involved are high.

Possible end land uses on RSA vary from an ecosystem that will blend with the surrounding eucalyptus forest at BGM, and a combination of commercial pasture and forest at HG.

A variety of rehabilitation strategies have been considered by the consulting engineers. Of those proposed, the scheme favored has the following basic features: (1) rehabilitation will follow initial consolidation and drying of the residue, 2–3 years after the termination of filling with slurry, (2) the outer edges will be reshaped and revegetated, and (3) the central areas will be covered with coarser refining residue from gold basement ore found below the present ore body at BGM. The overall revegetated residue area will have artificial drains and permanent mounds and ponds.

The main emphases of the engineering considerations for revegetation are on the methods of incorporating ameliorants into the residue, the water movement within the residue matrix, and the trafficability of the surface. Because of the nature of the constraints for revegetation, trials were directed at problems of salinity, salt migration and capillary rise, poor infiltration of water and waterlogging, surface cracking, and lack of residue structure.

Initial laboratory investigations showed that the residue consists of mainly clay and has a hydraulic conductivity of 1.9×10^{-2} m day^{-1}. When the residue was amended with 5% gypsum its hydraulic conductivity was increased by more than twofold to 5.4×10^{-2} m day^{-1}. Gypsum application increased the amount of sodium that was leached with a given volume of water and decreased sodicity. Gypsum also aided in pH adjustment by precipitating carbonate alkalinity as calcium carbonate, thus converting the residue to a calcareous material.

GREENHOUSE TRIALS

Greenhouse trials initially examined the growth of four species. The residue was amended with gypsum (rates equivalent to 0, 30, and 60 t ha^{-1}) and leached with deionized water equivalent to 75 mm or 150 mm of rain. Amounts of leaching water added were equivalent to 26% and 53% of annual May–June rainfall at Boddington. Four species, *Melaleuca halmaturorum*, *Atriplex amnicola*, *Agropyron elongatum*, and *Triticosecale* spp. cv Muir were grown. Water logging and salinity tolerances of test species were as follows:

Melaleuca	salt and waterlogging tolerant
Atriplex	salt tolerant
Agropyron	salt and waterlogging tolerant
Triticale	waterlogging tolerant

Gypsum was generally very beneficial to plant growth, as was leaching. *Atriplex* grew successfully under all conditions including the untreated residue, but the other three species did poorly in unamended residue. Gypsum also improved the residue chemical properties by neutralizing its alkalinity. With gypsum addition (2.5% by weight), the pH of the residue was decreased from 9.3 to 7.6, and subsequent leaching further decreased the pH to 7.2.

The next greenhouse trial was initiated to determine the value of topsoiling the residue and to evaluate the optimum gypsum level required to establish triticale, the most salt-sensitive plant species. Salt was also added to the residue to increase the salinity equivalent to 500 and 1,000 mS m^{-1} in a 1:5 soil:water extract. It was found that gypsum at 6% had little effect in alleviating growth at high salinities but at the lower salinity level there was a near linear growth response to increasing gypsum levels.

The combined effect of salinity and waterlogging on plant growth was examined because when both occur they can be more damaging to plants than when present singly. In saline waterlogged residue, triticale failed to germinate directly on the residue. A surface layer of coarse sand covering the residue allowed the plants to germinate and survive for some time before they died. Saltbush, however, fared much better under salt and water-logged conditions. An examination of triticale roots under saline conditions revealed that growth was limited to the topsoil layer and the adjacent 5 cm of the residue which had a lower salt content ($EC_{1:5}$ 137–373 mS m^{-1}) than the underlying residue ($EC_{1:5}$ 850 mS m^{-1}). This suggests that for successful growth, the rooting zone has to be sufficiently low in salt throughout the growing season.

Methods of nutrient addition were also examined by assessing the performance of *Agropyron* plants grown in the residue amended with 10% gypsum. Nutrients were supplied by one of the following ways: poultry manure; a basal chemical fertilizer mixture having the same nutrient content as that of the poultry manure; by regular application of a balanced liquid fertilizer mixture; or with an additive combination of poultry manure and the liquid fertilizer. The experiment was evaluated under both moderate ($EC_{1:5}$ 250 mS m^{-1}) and highly saline conditions (salt added to increase the $EC_{1:5}$ by 500 mS m^{-1}). The addition of salt depressed plant growth. The basal application of chemical fertilizer also had a negative effect on plant growth. Both the poultry manure and the liquid fertilizer significantly improved plant growth and their combined effects were additive. Poultry manure was superior to liquid fertilizer in relation to plant growth. The poultry manure not only provided nutrients but also improved physical, chemical, and biological properties of the residue.

These findings have very significant practical implications. First, since the wet residue is not trafficable by heavy equipment, the application of all fertilizers must be done prior to winter rain. Secondly, rehabilitation relies on the ability of the winter rain to leach salt out of the rooting zone into the deeper residue layers. If water soluble chemical fertilizers are applied in the summer with gypsum, then during the winter, most of the nitrogen and possibly sulfur and potassium will be leached from the rooting zone, making them unavailable to plants.

In another pot experiment, the effect of leaching on the availability of nutrients in poultry manure was examined. Pots filled with the residue containing 5% gypsum were mixed with varying amounts of poultry manure, up to a maximum of 100 g per kg, then leached with deionized water equivalent to 200 mm of rain (71% of May–June rainfall). A thin layer of topsoil was placed over the residue and triticale grown. While some pots received deionized water, the others received a complete liquid fertilizer mixture which supplied approximately the same rate of nutrients as supplied by poultry manure. The plants that received the combined mixture of poultry manure (100 g kg^{-1}) and liquid fertilizer showed significantly more growth than those that received poultry manure alone. The poultry manure added, 100 g pot^{-1}, would give a thickness of 2.3 cm. Since the amount of poultry manure added was equivalent to half the minimum thickness achievable with equipment used in the field, it would be very unlikely that any nutrient deficiencies would be experienced under field conditions with poultry manure added at 1% (w/w).

From these greenhouse experiments the beneficial effects of gypsum amendment, leaching of salts, incorporation of organic matter as a nutrient source, and placing a topsoil over the residue were evident. Field trials were conducted to further assess the efficacy of the practices and amendments used in the greenhouse studies.

FIELD TRIALS—GENERAL

The goal of developing land preparation procedures was to overcome the following constraints: waterlogging; high salt content in the rooting zone; capillary migration of salts onto the dried surface; and high pH of the residue. The findings of the greenhouse experiments showed that gypsum addition and leaching of salts were prerequisites for growth of salt-sensitive species in the residue and formed the basis for development of land preparation methodologies.

Breaking Down of the Massive Apedal Clay Structure

The volume loss during drying and consolidation of the slurry leads to the formation of large apedal clay blocks. These massive clay structures have a relatively small surface area to mass ratio and as the drying is limited to evaporative water loss from the exposed surfaces it may take several years for complete drying of the first 50–100 cm of the unamended residue surface. However, the process of drying is accelerated when the exposed surface area is increased.

To increase the surface area, the residue can be ripped, plowed, and hoed during the dry part of the late summer. The initial ripping reduces the size of the clay blocks and facilitates the plowing of the residue. The process of increasing the surface area and accelerating the capillary migration of salts onto the residue surface was further accelerated by breaking the large lumps into smaller pieces by rotary hoeing.

Application of Chemical Amendments and Nutrients

The incorporation of chemical amendments and fertilizers into the residue is facilitated by the ripping and rotary hoeing process. Gypsum amendment can be effectively carried out when the surface is cultivated. Nutrients required for plant growth can supplied either by chemical fertilizers or organic fertilizers. Since the latter has a longer residence time within the rooting zone, poultry manure was used as the source of nutrients and applied immediately after gypsum in the late autumn.

Field trials were conducted in specifically-built impoundments. Results of relevant trials that demonstrate the success of direct revegetation of residue surfaces are presented here.

FIELD TRIAL—CELL A

Cell A was available for experimentation in summer of 1991/92. It had an area of 5,500 m² and a depth of 1.25 m and it incorporated an underdrainage system. It therefore represented the well-drained perimeter of a decommissioned RSA with a shallow depth of residue and would provide valuable information relating to the revegetation of that part of an RSA.

Selection of Land Preparation Techniques and Planting Time

The suitability of several land preparation procedures were examined. Cracks were either filled with gravel (2,000 m^3 ha^{-1}) to facilitate rapid movement of excess rainwater, or left unfilled. Residue was either prepared by a sequence of ripping (to 50 cm depth), plowing, and hoeing or left without any preparation. Incorporation of gypsum (0, 30, 60 t ha^{-1}); incorporation of salt to achieve 10,000 mg salt kg^{-1} residue; addition of poultry manure (equivalent to a thickness of 0 or 4–6 cm); placement of topsoil (to a thickness of 0 or 8–10 cm); planting in winter or spring, and species (no plants, *Agropyron*, Triticale, saltbush) were considered.

Fourteen out of a possible 768 treatments were selected and they were biased toward favorable conditions for plant growth. Table 12.1 shows 10 of the treatments relevant to this study. The experiment was laid out in a split-plot design with blocks perpendicular to the moisture gradient. To minimize equipment movement, each block was divided longitudinally into many plots. Two major treatments (filling of cracks with gravel or not) were randomly assigned to each main plot. The other treatments were then randomly assigned to each of the fourteen 5 m by 5 m subplots within a block.

The total biomass production in winter planted plots (10.7 ± 1.2 t ha^{-1}) was significantly higher than those planted in spring (6.7 ± 0.6 t ha^{-1}). In winter, the residue was waterlogged and saline. At the beginning of spring, the residue contained less water and salts and hence the conditions were more favorable for plant growth. The net aboveground biomass accumulation rates for winter planting and spring planting were similar, 4.38 and 4.48 g m^{-2} day^{-1}, respectively. However, winter-planted species had the advantage of having a longer growth period than those planted in spring. Winter planting will not be effective if the soil conditions are not favorable. The factors that should be considered are the water content, pH and EC of the residue, and the subsurface drainage within the impoundments. Thus the selection of planting time must be based on the following two factors:

a. soil conditions: extent to which the amelioration treatments have been effective in alleviating salinity and waterlogging;
b. the length of the available growth period.

The total biomass production in gravel filled plots (6.6 ± 0.6 t ha^{-1}) was significantly lower than those in unfilled plots (9.0 ± 1.0 t ha^{-1}). This may be due to the decreased availability of water due to rapid drainage. The grasses, triticale, and *Agropyron* which have aerenchyma in their roots and hence can tolerate waterlogging grew significantly better in plots where gravel was not applied. For *Agropyron*, the total biomass was 3.9 ± 0.4 t ha^{-1} in plots with gravel filled cracks versus 12.0 ± 2.4 t ha^{-1} in unfilled plots. By contrast, saltbush which is not waterlogging tolerant, grew better in the gravel filled treatments (5.2 ± 0.8 t ha^{-1}) than in nontreated plots (1.9 ± 0.4 t ha^{-1}).

The effect of gypsum is also of interest. The grasses which were less salt tolerant performed significantly better under the high rate (60 t ha^{-1}) than the low rate (30 t ha^{-1}) of gypsum application (6.4 ± 0.7 t ha^{-1} versus 2.1 ± 0.7 t ha^{-1}). The effect of the higher gypsum rate was more prominent in the winter-planted plots (8.4 ± 2.3 t ha^{-1}) than those

Table 12.1. Selected Treatment Details and Results of Cell A Trial.

Treatment	Rotary Hoeing	Gypsum (t ha⁻¹)	Top Soil Placement	Poultry Manure	Planting Win	Planting Spr	Gravel in Cracks Agropyron (t ha⁻¹) Y	Agropyron N	Triticale (t ha⁻¹) Y	Triticale N	Salt Bush (t ha⁻¹) Y	Salt Bush N
1	N	60	Y	Y		SB	2.83		0.03		2.9	
2	Y	30	Y	Y	SB	Agro Trit	2.9	4.43	0.43	1.61	9.7	4.13
3	Y	60	N	N		SB Agro Trit	2.1	2.98	0	0.03	0.3	0
4	Y	60	Y	N		Agro Trit	3.03	4.28	0.6	0.17		
5	Y	60	N	Y	Agro Trit		13.0	17.6	1.38	0.26		
6	Y	60	Y	Y		Agro Trit		6.23		0.17		
7	Y	60	Y	Y		SB Agro Trit		6.10		0.36		0.94
8	Y	60	Y	Y		SB Agro Trit	3.90	12.0	0.23	0.28	5.15	1.94
9	Y	60	Y	Y	SB	Agro Trit	2.53	5.58	0.1	0.09	0	2.63
10	Y	60	Y	Y		Agro Trit	4.60		0.73			
Common for all	LSD=0.05%							4.98		2.06		4.34

Note: SB = Salt bush; Agro = *Agropyron*; Tri = Triticale; Win = winter planting; Spr = spring planting; Y = Yes; N = No.

Table 12.2. The Effect of Gypsum Application on the pH and the EC ($mS\ m^{-1}$) [1:5, soil:water extract] Measurements of Residue Sampled in February, 1993.

Rate of Gypsum Application (t ha^{-1})	pH (±SE)	EC ($mS\ m^{-1}$) (±SE)
0	7.93 (±0.14)	657 (±138)
30	6.53 (±0.17)	147 (±25)
60	6.58 (±0.15)	211 (±48)

planted in the spring (5.9 ± 0.7 t ha^{-1}). Gypsum requirements of spring planting may be less as the combined effect of autumn gypsum application and winter leaching had ameliorated the residue. The pH and the EC measurements of the residue sampled in February 1993 confirmed that gypsum application and subsequent leaching had ameliorated the residue, creating conditions favorable for plant growth (Table 12.2). On the other hand, saltbush consistently yielded better under lower gypsum applications (8.7 ± 0.9 t ha^{-1}) than at the higher rate (3.1 ± 0.6 t ha^{-1}). All species grew significantly better when treated with poultry manure (8.6 ± 0.6 t ha^{-1} vs. 3.4 ± 0.4 t ha^{-1}).

During the period May 1992–January 1993, the area received a total of 872 mm of rain and 83% of that was received during the wet season, May–September. Thus the climatic factors throughout the growing season were favorable for plant growth. The aboveground biomass production of *Agropyron* under optimum land preparations were 17.6 and 6.10 t $ha^{-1}\ yr^{-1}$ for winter planting and spring planting, respectively. These figures are higher than 1–2 t $ha^{-1}\ yr^{-1}$ previously reported for *Agropyron* in ameliorated bauxite refining residues (Wong and Ho, 1994). The saltbush production in winter and spring plantings of 9.7 and 5.15 t $ha^{-1}\ yr^{-1}$, respectively, were very much higher than the average production figures, 3–4 t $ha^{-1}\ yr^{-1}$ for Western Australia (Brian Warren, Department of Agriculture, Western Australia, personal communication).

Ripping, plowing, rotary hoeing, gypsum amendment, poultry manure amendment and leaching of salts during the winter prior to planting were found to be beneficial for the establishment and successful growth of the three pioneer species evaluated. Using this combination of land preparation technology and amelioration techniques to test the survival and growth of 27 species, Samaraweera et al. (1994) reported that establishment of additional pasture or crop species, and nonendemic native vegetation (including tree species) is feasible on the amended residue.

FIELD TRIAL—CELL B

Cell B had an area of 7,150 m^2 and a depth of 2.25 to 5.85 m. It had no underdrainage and was filled by residue at the end of March 1993. Cell B was prepared for trials by rotary hoeing, addition of gypsum (60 t ha^{-1}) and topsoil (10 cm layer). The salinity of residue in cell B was increased to resemble that which is anticipated in the final decommissioned RSA. Salt was added during the land preparation at a rate of 5 t ha^{-1} to give a soil content of about 10,000 mg kg^{-1} (EC 1:5 531 ± 109 $mS\ m^{-1}$).

The revegetation strategy was to introduce species in a planned succession: grasses, then shrubs, and finally trees. The rationale was to establish vigorous growth of the pioneer grasses and shrubs for 1–2 years before introducing trees. The delay in planting trees was to allow time for the amelioration of salinity to aid in the establishment of those species with lower salt tolerance. The hypothesis was that accumulation of soil organic matter and formation of an organic mulch on the surface would prevent capillary salt rise. It was further hypothesized that the land preparation, residue amelioration, and land reshaping practices, along with the penetration of the residue by the fibrous root systems of pioneer species would accelerate the decrease in salinity and waterlogging in the surface 0.5–1 m. The pioneer species in Cell A *Agropyron*, triticale, and saltbush all survived over two and a half years though density of the annual, triticale, declined each year. Success of this strategy requires continued investigation and examination over several years. Preliminary results from the first two and a half years growth are presented below.

Alleviating Waterlogging and Salinity

Our goal was to prevent flooding of the residue and waterlogging in the rooting zone during the rainy season. Another goal was to prevent drying of the residue surface during the summer and thereby stop the process of capillary migration of salt onto the residue surface. These were to be achieved by gravel filling the cracks, reshaping, and mounding of the residue surface.

Gravel Filling and Reshaping

The residue consolidated with a 0.5% slope. Water runoff and drainage in clayey residue at such gentle slopes is rather slow and was improved by increasing the slope to 2.5%. Four areas were created in Cell B in autumn 1993, with and without gravel in cracks and with and without reshaping.

During this exercise, large machinery, such as front end loaders and dozers, were used to spread gravel. The use of large machinery was to provide an indication of how large-scale land preparation could be carried out. Cells were further subdivided and some plots were planted with pioneer species (*Atriplex amnicola*, *Agropyron*, and triticale) and others were left as bare plots.

The mean total biomass production in all four parts of Cell B during 1993–1994 was 4.37 t ha^{-1} with a maximum of 11.58 t ha^{-1}: both values were lower than corresponding 1992–1993 production of 7.73 and 30.7 t ha^{-1} for Cell A. Indeed, the mean value for Cell A was an underestimate, as some parts were planted in spring whereas all the pioneer species in Cell B were planted in the winter. Rainfall for the two years was similar and close to average; therefore, precipitation does not explain yield differences. Analyses of Cell B residue showed that leaching over a year had not decreased its salinity to the level required for growth of nonhalophylic plants. The increased salinity of the residue in Cell B suppressed the yields of triticale, *Agropyron*, and saltbush. For example, in February 1994 while the EC$_{1.5}$ of the residue taken from unamended parts of Cell A was 121 mS m^{-1}, the corresponding value for Cell B was 1,800 mS m^{-1}. Plant growth may have also been adversely affected by compaction due to earthwork operations. There is, however, sufficient information after one year's field results to conclude that the highest biomass yields were obtained when the cracks were filled with gravel and the residue surface reshaped. Both the reshaping and gravel filling of cracks had significant positive effects on plant growth (Figure 12.1). The highest total biomass was recorded in the plots that were reshaped after gravel filling the cracks.

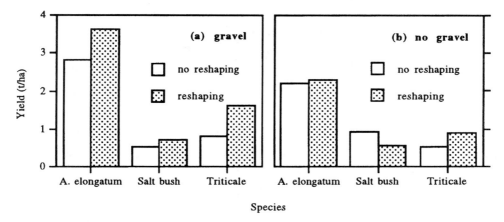

FIGURE 12.1. Effect of gravel filling of cracks and surface reshaping (no reshaping—0.5% ground slope; reshaping—2.5% ground slope) on yield (t ha^{-1}) of *Agropyron*, triticale, and saltbush in Cell B, December 1993. $LSD_{0.05}$ for comparison of means of *Agropyron*, triticale, and saltbush are 1.20, 0.52, and 0.26 t ha^{-1}, respectively.

Mounding

As the yield of pioneer species was relatively low and all tree species planted in the spring of 1993 died, niche mounding was investigated as an additional land preparation technique to sustain tree growth.

In each of the four areas of Cell B, plots were further subdivided so that one-half could be mounded. No new planting was carried out in the mounded or unmounded areas. At the time of preparation of mounds, the area was vegetated with *Agropyron* and saltbush. Triticale planted in winter 1993 had matured and died. The mounding operations destroyed aboveground shoots and damaged root systems of *Agropyron* and saltbush. The growth measurements made in the mounds in August 1994 showed that while there was no regeneration of *Agropyron* and saltbush, the production of Triticale was spectacular, in the range of 4–8 t ha^{-1} during the short growing season May–August 1994. This is particularly a significant achievement as the total rainfall for the period was about 500 mm. In contrast, the unmounded areas produced no regeneration of Triticale but production of *Agropyron* and saltbush for the period May 1993 to August 1994 was 4–5 t ha^{-1} and 1.5 t ha^{-1}, respectively. This clearly shows that biomass production can be optimized by having a mixture of mounded and unmounded areas which would enhance rapid growth of triticale during winter and *Agropyron elongatum* during the spring and summer periods.

Apart from the obvious avoidance of waterlogging and flooding, mounding decreased the salt content in the elevated portion of the mounds favoring plant growth and establishment. The residue/topsoil on the crests of the mounds in the reshaped area had $EC_{1.5}$ values of 300–900 mS m^{-1} and ESP of 2–10% in August 1994 after 480 mm of rain. The corresponding figures for unmounded reshaped areas were EC 700–1,100 mS m^{-1} and ESP 3.5–9.0%. The EC and ESP values are well within the ranges acceptable for plant growth. Yet the $EC_{1.5}$ and ESP of the residue near the base of the mounds were 26,000–46,000 mS m^{-1} and 8–21%, respectively.

The effects of mounding on plant growth are shown in Figure 12.2.

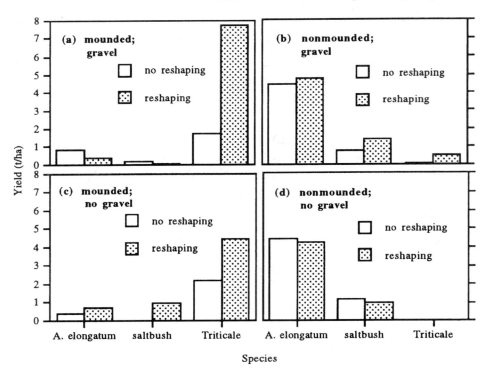

FIGURE 12.2. Effect of mounding, gravel filling of cracks and reshaping the ground surface (no reshaping—0.5% ground slope; reshaping—2.5% ground slope) on yield (t ha^{-1}) of *Agropyron*, triticale, and saltbush in Cell B, October 1994. LSD$_{0.05}$ for comparison of means of *Agropyron*, triticale, and saltbush are 1.44, 2.53, and 0.64 t ha^{-1}, respectively.

CONCLUSIONS

It is too early to conclude from research and monitoring carried out so far whether eucalyptus forest can be reestablished on the RSA, but establishment of nonendemic salt-tolerant native species and a variety of agricultural species appear to be useful rehabilitation species.

Based on research thus far, a revegetation prescription would appear to include:

- filling of cracks with gravel and reshaping of selected areas, to facilitate drainage
- ripping and plowing; to break up the massive blocks of residue into smaller blocks
- rotary hoeing: to break down the residue blocks to clods small enough to provide macropores, increase drying, accelerate surface salt migration, and facilitate subsequent leaching of salts from the surface
- gypsum addition: 60 t ha^{-1} to decrease pH, facilitate aggregation, and improve permeability and leaching of salts
- poultry manure: 2 to 4 cm thickness to give initial nutrients for plant growth and provide organic matter for aggregation of the soil, and
- topsoil placement: to provide a suitable medium for germination or establishment of seedlings.

TOPICS THAT REQUIRE FURTHER RESEARCH

A number of research questions remain unresolved but of crucial importance to the long-term effectiveness and sustainability of revegetation of gold ore refining residue. They are:

- effect of filling of cracks with gravel to resolve whether it is essential
- rate of the export of salt out of the system, as this will determine whether in the longer term, the soil will revert to surrounding conditions
- models for salt and water movement in the residue storage impoundments to determine how much leaching is required for effective removal of salt from the rooting zone
- screening of a wider range of plant species for their ability to withstand anticipated soil-related constraints in the amended residue
- effect of planting time on establishment of different species used in revegetation program
- long-term sustainability of the ecosystem including nutrient cycling, hydrological balance, and biodiversity
- finding low cost alternatives for treatments such as the fertilizers.

ACKNOWLEDGMENTS

The research described in this chapter was funded by Boddington Gold Mine and Hedges Gold. We gratefully acknowledge the support and assistance provided by the companies in carrying out the research, and we would like to particularly thank Messrs. Gerry Rayner and Ray Vergone and their staff.

REFERENCES

Department of Resources Development. Downstream Processing—An Overview of Resource Development Processing in Western Australia. Department of Resources Development, Perth, 1996.

Kaub, J. Gold Moves to Top of Resource Exports. *Prospects*, May, 10–11, 1995.

Samaraweera, M.K.S.A., S. Beaton, R.W. Bell, G.E. Ho, E.G. Barrett-Lennard, R. Vergone, and G. Rayner. Selection of Plants for Revegetation of Goldmine Residue in the Jarrah Forest of South-Western Australia, in *Proceedings of the 3rd International Conference on Environmental Issues and Waste Management in Energy and Mineral Production*. Brodie-Hall Research & Consultancy Centre Pty Ltd., Curtin University of Technology, Perth, 1994, pp. 619–626.

Wong, J.W.C. and G.E. Ho. Viable Technique for Direct Revegetation of Fine Bauxite Refining Residue, in *Proceedings of the International Bauxite Tailings Workshop*. Australian Bauxite and Alumina Producers, Perth, 1994, pp. 258–268.

Chapter 13

Overview of Acid Mine Drainage Issues and Control Strategies

S.D. Miller

INTRODUCTION

Management of mining and processing waste is one of the major physical environmental concerns associated with mining operations. This concern increases significantly where the mine waste materials contain reactive sulfide minerals such as pyrite. Reactive pyrite is the compound primarily associated with potential acid drainage. Where acid drainage occurs, it poses a major risk of adversely affecting surrounding water and soil resources. Acid drainage can be a long-term issue, and hence control of acid drainage is essential for minimizing the risk of any ongoing environmental liability from mining operations.

The process of sulfide oxidation is a natural weathering process which occurs when material containing reactive sulfides is exposed to oxygen. This in turn can lead to the production of acid drainage when water leaches away the products of sulfide oxidation from the site of formation to the receiving environment. These sites can be found within ore deposits and are formed by natural in situ sulfide oxidation. However, mining increases the rate of exposure of fresh rock to oxidizing conditions relative to that which occurs though natural weathering and the erosion process. If the rate of acid generation exceeds the capacity of the rocks and surrounding environment to assimilate this increased acid load, then low pH drainage containing elevated concentrations of soluble sulfate may occur.

Although the management of sulfide-containing mine waste is an important pollution issue faced by the mining industry, technologies are available, and experience has demonstrated that cost-effective control is feasible provided the appropriate strategies are implemented at an early stage of the mine life. It is always considerably more costly (in financial, environmental, and social terms) to remediate acid drainage problems after they have developed than to control the generation process through an integrated waste disposal operation.

IDENTIFICATION OF POTENTIALLY ACID-GENERATING MINE WASTE MATERIALS

The first step in evaluating the risk of acid drainage is to characterize the acid-forming and acid-neutralizing nature of all waste rock and process waste types that will be produced during the life of the mining operation.

Acid-drainage prediction investigations are aimed at classifying the acid-generating characteristics of individual samples of mine rock and process wastes. The steps involved include sampling (generally from drill core or drill cuttings), laboratory testing, and data interpretation. Experience shows that the simple acid-base procedures which calculate the net acid-producing potential (NAPP) of a sample is ideal for the initial screening. The net acid-generation (NAG) test has proved to be reliable for confirming the acid-generating nature of samples. The results of these two tests provide a sound basis for classifying rock types for operational management. The NAG test has been demonstrated to be ideally suited for field use, and as an operational monitoring tool for the in-pit identification and classification of waste types.

The results of the NAPP and NAG tests allow rocks to be classified into the following categories:

- Acid consuming (ACM)
- Nonacid forming (NAF)
- Potentially acid forming - low capacity (PAF-LC)
- Potentially acid forming - high capacity (PAF-HC)

Where further information is required on the kinetics of the acid generation and acid neutralizing processes, kinetic NAG and column tests are recommended.

Based on the results of these investigations, geochemical rock types and waste types are identified and waste quantities and mining schedules defined.

SULFIDE OXIDATION AND ACID DRAINAGE GENERATION

The rate of acid drainage generation from a waste emplacement is determined by the following key processes:

- sulfide oxidation rate;
- geochemical reactions including acid neutralization and equilibrium solubility with precipitated secondary minerals; and
- leaching and transport of oxidation and reaction products.

The sulfide oxidation rate is determined by the reactivity of the material (also known as the intrinsic oxidation rate) and the rate of supply of oxygen to the reactive sulfides contained in the material. Control of oxygen flux to the reactive material is the key factor in limiting the sulfide oxidation rate.

The acid-buffering capacity generally results from chemical reaction with inherent calcite and other carbonate minerals which may also occur within the waste material. Neutralization and other geochemical equilibrium reactions can provide a strong buffering capacity, and these geochemical controls will limit the concentration of sulfates and metals in solution. Depending on the mineral composition and reactivity, geochemical controls may persist for only a short period or may continue into the long term.

Leaching of oxidation and neutralization products from a waste emplacement is probably the least understood of the three key processes. Infiltrating water into a dump is known to move through preferential flow paths, and hence leaching will be intermittent and variable.

Consideration must be given to each of these processes when formulating a control or remediation plan. Although the availability of oxygen to reactive sulfides is generally the rate-limiting mechanism for sulfide oxidation, geochemical reactions and leaching efficiency are critical mechanisms determining the rate of release of acid, sulfate, and associated metal contaminants from a waste emplacement. Controlling the release rate to a level that can be assimilated by the receiving environment without adverse impacts is the primary purpose of any control or remediation plan.

CONTROL STRATEGIES

Strategies which reduce the sulfide oxidation rate, enhance geochemical solubility control reactions, and reduce the extent and rate of leaching can therefore be used to control acid drainage. Strategies which aim to meet these requirements and have been implemented or are proposed for existing mining operations, include:

- *Burial and containment of potentially acid forming rock types.*
  Designed to isolate potentially acid-forming materials from leaching or to control the oxygen flux to the reactive materials. Compacted layers can also be incorporated and designed as internal seepage barriers or oxygen diffusion barriers. This is the most commonly applied strategy for waste rock where the amount of potentially acid-forming material is less than about one-third of the total waste rock mined.
- *Compacted dump construction.*
  An extension of the burial and containment strategy is to construct a compacted dump from the bottom up in a similar manner to an earthen embankment. The dump is built in thin compacted layers (less than 1 m thick) with an overall permeability less than 10^{-8} m s^{-1}.
- *Segregation and controlled placement of potentially acid-forming (PAF) material.*
  This strategy segregates the potentially acid-forming materials into smaller dumps that can be contained and treated according to the specific needs. This strategy significantly reduces the surface area of placed potentially acid-forming material and hence can reduce the potential overall acid load.
- *Blending PAF with nonacid-forming (NAF) or acid-consuming (ACM) material.*
  This approach takes advantage of the inherent buffering capacity in the NAF material, but can only be practically applied to waste rock where sufficient high ANC material (e.g., limestone) occurs and the mining practice and stratigraphic location of the material types are suitable. Most effective with process tailings and conveyor/spreader constructed waste rock dumps.
- *Treatment with lime (CaCO₃).*
  Lime (as crushed limestone CaCO$_3$) can be used to supplement the inherent ANC and promote geochemical control during active placement and exposure. Generally it is not feasible to apply sufficient lime to buffer the total potential acidity except in materials that have a naturally low acid capacity (less than 0.5% sulfur). For operational control, lime addition should be sufficient to prevent the pH from falling below about 5.5.
- *Placement of an engineered earth material cover to minimize infiltration.*
  Earth material covers are commonly used to reduce the leaching rate and seepage volume. A compacted clay layer (permeability less than 10^{-8} m s^{-1}) is normally required in wet environments. In dryer climates an evapotranspiration layer can be used.
- *Placement of an engineered earth material cover to control oxygen diffusion.*
  This type of cover is generally multilayered, incorporating an oxygen diffusion barrier layer. This layer must be designed and constructed to retain an air voids content of less than 10% to be effective; i.e., more than 90% of the voids must be water filled. The lower the air void content and the thicker the layer, the more effective the cover will be.
- *Placement of an NAF oxygen-consuming outer layer.*
  This type of cover has recently been developed by EGi to take advantage of the oxygen-consuming capacity of contained sulfides within some nonacid-forming waste rock types. These materials consume oxygen and prevent oxidation of the underlying potentially acid-

forming materials. This strategy can be used on intermediate covers and incorporated into the final outer zone for long-term oxidation control.

- *Placement under a permanent water cover.*

    Placement of potentially acid-forming materials under a permanent water cover is the most secure long-term control strategy and essentially eliminates the risk of acid drainage. This is because diffusion of oxygen in water is about four orders of magnitude lower than in air, and the amount of dissolved oxygen in the water is not sufficient to be a concern. A water cover is the preferred option for tailings facilities where the hydrological regime is suitable. However, constructed flooded waste rock dumps and codisposal of tailings and waste rock in a single flooded impoundment are also feasible.

- *Pyrite removal or treatment.*

    This option can be applied to process tailings by incorporating pyrite flotation cells in the mill circuit to condition tailings prior to discharge.

In many situations where potentially acid-forming materials occur, a combination of the above strategies is used. The combination selected will depend on the geochemical and geotechnical characteristics of the rock, material production schedules, and site-specific conditions.

CONCLUSION

Techniques and strategies for prediction and control of acid drainage and prevention of revegetation problems from sulfide-generated acidity in mine wastes have been proven and are available to the industry. For effective implementation of these strategies, planning for acid-drainage control must parallel geological and mining studies. As a minimum, sulfide assays should be conducted on exploration drill hole samples representing material types from the ground surface to the base of the ore deposit (or maximum possible mining depth). All drill cuttings and core from the surface to the base of drilling should be retained and stored under cover. Failure to take due account of the significance of potentially acid-forming materials in the mine waste can have significant cost implications, increase the risk of long-term liability, and result in community dissatisfaction with the mining industry.

It is expected that pressure on the industry to demonstrate the environmental security of mining and waste disposal facilities will continue to increase. A key component of this will be the geochemistry of wastes and impact on water and soil resources. The long-term performance of earth material covers is likely to receive greater attention given the uncertainty surrounding long-term performance and the need for ongoing maintenance. Covers are perceived as having a risk of failure and therefore will be an ongoing liability to the industry. In wet environments the trend is likely to be toward total dump control such as blending for geochemical controls, placement of internal sealing layers and construction of outer zones for long-term oxidation control, and there will be greater use of water covers for tailings (this implies the construction of water-retaining tailings impoundments). Simple soil covers are unlikely to provide adequate long-term security.

Codisposal of tailings and waste rock will also be further developed. Placement of sulfide-containing mine wastes below permanent natural waters such as the ocean and lakes can be an environmentally acceptable option and should not be excluded.

Site-specific characteristics are critical for determining the best and most practical acid mine drainage control strategy. Guidelines are available, but regulations must acknowledge site variability and promote innovative and practical developments for control of acid mine drainage.

Chapter 14

Principles of Economic Mine Closure, Reclamation and Cost Management

D.R. Morrey

INTRODUCTION

Within the United States, the regulatory environment at both state and federal levels has exerted significant economic pressure on mine operators. In particular, increased stringency in standards for environmental protection on and around mines has a general tendency to increase operating costs during production, and capital costs at closure. Also, as a result of increasing environmental awareness in the public sector, mine operators must be aware of contingent liabilities related to regulatory compliance and third party, common-law action. This awareness applies as much to the Asian environment, as it does to the USA.

Hutchison and Ellison (1992) presented evidence of the cost sensitivity of mining operations to changes in environmental regulation in California. For example, more stringent requirements for precious metal waste rock containment were estimated to increase rock dump management costs threefold, if it should be necessary to introduce synthetic liner systems for groundwater protection. Undoubtedly, the economics of mining projects are sensitive to regulatory requirements, and can incur reduced financial rates of return resulting from increased costs in environmental management during operations and at mine closure.

The most expensive closure components relate to physical rehabilitation of mine disturbance, and the elimination of long-term maintenance requirements and liabilities, beyond closure. However, with careful planning, design and technology selection, rehabilitation and closure costs can be significantly reduced, particularly if reclamation is performed concurrently with mining operations, and partial closures of exhausted mine components are effected (Dahlstrand, 1995).

AIMS AND OBJECTIVES OF MINE CLOSURE

The aims of mine closure are generic, regardless of the regulatory arena within which the mine has operated. In general, the closure of each mine component should comply with both national and local regulations, where appropriate, and should fulfill specific permit requirements. The mine operator also aims to rehabilitate and close mine sites at low cost, while minimizing exposure to environmental liability in the long term. The generic aims and objectives of mine rehabilitation and closure planning are described in the Ontario Closure Plan Technology Guidelines (Ontario Ministry of Northern Development and Mines, 1991). The Ontario Guidelines summarize the main objectives of mine site rehabilitation as:

- protection of public health and safety;
- elimination or reduction of environmental impact; and
- restoration of disturbed land to its premining land use potential, or an acceptable alternative.

In so doing, maintenance of air quality, the protection of surface and groundwater supplies, and the restoration of agricultural or ecological resources are accommodated. Each of these objectives may be considered in terms of designing for physical stability, waste management, and acceptable land use. In terms of physical stability, all facilities and structures which remain after closure should not pose any risk to human health and safety, or to the environment, as a consequence of failure or deterioration. In this context, structures should remain stable under the influence of extreme events and perpetually disruptive forces. Similarly, health, safety, and environment should not be adversely affected by wastes which contain potentially toxic or corrosive chemical components, which could include acid-rock drainage, soluble metal salts, cyanide, and miscellaneous solvents and hydrocarbons. Mechanisms of release of these components to the environment include leaching, runoff, spillage, and erosion.

These concepts are widely accepted in the mining industry and have become standard frames of reference for rehabilitation and closure planning internationally. However, the author wishes to introduce the principles of land use capability, feasibility, and biological stability as additional criteria upon which rehabilitation and closure designs should be based.

The rehabilitation and closure plan should take cognizance of postclosure land use alternatives. Often, restoration of premining land use potential is the only statutory requirement. However, land use planning must also include assessments of land capability and land use feasibility. In this context, potentially limiting factors such as severely altered topography or surficial contamination, which could preclude certain types of land use, must be identified. Therefore, the application of United States Department of Agriculture criteria for land use capability (Klingbeal and Montgomery, 1961), its British modification (Bibby and MacKney, 1969), and FAO guidelines for assessing land suitability (Riquier et. al., 1970; and FAO, 1976) or other internationally accepted protocols, are recommended in the review of reclaimed mined land afteruse.

In cases where an ecological land use is appropriate, surface rehabilitation is considered successful when acceptable vegetation cover has been established upon land disturbed by mining activities. Usually, extent of cover and vegetation production are the only performance parameters of concern. However, the long-term stability of the surface systems depends upon the restoration of a complex of interactive components which allow ecologically rehabilitated land to be self-sustaining. As a general rule, the biological stability of reclaimed land, including vegetated mine wastes, increases with species diversity and habitat diversity. Long-term sustainability also depends upon the restoration of microbiological processes within the rooting zone. Therefore, the approach to successful rehabilitation should be holistic, and aimed at ecosystem restoration and soil development, rather than vegetation establishment alone.

CLOSURE PLANNING AND DESIGN

Conventionally, closure planning progresses in three phases, which may be summarized as:

- environmental characterization
- identification and description of mining components planned for closure
- selection of closure technology alternatives.

Descriptions of environmental and mining components are usually provided as integral parts of environmental impact assessments and conceptual reclamation plans, submitted to regulatory authorities and funding agencies during the feasibility and permitting processes. However, selection of closure technologies and final designs for closure of each mine component are often based upon a systematic procedure implemented later in mine development, as described in Figure 14.1. The evaluation of closure options is based upon assessments of effectiveness, risk of failure, long-term stability, cost, and liability.

Current trends in closure planning and design involve technical review and analyses of risk and cost benefit in both engineering and environmental terms. Conceptual designs may be screened by means of quantitative performance assessments which are relevant to design evaluations during operational and closure phases. Such analyses assist in the selection of the most appropriate design alternatives. Selection can be facilitated further by way of objective decision analysis.

Performance assessments incorporate engineering and environmental risk assessments, and decision analysis. Environmental risk assessments can be both human and ecological. The advantages of a risk-based approach to closure planning lie in the quantification of subjective factors and the analysis of uncertainty related to engineering design performance and cost. The outcome of this type of assessment is either the selection and implementation of an appropriate technology, or risk management. Risk management may involve redesign, additional data collection and analysis, or more detailed modeling, leading to final selection. In doing so, mine management aims to reduce risk and uncertainty. A schematic flow diagram which describes a performance assessment protocol is shown in Figure 14.2.

ENGINEERING PERFORMANCE ASSESSMENT

An engineering assessment should be performed for all structures remaining on the mine site to evaluate their postclosure performance. The performance assessment should include an analysis of stability under extreme events, such as seismic loading, and under the influence of perpetually disruptive forces, such as erosion or expansive clay movement. At a minimum, the stability of tailings impoundments and waste rock facilities must be determined. The forces such as seismic loading or impacts from hydrological events used in this analysis are typically more conservative than those used for the operational life of the structures. Bedrock acceleration as a result of the maximum credible earthquake or the erosional forces and storage requirements which result from runoff following probable maximum precipitation are typical examples of conservative design parameters.

The performance analysis includes an evaluation of failure scenarios of engineering structures such as releases of large volumes of tailings as a result of an impoundment failure, chronic releases of contaminants through wind and water erosion, and the performance of the final reclaimed landscape. These analyses are usually quantitative but can also be qualitative. Deterministic analyses are commonly used to evaluate the relative stability of structures, and the results are expressed as factors of safety. Probabilistic analyses can also be used to express the reliability of structures (Harr, 1967). In the latter case, the results are expressed as a probability of failure. Because probabilities are linearly scaled

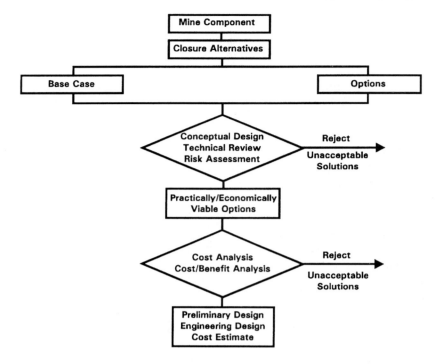

FIGURE 14.1. Selection of closure options (Adapted from BC AMD Task Force, 1989).

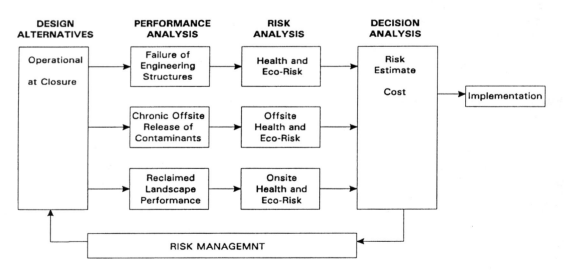

FIGURE 14.2. Performance assessment protocol.

(factors of safety are typically not linearly scaled), the results are well suited as input to economic analyses.

HEALTH AND ECOLOGICAL RISK ANALYSIS

This type of analysis is a necessary component of the selection procedure if the significance of environmental impacts of engineering failure, both during operations and beyond closure, is a concern. The analysis may be qualitative or quantitative, and relies upon the fundamental questions of risk assessment: What can happen? How likely is it that it will happen? and, If it does happen, what are the consequences? (Kaplan and Garrick, 1981). The four general steps involved in both qualitative and quantitative assessments are: hazard identification and assessment; exposure assessment; consequence assessment; and risk characterization. The result of the analysis includes an estimate of the magnitude of risk associated with selected engineering failure modes and exposure/consequence scenarios. Additionally, contaminant-specific probabilities of exceeding prescribed environmental standards are given for each closure design alternative and failure mode.

COST-BENEFIT ANALYSIS

The relative cost benefit of closure options can also be assessed with the aid of models which may be subjective or objective, qualitative or quantitative, and deterministic or stochastic. Each type of model may include evaluations of performance criteria which may be summarized as:

- practicality (of implementation)
- durability
- risk of failure
- environmental impact
- complexity of installation
- maintenance requirement
- effectiveness.

These criteria are usually assessed in a subjective and semiquantitative manner, to provide a cost per unit benefit ratio, for each closure option. However, the process can be made more rigorous if probabilistic analyses of performance are applied. Difficulty arises, though, when reliable cost estimates for personnel, materials, and equipment are difficult to obtain, either due to significant variance in quotes, uncertainties of mass/volume estimates, or the inflationary variables associated with long-term planning. This type of uncertainty in closure cost analysis can be managed using relatively simple statistical tools (Morrey and Van Zyl, 1994).

ADDRESSING CLOSURE COST UNCERTAINTIES

The uncertainty or variability of parameters can be quantified using objective statistical methods, or subjective decisions can be used to identify their variability. Typically, closure costs are calculated with a spreadsheet in which the quantities and the unit cost for the various items are listed and assumed to be deterministic. Two spreadsheet supplements are available which can describe any of the cells in a spreadsheet as a variable. These additional software components are @RISK (Palisade, 1991) and CrystalBall (Decisioneering, 1993). @RISK is compatible with the Lotus spreadsheet, while CrystalBall can be used with Lotus as well as Excel software.

By defining the variable in any cell as a random variable, Monte Carlo or Latin Hypercube modeling can be performed to generate results expressed as a distribution curve. This will allow a large number of probability distributions to be simulated for the random variables. Usually, a mean and standard deviation are necessary to define the parameters for the probability density function. In the case of some distributions, such as the β and γ distributions, it is also necessary to identify scale and/or shape parameters.

EXAMPLE

Consider a simplified example for closure costs of a mine waste dump. Assume that there is no acid drainage present and that the waste dump can be reclaimed in place by resloping the outer surface, placing growth medium, and vegetating the covered surface. Table 14.1 provides a typical cost estimate to perform the work. By assuming that all quantities and unit costs are random variables, the resulting closure cost based on expected values (mean values) can be estimated.

Table 14.2 lists the distributions and parameters for quantities as well as unit rates. For recontouring, a triangular distribution is assumed for the quantity of earthworks. The triangular distribution has a minimum value of 180,000 m^3, a maximum value of 220,000 m^3, and a mean value of 200,000 m^3. The unit cost is assumed to follow a normal distribution with a mean value of \$0.38 and a standard deviation of \$0.08. After performing the Monte Carlo simulation, the probabilistic results for the cost estimate shown in Figure 14.3 are obtained. These results show that when using expected values for all parameters, the predicted value for closure costs is \$139,700 (the results shown in Table 14.1). However, there is a 10% probability that the closure cost will exceed \$163,700 or that it will be smaller than \$116,900. By using this approach, the potential high and low ends of the cost estimate can be identified, and this can be used by management as a decision-making tool for funds to be set aside for closure.

CLOSURE FUND ESTIMATION

Financial provision for closure may be planned for by considering the following tasks:

- estimation of time to closure
- estimation of period of postclosure reclamation and monitoring
- determination of annual costs of postclosure reclamation
- determination of value of fund required at closure
- forecast of rates of return on fund
- calculation of periodic payments.

Generally, closure costs may be reduced significantly if concurrent reclamation and partial closures of mine components are effected during the operational life of the mine. Such costs may be accounted for as ongoing operational costs and can be easily planned. However, the capital cost of final closure has several uncertainties attached to it.

Apart from the element of specific reclamation cost as illustrated above, mine management may wish to consider contingent costs associated with unforeseen environmental liabilities within the context of closure fund estimation. Exposure to such liability at and beyond closure is an uncertain variable, involving an uncertain cost. Often, the exposure arises as a consequence of design failure, changes in regulations, or litigation.

Table 14.1. Reclamation Cost Example.

Cost Item	Units	Quantity	Unit Cost	Cost
Recontouring	m³	200,000	$0.38	$76,000
Placement of growth medium	m³	28,000	$1.65	$46,200
Vegetative cover	hectares (ha)	35	$500	$17,500
			TOTAL COST	$139,700

Table 14.2. Distribution Functions for Reclamation Example.

Variable	Units	Distribution	Mean	Standard Deviation	Range
Recontouring	m³	triangular	200,000		180,000–
	$/m³	normal	$0.38	$0.08	220,000
Growth medium	m³	triangular	28,000		25,000–
placement	$/m³	normal	$1.65	$0.17	31,000
Vegetative cover	ha	triangular	35	$75	30–40
	$/ha	normal	$500		

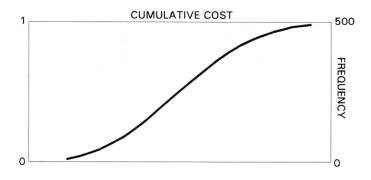

PERCENTILE	COST ($)
0%	92,462
10%	116,882
20%	123,567
30%	128,952
40%	133,688
50%	139,073
60%	144,087
70%	150,153
80%	155,229
90%	163,771
100%	203,882

FIGURE 14.3. Cost estimates.

OVERALL CLOSURE COST CALCULATION

Management would be able to include uncertain or risk-costs in the overall closure cost calculation, as follows:

$$Q = \sum \frac{1}{(1+i)} \bullet (B_t - C_t - R_t) \tag{14.1}$$

Where Q = net closure cost
        B = financial benefits (e.g., asset recovery)
        C = nonoperational reclamation and closure costs (including postclosure monitoring and maintenance)
        R = risk costs (costs x uncertainty)
        i = annual interest rate
        t = time in years.

Note that financial benefits (B), reclamation and closure costs (C), and risk cost (R) can be expressed as probabilistic quantities. Risk cost (R) is usually expressed as a function of real cost and uncertainty, as follows:

$$R = p \times c \times \gamma \tag{14.2}$$

Where p = probability of cost occurring
        c = value of cost
        γ = management function.

The management function is a parameter which describes the overall behavior of a system and/or its management, in terms of risk. Conservative, risk-averse management sets $\gamma < 1.0$; whereas in the case of poor management (whether identified by the company or during due diligence or regulatory review), the value of the management function is typically $\gamma > 1.0$.

By using Equation 14.1, it is possible to obtain the total expected net cost Q on an annual or cumulative basis by using the expected costs from Monte Carlo analyses and by estimating the risk cost.

It should be noted that calculations based on expected values only are not fully "probabilistic." In reality, it is preferable to apply distributions of input data, to provide a distribution of contingency fund estimates. The contingency fund would be calculated in the same way, but using a statistical distribution of values for each variable.

CORPORATE PREFERENCES

Not all corporations have a clear philosophy on what closure objectives they should follow as a principle for closure of existing mines. The three objectives typically available are:

- final closure (not requiring ongoing maintenance)
- passive care following closure
- active care following closure.

The period of postclosure maintenance may be short- or long-term. In the case of new mines, the choice should always be to develop a design that will allow for a final closure solution. There are a number of jurisdictions in the world where permits for new mines will not be issued if anything other than a final closure solution is proposed.

As a consequence of poor experiences with active care programs, such as ongoing treatment of effluent from waste dumps, corporate philosophy may be that only final solutions will be acceptable at existing mines. However, when such a solution is developed, and the initial capital and operating cost estimates become exceedingly high, it may be necessary to review the strategy and consider closure designs with alternative objectives in mind. The uncertainties associated with corporate preferences may also change with time.

CONCLUSIONS

Globally, the mining industry is tending toward an integrated approach to environmental management, operations, and economic planning. Also, environmental impact mitigation and concurrent reclamation are now implemented more frequently as integral components of the mining process, at least for economic reasons. This holistic approach may also accommodate sequential closure of selected mine components during operations. The aim of partial closure is to reduce effort and capital costs during the final decommissioning phase, while reducing environmental liabilities during operations. The concept of integrated operations and reclamation planning at an early stage in mine feasibility assessment and during the permitting process is now frequently adopted in the development of new mining projects, and expansions of existing facilities. Significant benefits derived from integrated operations and environmental planning, and concurrent implementation, are evident. Examples of such benefits include enhanced operational cost benefit, reduced capital costs at closure, and reduced regulatory pressure, which is frequently associated with poor environmental compliance during mine development and production.

REFERENCES

Bibby, J.S. and D. MacKney. Land Use Capability Classification, Soil Survey Technical Monograph No. 1, 1969. HMSO, London, 1969.

British Columbia Acid Mine Drainage Task Force. Draft Acid Rock Drainage Technical Guide, Vol. 1. Steffen, Robertson and Kirsten (BC) Inc., Canada, 1989.

Dahlstrand, A. Closure Concerns at Sonora Mining's Jamestown Mine. Mining Engineering. March 1995, pp. 236–239, 1995.

Decisioneering. CrystalBall Version 3.0, Forecasting and Risk Analysis, Decisioneering Inc., Denver, CO, 1993.

FAO. FAI Soils Bulletin No. 32. FAO, Rome, 1976.

Harr, M.E. *Reliability-Based Design in Civil Engineering*. McGraw-Hill, New York, 1987.

Hutchison, I.P. and R.D. Ellison, Eds. *Mine Waste Management*. Lewis Publishers, Boca Raton, FL, 1992.

Kaplan, S. and B.J. Garrick. On the Quantitative Definition of Risk. *Risk Anal.* (I), 1, pp. 11–27, 1981.

Klingbeal, A.A. and P.H. Montgomery. Land Capability Classification, USDA Agricultural Handbook No. 210, Washington, DC, 1961.

Morrey, D.R. and D. Van Zyl. Including Uncertainty in Mine Closure. *Proc. 5th Western Regional Conference on Precious Metals, Coal and the Environment*. Society for Mining, Metallurgy and Exploration. Black Hills, SD, 1994.

NEPA. National Environmental Policy Act. United States Federal Government, Washington, DC, 1969.

Palisade. Risk Analysis and Modelling, @RISK, Palisade Corporation, 1991.

Riquier, J. et al. A New System of Soil Appraisals in Terms of Actual and Potential Production. FAO, Rome, 1970.

MANAGEMENT OF DERELICT LANDS

Chapter 15

The Importance of Nitrogen in the Remediation of Degraded Land

A.D. Bradshaw

The ultimate requirement in remediation of degraded land, unless it is to be built over, is the establishment of a vegetation cover. This ensures stability, reduces potential pollution, provides amenity, and in many cases produces an economic product. Because plants have the capacity to grow and accumulate organic matter without outside support, they provide a self-sustaining and genuinely economic solution.

But this is only possible if the plants have the right environmental conditions and sufficient soil resources to enable this growth to take place. It is relatively easy to recognize when environmental conditions are inappropriate for growth due to soil problems such as extreme acidity or metal toxicity. It is less easy to recognize when the problem is lack of soil resources, particularly lack of mineral resources.

THE IMPORTANCE OF THE NITROGEN CAPITAL

In practice, when land is being remediated, experience has shown that nitrogen is commonly an important limiting factor (Bloomfield et al., 1982). This is understandable. Of all the nutrients that plants obtain from the soil, nitrogen is the nutrient required in the greatest amount by living tissue. This is demonstrated by the amounts actually taken up by plants annually and by the amounts stored in capital, particularly in grassland ecosystems. In woodland ecosystems, because of the larger fraction of woody material, calcium may be accumulated in somewhat larger total amounts, but nitrogen uptake is still highest (Table 15.1). Adequate amounts of nitrogen must therefore be available in soils if plants are to grow properly.

Nitrogen, however, is unlike other nutrients. It does not occur in soil minerals but is accumulated in soil almost entirely in the surface layers, mostly by the process of biological fixation. Apart from the small amount occurring in a soluble form, it is stored in the soil in organic matter. This nitrogen becomes available to plants only when the organic matter breaks down as a result of the activities of microorganisms.

When land is degraded, the surface soil containing the organic matter and the store of nitrogen is usually lost. Successful remediation therefore depends on replacing that nitrogen. Because the nitrogen used by plants is that released by breakdown of the organic matter, and because this occurs at a slow rate in most conditions (less than 10% yr^{-1}), the amount that must be replaced is considerable. In temperate regions plants require about 100 kg N ha^{-1} yr^{-1}. If the rate of decomposition of organic matter and therefore release of mineral nitrogen is only about 10% yr^{-1}, the soil must contain about 1,000 kg N ha^{-1} if

Table 15.1. Comparison of Capital and Uptake of Nitrogen and Other Elements in a Temperate Forest and Grassland (in italics).[a]

Component	Chemical Element									
	N		P		K		Ca		Mg	
Capital (kg ha⁻¹)										
Aboveground biomass	351	*43*	34	*4.0*	155	*25.2*	383	*5.8*	36	*11.3*
Belowground biomass	181	*54*	53	*3.7*	63	*3.1*	101	*9.7*	13	*32.0*
Forest floor	1256	*78*	66	*372*	38					
Uptake (kg ha⁻¹ yr⁻¹)										
Vegetation uptake	79.6	*162*	8.9	*16*	64.3	*73.7*	62.2	*30*	9.3	*7.0*
Vegetation accumulation	9.0	*2.3*	5.8	*8.1*	0.7					

[a] From Bormann and Likens, 1979 and Perkins, 1978.

Table 15.2. The Organic Soil Nitrogen Capital Needed (kg N ha^{-1}) to Satisfy Different Nitrogen Requirements, Assuming Various Decomposition Rates.[a]

Annual Requirement	Decomposition Rate				Type of Ecosystem
	2%	10%	20%	100%	
200	10,000	2,000	1000	200	warm
100	5,000	1,000	500	100	temperate
50	2,500	500	250	50	cool/arid
Type of ecosystem	montane	cool temperate	warm temperate	tropical	

[a] From Bradshaw, 1983.

100 kg N ha^{-1} yr^{-1} is to be released and the soil/vegetation system is to be self-sustaining. The detailed amounts required will depend on plant growth and rates of organic matter decomposition. In tropical conditions a smaller soil nitrogen capital may be adequate (Table 15.2).

But since the vegetation itself contains nitrogen, this must also be replaced. Here the amount required differs markedly depending on the nature of the vegetation; in grassland the vegetation may contain no more than 100 kg N ha^{-1}; in woodland there can be more than 500 kg N ha^{-1} (Table 15.1). All this suggests that for practical purposes we can assume that to establish a satisfactory self-sustaining vegetation cover at least 1000 kg N ha^{-1} must be found or provided. Yet in most types of degraded land no more than 100–200 kg N ha^{-1} is likely to be present at the outset.

To cope with this overriding problem of inadequate nitrogen supply, it may be possible in some situations to choose species with low nitrogen requirements and tolerant of low levels of soil nitrogen. But these typically have low growth rates and will not be very successful at providing a stabilizing plant cover.

Successful remediation therefore normally requires:

(a) the accumulation of nitrogen to provide the soil capital
(b) the accumulation of nitrogen to provide the vegetation capital.

To ensure success, accumulation must be achieved as rapidly as possible, otherwise nitrogen deficiency may lead to poor growth and even plant death, and in many situations erosion and other problems connected with degraded land can return.

THE PROVISION OF NITROGEN CAPITAL

How can nitrogen be provided? Several contrasting methods are available:

1. By storing and returning the original topsoil. Surface soil usually contains about 0.2% N. If this soil can be stripped before the disturbing/degrading operation begins and be stored suitably for a short time in a way that its qualities are not destroyed, usually in low heaps, it can be respread later and the nitrogen capital restored (Harris et al., 1996). Such soil conservation usually has many other benefits since the organic matter is conserved and therefore the soil physical structure. Accumulations of other nutrients besides nitrogen, in an available form, will also be retained.

Table 15.3. Total Nitrogen Contents and Nitrogen Released by Top Soil Layers of Different Depths, with Different N Contents, and with Different Annual Rates of Mineralization.[a] (Compare with suggested requirements given in Table 15.2.)

N content		Depth of Top Soil Layer			
		100mm	200mm	300mm	
0.02%	total N	200	400	600	kg N ha^{-1}
	N released				
	2% mineralized	4	8	12	kg N ha^{-1} yr^{-1}
	5%	10	20	30	
0.05%	total N	500	1000	1500	kg N ha^{-1}
	N released				
	2% mineralized	10	20	30	kg N ha^{-1} yr^{-1}
	5%	25	50	75	
0.1%	total N	1000	2000	3000	kg N ha^{-1}
	N released				
	2% mineralized	20	40	60	kg N ha^{-1} yr^{-1}
	5%	50	100	150	
0.2%	total N	2000	4000	6000	kg N ha^{-1}
	N released				
	2% mineralized	40	80	120	kg N ha^{-1} yr^{-1}
	5%	100	200	300	

[a] From Bradshaw, 1989.

It is not always possible, however, to retain all the original topsoil, or it may become diluted by subsoil. Either will diminish the nitrogen capital. The presence of subsoil may also reduce rates of mineralization. The effects of these on the annual supply of mineral nitrogen are shown in Table 15.3. Nevertheless, since careful replacement of topsoil allows instant restoration of nitrogen capital and other soil qualities, it is now mandatory in many countries. It is often possible, particularly in surface mining, to overcome storage problems by organizing a system of continuous removal and immediate replacement, which reduces costs and enables many original species to be carried over.

2. By fertilizer. Simple or compound mineral fertilizers are readily available in most countries, although they are usually expensive. They can be applied easily by normal agricultural machinery. Their most serious drawback is that considerable leaching losses can occur at the outset, especially in the period before the vegetation cover is properly established (Dancer, 1975; Marrs and Bradshaw, 1980). As a result, repeat applications are essential, of about 50–100 kg N ha^{-1} each time. These will then need to be continued for about 10 years to build up the required capital. This long-continued aftercare is often difficult to arrange or pay for.

3. By organic wastes. Organic materials have the advantage that the nitrogen is already held in an organic form and so will be held and released at an appropriate rate. Because the nitrogen is not released immediately, large amounts can be supplied in a single application, even as much as 1000 kg N ha^{-1}. Sewage sludge is the most common and suitable material (Hall, 1991). It has the advantage for many purposes that it contains high levels of phosphorus. It can be used in the initial treatment or later, as an aftercare treatment, when nitrogen deficiencies may become manifest (Figure 15.1). In industrial areas, sewage sludge may contain levels of toxic metals too high to allow it to be used. But it must be remembered that the levels that will be permissible must be calculated in relation to a single

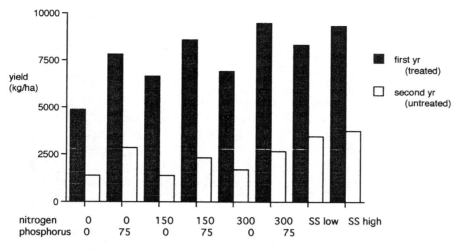

FIGURE 15.1. A comparison of the effects of addition of fertilizer or sewage sludge containing an equivalent amount of available nitrogen to a pasture newly established on reclaimed colliery spoil and suffering from regression (redrawn from Michael et al., 1991). The fertilizer treatments show that the pasture is deficient in both nitrogen and phosphorus; these are effectively relieved by the sewage sludge, the effects of which persist into the second year.

Table 15.4. Nutrient Content and Organic Matter Content of Some Different Organic Materials. (% dry solids).[a]

Material	Nitrogen	Phosphorus	Potassium	Organic Matter
Farmyard manure	0.6	0.1	0.5	24
Poultry manure	2.3	0.9	1.6	68
Sewage sludge (air dry)	2.0	0.3	0.2	45
Domestic refuse	0.5	0.2	0.3	65
Straw	0.5	0.1	0.8	95

[a] From Bradshaw and Chadwick, 1980.

application, and not the repeated applications that are usual when it is applied to agricultural land. In many areas other organic wastes may be available (Table 15.4).

4. By atmospheric precipitation. The importance of atmospheric contributions must not be forgotten. In nonindustrial areas throughout the world the contribution is usually not more than 10 kg N ha^{-1} yr^{-1}. This can help maintain nitrogen levels but will hardly help to build up capital. In industrial areas, however, due to the burning of fossil fuels, atmospheric precipitation of nitrogen can reach 50 kg N ha^{-1} yr^{-1}, or even 100 kg N when dry deposition is included. This is sufficient to provide for immediate annual needs of the vegetation and will subsequently become part of the soil capital. It is noticeable that in areas where there is considerable burning of fossil fuels, nitrogen deficiencies in remediation work are not so noticeable, especially after the first year. Because the nitrogen is spread in a diffuse manner, species with extensive root systems, scavenging species, will be particularly able to take advantage of this source.

5. By free-living nitrogen-fixing microorganisms. The importance of this source of nitrogen is often stressed. In certain conditions, notably in rice fields, free-living microorganisms, mostly blue-green algae, can contribute 50–100 kg N ha^{-1} yr^{-1}. But usually the amount is less than 10 kg N, and therefore not very significant.

6. By symbiotic nitrogen-fixing microorganisms. The activities of the symbiotic nitrogen fixers, where there is a highly developed relationship between microorganisms and host, as in most members of the Leguminosae, is usually the most valuable source of nitrogen in the remediation of degraded land. Normal rates of net accumulation of nitrogen by such species is commonly at least 50 kg N ha^{-1} yr^{-1}, and can readily reach over 100 kg N even on degraded soils (Dancer et al., 1977). The advantage of the close relationship between the nitrogen-fixing microorganisms and their plant hosts is that the plant obtains the nitrogen directly and not via the soil. But at the same time, as the plant sheds its older parts both above and below ground, nitrogen is soon accumulated in the soil, in an organic form. Because of the low C/N ratio of this material it decomposes rapidly, so that the nitrogen from the nitrogen-fixing species soon becomes available to other, nonfixing, species.

This can be well seen in the effects of alder (*Alnus* sp.), which has root nodules containing an N-fixing actinomycete. Planted on kaolin waste, it has a very significant effect on soil nitrogen (Figure 15.2). This translates both into a very high rate of growth itself and a remarkable effect on the growth of adjacent nonnitrogen-fixing species (Figure 15.3).

The Best Option

Which is the best option of these different methods of accumulating nitrogen? It will depend on circumstances and the desired end use. The governments of developed countries are now requiring Method 1, because it leads to instant restoration of all soil qualities, not only the nitrogen capital. However, it requires costly earthmoving equipment which must be carefully organized. Method 2 is possible in almost all situations, but it requires repeated applications and is, ultimately, expensive. Method 3 depends on circumstances, particularly on the availability of N-rich organic material within reasonable distance of the site being remediated. Method 4, to be effective, depends entirely on the site being situated in an area where there is high atmospheric pollution. Method 5 is never more than a contributor to other methods.

There is no doubt that Method 6 is universally the most valuable, because once the nitrogen-fixing species have become established they provide large amounts of nitrogen at no cost. A wide variety of species are available, of different life-forms and soil and climatic adaptations (Table 15.5). These can either be used directly as fodder or timber, or as a contributor by their nitrogen to the growth of other desirable nonfixing species. The latter approach has been conspicuously successful in the rehabilitation of jarrah forest (*Eucalyptus marginata*) in Western Australia after mining for bauxite; the *Acacia* species used lead to an accumulation of 112 kg N ha^{-1} yr^{-1} (Ward and Koch, 1996). But work is nearly always required to determine the best species and techniques of use. The choice of species will be related to the proposed end-use, but also to the desired associated species and to the soil conditions. A systematic development program is required (Table 15.6).

In practice, several of these methods may be combined together. In soils very deficient in nitrogen, fertilizers provide an immediate source to help the vegetation to become established, even when nitrogen-fixing species are being used. Organic wastes are a good alternative, especially because their controlled release of mineral nitrogen ensures that the

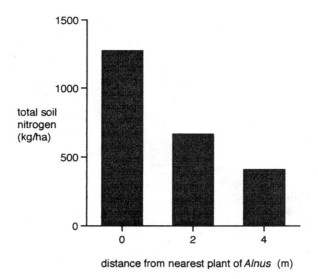

FIGURE 15.2. Total soil nitrogen accumulated on kaolin mine waste after 11 years at different distances from stands of nitrogen-fixing alder (*Alnus* sp.) (derived from Kendle and Bradshaw, 1992). In this period the alder has raised the nitrogen content of the waste in the vicinity of its roots to well over 1000 kg N ha^{-1}.

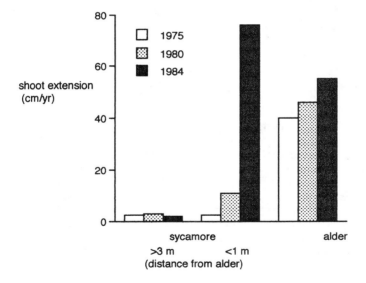

FIGURE 15.3. Extension growth of nonnitrogen-fixing sycamore trees (*Acer pseudoplatanus*) on kaolin mine waste at different distances from nitrogen-fixing alder (*Alnus* sp.) (derived from Kendle and Bradshaw, 1992). The progressive effects on the growth of the sycamore of the nitrogen accumulation shown in Figure 15.2 are very apparent.

growth of nitrogen-fixing species is not inhibited, but their use is dependent on their availability. In all situations, atmospheric precipitation and free-living nitrogen fixers can be expected to contribute to some extent.

Table 15.5. Legumes Particularly Suitable for Degraded Land.[a]

	Soil	Climate
Herbaceous		
Amorpha fruticosa (indigo bush)	NC[b]	W[c]
Centrosema fruticosa (centro)	AN	W
Coronilla varia (crown vetch)	AN	CW
Desmodium uncinatum (silver desmodium)	AN	W
Lathyrus sylvestris (mat peavine)	NC	W
Lezpedeza bicolor (lezpedeza)	AN	W
Lezpedeza cuneata (sericea)	AN	W
Lezpedeza japonica (Japan lezpedeza)	AN	W
Lotus corniculatus (birdsfoot trefoil)	NC	CW
Lupinus arboreus (tree lupin)	ANC	CW
Medicago sativa (alfalfa)	NC	CW
Melilotus alba (white sweet clover)	ANC	CW
Melilotus officinalis (yellow sweet clover)	ANC	CW
Phaseolus atropurpureus (siratro)	ANC	W
Stylosanthes humilis (Townsville stylo)	AN	W
Trifolium hybridum (alsike clover)	ANC	C
Trifolium pratense (red clover)	NC	C
Trifolium repens (white clover)	NC	C
Woody		
Acacia albida (winter thorn)	NC	WA
Acacia auriculiformis	ANC	WA
Acacia crassiocarpa (golden wattle)	ANC	WA
Acacia saligna	NC	WA
Acacia tortilis (umbrella thorn)	NC	WA
Albizia lebbek (lebbek)	NC	CW
Calliandra callothyrsus	NC	W
Dalbergia sissoo (sissoo)	ANC	CWA
Gliciridia sepium	ANC	W
Gleditsia triacanthos (honey locust)	ANC	CW
Leucaena leucocephala (subabul)	ANC	W
Peltophorum pterocarpum (copperpod)	ANC	W
Prosopis glandulosa (mesquite)	NC	WA
Prosopis tamarugo (tamarugo)	NC	WA
Sesbania grandiflora (agati)	ANC	W

[a] From Bradshaw and Chadwick, 1980 and National Research Council, 1979.
[b] A=acid; N=neutral; C=calcareous.
[c] C=cool; W=warm; A=arid.

DEVELOPMENT OF A SUCCESSFUL PROGRAM

In practice, each of the options requires investigation so that their potential value for the site in question can be assessed. Successful use of nitrogen-fixers (Method 6) requires this particularly, because it is dependent on biological processes and successful growth of particular species.

Some of this development must be done at the outset. But some of it can usefully be done within the remediation operation itself, using the operation to provide the appropriate experimental conditions. But any experiments must be laid out critically—with replica-

Table 15.6. Essential Features of a Development Program for the Use of Nitrogen-Fixing Species in Remediation of Degraded Land.

1. Decide end use/requirement (e.g., stabilization, agriculture, forestry, nature conservation, etc.)
2. Determine likely range of species to be used
   (a) native
   (b) others
3. Identify N-fixers within these
4. Examine literature for other possible N-fixers
5. Establish trials of N-fixers on relevant soils
   (a) unamended
   (b) with various amendments
6. Establish trials to assess importance of inoculation by microorganisms and of use of particular strains
7. Establish trials of N-fixers
   (a) with other species
   (b) with different management regimes

tion, randomization, and factorial designs where appropriate. Odd plots containing treatments without proper controls are not satisfactory.

However, in remediation work it can be valuable to have a simple experiment which is repeated in several different places and conditions rather than one complex experiment only carried out in one place on one lot of material. In this way the universality of the method, and the importance, or otherwise, of local conditions can be assessed.

Within the accepted flowchart for restoration (Figure 15.4) the critical importance of nitrogen means that there is, therefore, an important subset of work which, if successfully carried out, can add considerably to the success of all remediation projects.

REFERENCES

Bloomfield, H.E., J.F. Handley, and A.D. Bradshaw. Nutrient Deficiencies and the Aftercare of Derelict Land. *J. Appl. Ecol.*, 19, pp. 151–158, 1982.

Bormann, F.H. and G.E. Likens. *Pattern and Process in a Forested Ecosystem*. Springer-Verlag, New York, 1979.

Bradshaw, A.D. The Reconstruction of Ecosystems. *J. Appl. Ecol.*, 20, pp. 1–27, 1983.

Bradshaw, A.D. The Quality of Topsoil. *Soil Use Manage.*, 5, pp. 101–108, 1989.

Bradshaw, A.D. Alternative Endpoints for Reclamation. In *Rehabilitating Damaged Ecosystems*, 2nd ed., Cairns, J., Jr., Ed., Lewis Publishers, Boca Raton, FL, 1995, pp. 165–186.

Bradshaw, A.D. and M.J. Chadwick. *The Restoration of Land*. Blackwell Scientific Publications, Oxford, 1980.

Dancer, W.D. Leaching Losses of Ammonium and Nitrate in the Reclamation of Sand Spoils in Cornwall. *J. Environ. Qual.*, 4, pp. 499–504, 1975.

Dancer, W.D., J.F. Handley, and A.D. Bradshaw. Nitrogen Accumulation in Kaolin Mining Wastes in Cornwall. II. Forage Legumes. *Plant Soil*, 48, pp. 303–314, 1977.

Hall, J.E., Ed. *Alternative Uses for Sewage Sludge*. Pergamon, Oxford, 1991.

Harris, J.A., P. Birch, and J. Palmer. *Land Restoration and Reclamation: Principles and Practice*. Addison, Wesley, Longman, Harlow, Essex, 1996.

Kendle, A.D. and A.D. Bradshaw. The Role of Soil Nitrogen in the Growth of Trees on Derelict Land. *Arboricultural J.*, 16, pp. 103–122, 1992.

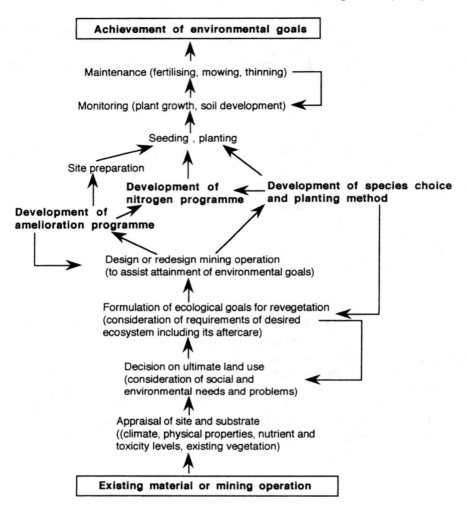

FIGURE 15.4. The normal steps to be taken in the development of a program for the remediation of degraded land (Bradshaw, 1995) should be extended to include a specific program for the remediation of nitrogen deficiency (see Table 15.6).

Marrs, R.H. and A.D. Bradshaw. Ecosystem Development on Reclaimed China Clay Wastes. III. Leaching of Nutrients. *J. Appl. Ecol.*, 17, pp. 803–813, 1980.

Michael, N., A.D. Bradshaw, and J.E. Hall. The Value of Fertilizer, Surface Applied and Injected Sewage Sludge to Vegetation Established on Reclaimed Colliery Spoil Suffering from Regression. *Soil Use Manage.*, 7, pp. 233–239, 1991.

National Research Council, Advisory Committee on Technological Innovation. Tropical Legumes: Resources for the Future. National Academy of Sciences, Washington, DC, 1979.

Perkins, D.F. The Distribution and Transfer of Nutrients in the *Agrostis-Festuca* Grassland Ecosystem. In *Production Ecology of British Moors and Montane Grasslands*, Heal, O.W. and D.F. Perkins, Eds., Springer-Verlag, Berlin, 1978, pp. 375–396.

Ward, S.C. and J.M. Koch. Biomass and Nutrient Distribution in a 15.5 Year Old Forest Growing on a Rehabilitated Bauxite Mine. *Australian J. Ecol.*, 21, pp. 309–315, 1996.

Chapter 16

Diagnosis and Prognosis of Soil Fertility Constraints for Land Restoration

R.W. Bell

INTRODUCTION

The finite nature of existing global land resources means that degraded land must be restored if the demand for increased food, fiber, and timber production is to be satisfied. The agents that threaten land quality and productivity are varied in nature, however; many of them have a nutritional component and it is the methodology behind the diagnosis and prognosis of such nutritional disorders that is the main emphasis of this paper.

Soil fertility constraints for land restoration are diverse in their origin and nature and can be grouped under the following broad categories: acidity, alkalinity, salinity, sodicity, nutrient deficiency, and elemental toxicity. All of the preceding chemical constraints impact on plant nutrition. Nutritional constraints in pot experiments with rehabilitation substrates, in field experiments, and in post-revegetation monitoring and management can be diagnosed using plant analysis, plant symptoms, and plant response. In addition, plant and soil analysis can be used for the prognosis of possible soil fertility constraints which may limit plant growth after restoration, but their use depends on properly calibrated standards for the interpretation of the plant and soil analysis. Principles for the proper calibration of the soil and plant analysis standards will be discussed with particular reference to nutrient deficiencies. Long-term success in alleviating soil fertility and plant nutrient constraints also needs to be assessed in land restoration since nutrient cycling is one of the key ecosystem functions that restoration seeks to reestablish (Hobbs and Norton, 1996). Though soil and plant analysis are used in nutrient cycling studies, their application and the types of standards used for managed agricultural postmine use differs from the cases where restoration is aimed at a native bushland, grassland, or forest. These differences are discussed below.

The term "restoration" is used in the present review as a generic term after Hobbs and Norton (1996), who suggest "that restoration occurs along a continuum and that different activities are simply different forms of restoration." Given that most of the study concerns restoration of land degraded by mining, the term rehabilitation as defined by Aronson et al. (1993) is also appropriate.

DIAGNOSIS OF MINERAL DISORDERS

Alleviation of soil fertility constraints depends on being able to correctly identify the various limiting disorders which are already causing a yield constraint. This process of identifying a disorder is known as diagnosis (Smith, 1986). The most convincing diagnosis of a disorder is to apply a corrective treatment in the field and demonstrate alleviation of the constraint by the plant response (Craswell et al., 1985). However, this approach is rather costly when many sites are involved, where inappropriate selection of treatments or unseasonal conditions cause total failure of the experiment, or multiple nutrient constraints causes the process of identifying all the key constraints to take several years. Indeed, in the case of restoration, responses may require many years to fully assess. Hence this approach to diagnosis, although indispensable, must be used selectively. When completed, field response experiments serve as the benchmark against which most other forms of diagnosis are calibrated (Peck and Soltanpour, 1990). On mine sites where many of the constraints are rather site-specific, field response experiments are indispensable as the indicator of restoration success.

On many occasions in revegetated areas, poor growth or variable growth will prompt the need for diagnosis of the cause so that corrective actions can be initiated. However, it would be preferable to predict the occurrence of such disorders before revegetation begins so that appropriate amelioration can be applied initially. The use of soil and plant analysis in this way is known as prognosis because the aim is to predict those mineral disorders that will in the future limit plant growth (Craswell et al., 1985).

Field Response

Several precautions are necessary for field experiments to accurately diagnose the actual mineral disorder constraining growth. The basis principle is that growth will be limited by the most limiting factor. Unless the most limiting factor has been identified or added to the experiment then the experiment can be expected to achieve little. This is especially important on degraded sites such as mine sites which suffer from multiple deficiencies because the alleviation of only one or some of them will fail to give a true indication of the responsiveness of plant growth at the site. In another example, extreme salinity in mine residues may limit growth to such an extent that nutrients fail to cause growth responses even though the growth substrate is in fact very low in many nutrients (Ho et al., this volume). Field experiments can also fail to diagnose a deficiency if nutrients are added at rates which are substantially less than optimal levels. In such cases, a preliminary subtractive design experiment comprising a complete nutrient treatment and the omission of single nutrients can be an effective approach (Craswell et al., 1985; Bell et al., 1989). Preliminary pot experiments are a particularly effective approach when a multiplicity of nutrient constraints exist and the interactions between them needs to be defined before proceeding to field experiments (Bell, 1982).

In cases of multiple deficiency, care must be taken with the source of nutrients added. Sulfate forms of nitrogen (N) or potassium (K) often give responses because of their supply of sulfur (S) on sites that are not only deficient in N or K but also in S. Less well recognized is the variable content of S in compound fertilizers. In Thailand, for example, compound fertilizers marketed with the same NPK content varied in S from 2 to 20% (Bell et al., 1989). Other examples exist for micronutrients which occur in some macronutrient fertilizers in significant amounts either as a contaminant (e.g., B in triple superphosphate; Bell et al., 1989), or as an additive. For the correction of deficiencies, these additives or contaminants may cause erroneous conclusions about the most limiting factors, or variable responses from site to site and year to year that are difficult to explain.

In degraded sites, there is usually substantial depletion of macronutrients, especially N, due to their removal in aboveground biomass, surface litter and topsoil, or due to losses during stockpiling of topsoil. Hence for land restoration very large additions of N fertilizer may be required (Bradshaw, 1983). In native eucalyptus forests of southwest Australia, Ward and Koch (1996) report that the surface 1 m of soil contains 1800 kg Nha^{-1}. Most of this N is incorporated into the soil organic matter pool where it mineralizes at a rate which supplies enough N for 5 t dry matter ha^{-1} yr^{-1}. The N requirement will obviously vary with biomass production and climate but clearly amounts of N fertilizer in the order of 1,000–2,000 kg Nha^{-1} cannot be added in a single dose without being toxic to plants or causing huge N losses by leaching. Fewer such problems are experienced with P or micronutrients, although the possibility of inducing toxicities from high fertilizer rates on sandy substrates should not be overlooked.

Creating the conditions which favor high legume abundance in revegetated plant communities and high rates of N_2 fixation is an obvious solution to the problem of supplying high amounts of N to the land being restored, and of doing so without costly repeated annual fertilizer additions (Ward and Koch, 1996). In environments where there are no native legumes, the problem of selecting suitable species is an added constraint to restoration because not only must the introduced species be able to nodulate and fix nitrogen with native or introduced *Rhizobia* but they must also be productive without becoming a potential weed.

Plant Symptoms

Plant symptoms may be a useful indication of the nature of a mineral disorder. Most nutrient deficiencies and elemental toxicities impair plant metabolism in a characteristic way which is manifest in distinctive symptoms on the plant (Robson and Snowball, 1986). An increasing number of publications have color prints of the distinctive symptoms of mineral disorders (Robson and Snowball, 1986; Grundon, 1987; Bennett, 1993; Dell et al., 1995), the most comprehensive of them being Bergmann (1992). In addition, keys for diagnosing symptoms have been prepared for various crops (Sprague, 1964; Robson and Snowball, 1986; Grundon, 1987; Bergmann, 1992). For the inexperienced diagnostician, these keys may be the most reliable approach to correct identification of a mineral disorder using plant symptoms. Most of the publications also include photographs of common nonnutritional disorders which may cause misdiagnosis.

With familiarity of a particular plant species and its symptoms of nutrient disorders, and of the local environment and soils in which it is grown, reliable diagnosis of the disorder can be achieved. For example, hollow heart in peanut kernels is such a distinctive symptom of B deficiency that it was used as an indicator of low B soils for a survey of over 3700 sites in Thailand (Bell et al., 1989). However, even without a high level of familiarity of the symptoms, they can still be used to narrow down the range of possible disorders to a small number that can be tested further by other means, particularly plant and soil analysis.

However, reliance on plant symptoms has several limitations (Robson and Snowball, 1986). First, symptoms generally develop only in cases of severe to very severe disorders. Thus mild to moderate deficiencies or toxicities may significantly depress plant growth without producing any symptoms that can be diagnosed. In these cases, plant analysis is required for diagnosis. Secondly, the symptom expressed may not indicate the direct cause of the mineral disorder. Many symptoms of Al toxicity are expressed as induced deficiencies of other elements such as P, Ca, or Mg. Similarly, Fe chlorosis symptoms are often induced in response to a toxicity of a range of heavy metals including Mn, Cu, Zn, and Cr (Bergmann, 1992). Thirdly, symptoms of a mineral disorder may vary quite substantially between cultivars of the same species. Fourthly, some symptoms are difficult to distinguish from one another. In legumes reliant in symbiotic N fixation, N and molybdenum deficiencies are indistinguishable. Finally, many nonnutritional disorders cause plant symptoms that can be misdiagnosed as a nutrient deficiency or toxicity. For example, frost and drought can cause sterility in wheat heads that closely resembles the symptoms of Cu deficiency (Robson and Snowball, 1986).

CALIBRATION OF SOIL AND PLANT TESTS

For use in diagnosis and prognosis of nutrient disorders, both soil and plant tests must be calibrated. Standards are developed by correlation of plant response with either nutrient concentrations in a part of the plant (plant analysis), or with the concentration of element extracted from the soil with a chemical solution (soil analysis).

A soil test uses a chemical solution, usually a dilute acid, a dilute base, or a chelate to extract a fraction of an element from the soil that is generally thought to represent that fraction of the element available to the plant. Its purpose is to simulate the extraction of the nutrient from the soil by plant roots. However, by contrast with the chemical extraction which occurs in 5 min to 24 h, plant roots extract nutrients continuously over a period of months or years. Clearly the soil test will not extract all the nutrients accessed by plant roots, but for the test to be useful, the nutrients it extracts must be correlated with plant response. The most effective soil tests will be those that remove nutrients from similar pools to those which supply plant roots. A good example of the development of a soil test which is based on a thorough knowledge of the forms of the element in soils and their relative importance as pools for plant uptake is that by Blair et al. (1991) for soil S testing. Prior research showing that S-esters in organic matter were a major pool of S absorbed by plants led to the development of an extraction of soils with KCl at 40°C because this most effectively removed the ester forms of S. By contrast, other tests which extracted S primarily from pool sorbed on the soil mineral colloids were less effective in predicting plant response to S. Tests which are not based on such a thorough understanding of the pools and chemical equilibria of nutrients which supply plants, are empirical tests which can be expected to vary in effectiveness from soil to soil.

The standards used for the interpretation of soil and plant tests are obtained from a correlation of the soil or plant test value with plant growth, crop yield, or the quality of the harvested product. In the generalized relationship shown in Figure 16.1, the curve exhibits four distinctive ranges. In the deficient range (D in Figure 16.1), low concentrations are associated with depressed yield, and small increases in soil or plant test values are associated with large increases in yield. In the marginal or transition range (C), both yield and nutrient concentration increase through a transition to the third range where growth is at a maximum. Increases in soil or plant test values do not increase yield in the range known as the adequate or sufficient range (S). At excessive levels of a nutrient in the soil or plant, growth may be depressed by toxicity (T).

Soil and plant test data may be calibrated as deficient, marginal, adequate, and toxic. For interpretation, the end points of the ranges are of particular importance. In other definitions, the marginal range is referred to as a critical range (Dow and Roberts, 1982). An alternative approach is to define a single concentration in the critical range which corresponds with 90 or 95% of maximum yield as the critical concentration (Ulrich and Hills,

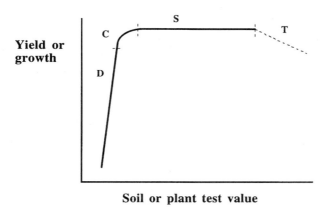

Yield or growth

Soil or plant test value

FIGURE 16.1. Generalized relationship between soil or plant test concentration and yield. From Smith, 1986. D-deficient range; C-critical or marginal range; S-sufficiency or adequate range; T-toxic range.

1967). In practice, a critical range better represents the inherent variability with all soil and plant tests, especially when field calibrated (Smith, 1986).

When the soil or plant test concentration is in the deficient range, the recommendation is generally for fertilizer treatment. For concentrations in the critical range, the decision to fertilize or not is a matter for the land manager to decide based on the value of the crop, the age of the crop, and the biomass expected. In the adequate range, fertilizer might still be recommended in order to replace nutrients removed in harvested products so as to maintain soil nutrient levels (Dahnke and Olson, 1990). Similarly for land restoration, fertilizer will often be needed even when the soil test concentration is adequate because of the need to establish a pool of nutrients for nutrient cycling.

Limitations of Plant and Soil Testing

The most satisfactory results from soil and plant testing will be obtained when users recognize the inherent strengths and weaknesses of each test so that they do not use the test inappropriately, or overextend its interpretation. Plant analysis cannot diagnose the underlying cause of the disorder detected in the plants. It merely confirms that concentrations of an element in a plant part are above or below the critical concentration. For example, while low P in the soil is usually the cause of low P in the plant, environmental factors like low soil water that restrict P uptake can also induce P deficiency in the plant (Fisher, 1980). Similar interactions of low soil water on B and Mn concentrations have been reported (Kluge, 1971; Hannam et al., 1985). In the case of Al toxicity, referred to earlier, plant analysis may diagnose the deficiency induced by excess Al but not the underlying cause. Misdiagnosis of the cause will likely lead to an ineffective treatment being prescribed.

Plant analysis conducted on a single occasion cannot predict the amount of fertilizer required in the current season to correct a disorder. By contrast, soil tests can be calibrated to estimate, before sowing, the fertilizer requirements for the present year's crop (Dahnke and Olson, 1990).

Soil analysis suffers from some limitations that plant analysis does not. Soil analysis will usually not diagnose the existing cause of a mineral disorder whereas plant analysis does because it is an integrated measure of soil, plant, and environmental variables that affect growth and nutrient uptake. By contrast, soil test results only provide a measure of the extractable nutrient concentration in soil at planting. Soil analysis may be a poor indicator of the growth response of deep rooted perennial plants to nutrients. First, whereas soil samples are mostly taken from a surface layer, plant roots may absorb nutrients from a considerable depth. Second, perennial species usually retain substantial storage pools of nutrients in roots, bark, stems, and buds which are recycled to stimulate annual flushes of shoot and root growth (Smith and Loneragan, 1997). Plant analysis is in these cases a better indicator of crop nutrient status.

The limitations outlined in the uses of plant and soil analysis can usually be avoided by the complementary use of both soil and plant analysis. The major limitation for the use of both soil and plant analysis arises from the lack of tests calibrated for the particular soil types, crop species, and environmental conditions under which the interpretation is to be made. Thus restoration after mining which often deals with unusual, extreme soil conditions and revegetation of native plant communities suffers from a scarcity of relevant calibrated soil and plant tests (Bell, 1981). However, with good planning and management during mining operations, it is often possible to avoid some of the worst fertility constraints for revegetation.

Factors Affecting Interpretation

Plant Analysis

Defined parts of plants are usually more appropriate for prognosis or diagnosis of mineral disorders that whole shoots (Smith, 1986). For elements which are phloem mobile (N, P, K), older plant parts better reflect nutrient status in the plant than young growing plant parts. For phloem immobile elements (Ca, Mn, B), the opposite is the case. For the variably mobile elements (Cu, Fe, Mo, Zn, Mg, S), the choice of plant part varies from expanding leaves to recently matured leaves, depending on the element and the plant species. The youngest mature leaf (approximately equivalent to the youngest fully expanded leaf) often represents a suitable compromise for most elements where specific guidelines are lacking, and has the advantage that an interpretation can be made for a wide range of elements in the single plant part.

The consequence of plant age differs for diagnosis compared to prognosis. Critical ranges of nutrients in defined plant parts are relatively constant with plant age when calibrated for diagnosis of disorders. By contrast, the critical range for prognosis will often vary with plant age. The distinction is illustrated in the following example of N deficiency diagnosis and prognosis in sugar beets (Ulrich and Hills, 1967). For diagnosis, 1,000 mg NO_3 kg^{-1} in sugar beet petioles appears to be the critical concentration and remained stable for an extended period of crop growth from 4 to 12 weeks after sowing (Figure 16.2). Figure 16.2a shows the relationship between NO_3-N in petioles of sugar beets and time for plants treated with three rates of N fertilizer (levels 1, 2, and 3) and Figure 16.2b the relationship between critical NO_3-N concentration for diagnosis which remained stable with time and the critical concentration for prognosis which declined with time. Whenever petiole NO_3 level fell below 1,000 mg kg^{-1}, a deficiency in the plant can be expected, except when it happens very late in the season close to harvest. The N treatment which ensures that petiole NO_3 remains continuously above the critical value for diagnosis was referred to as the minimum safe level by Ulrich and Hills (1967). The curve of NO_3 concentration over

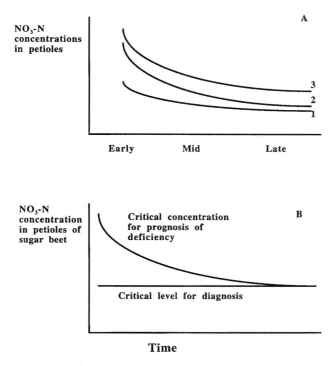

FIGURE 16.2. Generalized relationship between critical concentrations for diagnosis and prognosis based on petiole nitrate nitrogen (NO_3-N) concentrations in sugarbeet and their response to fertilizer N and time. From Ulrich and Hills, 1967.

time at the safe level defines the critical values for prognosis of N deficiency in sugar beets. At suboptimal N fertilizer application rates, petiole NO_3-N declined from adequate concentrations early in crop growth to deficient concentration during the growing season (Figure 16.2). Thus, in order to interpret a critical value for prognosis, the time of sampling is usually very important. However, there are other cases where the critical value for prognosis does not decline with time [e.g., Cu in wheat (Robson et al., 1984)]. When the nutrient supply is strongly buffered over time, the critical values for prognosis will show less change over time than for N, which is very labile in the soil. Thus, the effect of time of sampling on the critical value for prognosis will vary with soil properties as well the element concerned.

Soil Analysis

The amounts of an element extracted will vary greatly with different extractants and for different soil types. Consequently, the critical concentrations or critical ranges will also vary. For soil P extractions, Fixen and Grove (1990) cite 12 different methods which are being used or have been used in the USA. The use of a particular extractant is governed by whether the test values are correlated with response of the plant of interest in the soils of interest. The development of soil tests is now increasingly based on a prior fractionation of the element in the soil so as to identify the extractant which assesses the same fractions available to the plant roots (Fixen and Grove, 1990). When a diverse range of soils is evaluated, grouping of the soils may be necessary before any meaningful relationship between plant nutrient uptake, soil extractions, and soil fractions can be established.

Ideally, the soil samples collected for testing should be taken from the layers where plants absorb most of their nutrients. For shallow rooted annual crops, samples to 10, 15, or 20 cm are usually sufficient since this represents the layer where most roots grow (Brown, 1993). For deeper rooted annual crops like maize, especially in dryland environments, soil sampling to depths of 60 cm may be necessary to improve the prediction of deficiencies of elements such as N or S which are readily mobile in the soil (Dahnke and Olson, 1990).

Species may differ markedly in the critical range of extractable nutrients required for predicting deficiency. Differences in internal requirements, rates of absorption, rhizosphere modification, symbiotic associations, root morphology, yield potential, and growth rate all contribute to differences. Among cultivars of the same species, differences in critical soil test values and in fertilizer requirements have also been reported (e.g., Reuter et al., 1983).

Even with adequate nutrients supplied, yield potential varies considerably from year to year. Two factors operate to determine the effects of seasonal weather on the prediction of deficiency by soil tests. First, soil conditions such as low soil water can induce deficiency in one season (e.g., P, Mn, B) but not in another, even though the soil test value may not change. Secondly, seasonal weather determines the actual yield in a particular year by controlling water availability, radiation levels, disease incidence, and pest populations. The latter indirectly affects the accuracy of prediction of nutrient deficiency by altering the total nutrient demand. In years of severe drought, for example, nutrient responses are less likely because of low overall dry matter production even though the soil level is low enough under normal seasonal conditions to cause a nutrient induced yield depression (Figure 16.3).

In addition, the decisions on whether to fertilize in a particular year will be determined by the economic value of the crop. In very high value crops, the potential loss of income from even marginal deficiency is so high relative to the cost of the fertilizer that fertilizer will be added when even a slight risk of deficiency exists.

PURPOSES OF SOIL AND PLANT TESTING

When the postmine land use after restoration is a managed agricultural system, then the uses of plant and soil testing are identical to those in agriculture and horticulture. For such applications, there is a substantial body of published information on soil and plant test concentrations that can be used for interpretation (e.g., Chapman, 1966; Martin-Prevot et al., 1984; Reuter and Robinson, 1986; Reuter et al., 1997; Westermann, 1990; Baigent, 1993; Havelin and Jacobsen, 1994). For a wide range of species, soil and plant analysis tests have been correlated with plant response, and the results calibrated so that they can be formulated into fertilizer recommendations.

By contrast, much less information is available for the restoration of native or near-native ecosystems for forestry or for nature conservation. The main difference between the agricultural applications of these standards and those in degraded sites is the fact that more extreme conditions may exist for plant growth in degraded sites than in agricultural land. In addition, restoration of ecosystems is inherently more complex because it involves many species rather than a single species. Not only is there the need for a much larger database to interpret mineral nutrient constraints, but the interaction of species with one another and the dynamic change of the ecosystem over time is outside the experience of most soil and plant analysis studies.

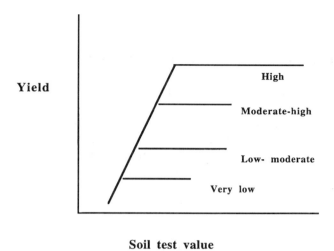

Soil test value

FIGURE 16.3. Generalized relationship between soil test concentration for nitrogen or phosphorus and yield. The critical concentration or critical range varies with the yield in a particular site or season. From Dahnke and Olson, 1990.

By contrast, with restoration of managed agricultural systems where maximum yield is the goal, restoration of native vegetation may place a greater premium on reducing the risks of failure or on protecting the natural resource base than on yield per se. Many native ecosystems are nutrient limited (Chapin, 1980), so that plant and soil tests will usually generate values in the deficient range of Figure 16.1. Consequently neither plant nor soil analyses may be very sensitive for discriminating between a severe and a moderate deficiency because while their productivity can be very different, their soil and plant test values are not. A possible solution to this problem is to monitor changes in plant and soil test values at a site over time. In the restored area, the time trend in relation to an undisturbed benchmark system may give a satisfactory indication of whether restoration is tending toward processes similar to the undisturbed ecosystem. Similarly, comparisons of soil and plant test values between a restored area and an undisturbed site can be useful in diagnosing low nutrient status (Ward et al., 1985).

CONCLUSIONS

Without representative samples, little realistic hope of accurate diagnosis or prognosis can be held. A clearly defined purpose for sampling should ensure that the samples represent the correct depth and time of year for the species or ecosystem being evaluated. Spatial and temporal variation in soil and plant samples is much greater than most nonsoils, specialists realize (Brown, 1993). Unless sufficient numbers of subsamples are collected, the risks of forming an erroneous conclusion about nutrient status are great. In restored mine sites, variability spatially and vertically may be much higher than that in undisturbed soils. Hence, a greater number of samples or increased stratification of sampling may be necessary. Avoidance of contamination of samples during collection and handling is also necessary for reliable results. On a mine site, the potential for contamination of samples is relatively high. Detailed reviews of sampling methodology and theory for plants (e.g.,

Reuter et al., 1986; Robinson, 1993) and soils (e.g., Brown, 1993) are available and should be consulted by all users of plant and soil testing.

Several interlaboratory comparisons suggest that considerable variation exists between them when using the same published methods with subsamples of the same thoroughly homogenized sample (Peverill, 1993). Thus, quality assurance is a necessary part of any laboratory analysis procedure if interpretation of soil and plant analysis data is to be meaningful.

Calibrated critical concentrations are often lacking for specific crop-soil combinations. The site-specific nature of many mine restoration projects means that applicable data may be lacking for interpretation. Inevitably, information which represents the best approximation will selected from the published literature for the purposes of interpretation. Users therefore need to be fully aware of the factors affecting interpretation and the interpretation limits of published information.

REFERENCES

Aronson, J., E. LeFoc'h, C. Floret, C. Ovalle, and R. Pontanier. Restoration and Rehabilitation of Degraded Ecosystems in Arid and Semiarid Regions. II. Case Studies in Chile, Tunisia and Cameroon. *Restor. Ecol.*, 1, pp. 168–187, 1993.

Baigent, R.A., Ed. Special Issue Soil and Plant Analysis. *Aust. J. Exp. Agric.*, 33, pp. 963–1086, 1993.

Bell, L.C. A Systematic Approach to the Assessment of Fertiliser Requirements for the Rehabilitation of Mine Wastes, in *Australian Mining Industry Council (AMIC), Environmental Workshop Papers*. 20–24 September, Canberra. AMIC, Canberra, 1981, pp. 163–175.

Bell, R.W., B. Rerkasem, P. Keerati-Kasikorn, S. Phetchawee, S. Ratanarat, P. Pongsakul, N. Hirunburana, and J.F. Loneragan. Mineral Nutrition of Food Legumes in Thailand with Particular Reference to Micronutrients. ACIAR Tech. Rep. 16, 1989, p. 52.

Bennett, W.F., Ed. *Nutrient Deficiencies and Toxicities in Crop Plants*. APS Press, MN, 1993, p. 202.

Bergmann, W., Ed. *Nutritional Disorders of Plants. Colour Atlas*. Gustav Fischer Verlag Jena, Stuttgart, 1992, p. 386.

Blair, G.J., N. Chiniom, R.D.B. Lefroy, G.C. Anderson, and G.J. Crocker. A Sulfur Soil Test for Pastures and Crops. *Aust. J. Agric. Res.*, 29, pp. 619–626, 1991.

Bradshaw, A.D. The Reconstruction of Ecosystems. *J. Appl. Ecol.*, 20, pp. 1–17, 1983.

Brown, A.J. A Review of Soil Sampling for Chemical Analysis. *Aust. J. Expt. Agric.*, 33, pp. 983–1006, 1993.

Chapin, F.S. The Mineral Nutrition of Wild Plants. *Ann. Rev. Ecol. Syst.*, 11, pp. 233–260, 1980.

Chapman, H.D., Ed. *Diagnostic Criteria for Plants and Soils*. University of California, Riverside, CA, 1966.

Craswell, E.T., J.F. Loneragan, and P. Keerati-Kasikorn. Mineral Constraints to Food Legume Crop Production in Asia, in *Food Legume Improvement in Asian Farming Systems*, Wallis, E.S. and D.E. Byth, Eds. ACIAR Proc. Series No. 18. Ramsay Ware Printing, Melbourne, 1985, pp. 99–111.

Dahnke, W.C. and R.A. Olson. Soil Test Correlation, Calibration, and Recommendation, in *Soil Testing and Plant Analysis*, 3rd ed. Westermann, R.L., Ed. Soil Sci. Soc. Amer. Book Series No. 3, ASA CSSA SSSA, WI, 1990, pp. 45–71.

Dell, B., N. Malajczuk, and T.S. Grove. *Nutrient Disorders in Plantation Eucalypts*, ACIAR Monograph No. 31, 1995, p. 110.

Dow, A.I. and S. Roberts. Proposal: Critical Nutrient Ranges for Crop Diagnosis. *Agron. J.*, 74, pp. 401–403, 1982.

Fisher, M.H. The Influence of Water Stress on Nitrogen and Phosphorus Uptake and Concentration in Townsville Stylo (*Stylosanthes humilis*). *Aust. J. Exp. Agric. Anim. Husb.*, 20, pp. 175–180, 1980.

Fixen, P.E. and J.H. Grove. Testing Soils for Phosphorus, in *Soil Testing and Plant Analysis*, 3rd ed., Westermann, R.L., Ed. Soil Sci. Soc. Amer. Book Series No. 3, 1990, pp. 141–180.

Grundon, N. *Hungry Crops: A Guide to Nutrient Deficiencies in Field Crops*, Queensland Dept. of Primary Industries, Brisbane, Australia, 1987, p. 246.

Hannam, R.J., R.D. Graham, and J.L. Riggs. Diagnosis and Prognosis of Manganese Deficiency in *Lupinus angustifolius* L. *Aust J. Agric. Res.*, 36, pp. 765–777, 1985.

Havelin, J.L. and J.S. Jacobsen. *Soil Testing: Prospects for Improving Nutrient Recommendations*. Soil Sci. Soc. Amer. Spec. Publ. No. 40, 1994.

Ho, G.E., M.K.S.A. Samaraweera, and R.W. Bell. Revegetation Directly on Salt-Affected Gold Ore Refining Residue: A Case Study, in *Remediation and Management of Degraded Lands*, held in Hong Kong, 3–6 December, 1996. Ann Arbor Press, Chelsea, MI, 1998.

Hobbs, R.J. and D.A. Norton. Towards a Conceptual Framework for Restoration Ecology. *Restor. Ecol.*, 4, pp. 93–110, 1996.

Kluge, R. Contribution to the Problem of Drought-Induced Boron Deficiency of Agricultural Crops. *Arch. Acker. Pflanzenbau Bodenkd.*, 15, pp. 749–754, 1971.

Martin-Prevel, P., J. Gagnard, and P. Gautier, Eds. *Plant Analysis as a Guide to the Nutrient Requirements of Temperate and Tropical Crops*. Lavoisier Publishing, New York, 1987, p. 722.

Peck, T.R. and P.N. Soltanpour. The Principles of Soil Testing, in *Soil Testing and Plant Analysis*, 3rd ed., Westermann, R.L., Ed., Soil Sci. Soc. Amer. Book Series No. 3, ASA CSSA SSSA, WI, 1990, pp. 1–9.

Peverill, K.I. Soil Testing and Plant Analysis in Australia. *Aust. J. Exp. Agric.*, 33, pp. 963–970, 1993.

Reuter, D.J. and J.B. Robinson, Eds. *Plant Analysis: An Interpretation Manual*. Inkata Press, Melbourne, Australia, 1986, p. 218.

Reuter, D.J. and J.B. Robinson, Eds. *Plant Analysis: An Interpretation Manual*, 2nd ed. CSIRO, Melbourne, Australia, 1997, 572 p.

Reuter, D.J., J.B. Robinson, K.I. Peverill, and G.H. Price. Guidelines for Collecting, Handling and Analysing Plant Materials, in *Plant Analysis: An Interpretation Manual*, Reuter, D.J. and J.B. Robinson, Eds. Inkata Press, Melbourne, Australia, 1986, pp. 20–33.

Robinson, J.B. Plant Sampling: A Review. *Aust. J. Exp. Agric.*, 33, pp. 1007–1014, 1993.

Robson, A.D. and K. Snowball. Nutrient Deficiency and Toxicity Symptoms, in *Plant Analysis: An Interpretation Manual*, Reuter, D.J. and J.B. Robinson, Eds. Inkata Press, Melbourne, Australia, 1986, pp. 13–19.

Robson, A.D., J.F. Loneragan, J.W. Gartrell, and K. Snowball. Diagnosis of Copper Deficiency in Wheat by Plant Analysis. *Aust. J. Agric. Res.*, 35, pp. 347–358, 1984.

Smith, F.W. Interpretation of Plant Analysis: Concepts and Principles, in *Plant Analysis: An Interpretation Manual*, Reuter, D.J. and J.B. Robinson, Eds. Inkata Press, Melbourne, Australia, 1986, pp. 1–12.

Smith, F.W. and J.F. Loneragan. Interpretation of Plant Analysis: Concepts and Principles, in *Plant Analysis: An Interpretation Manual*, 2nd ed. Reuter, D.J. and J.B. Robinson, Eds. CSIRO, Melbourne, Australia, 1997, pp. 1–33.

Sprague, H.B., Ed. *Hunger Signs in Crops*. David McKay Company, New York. 1964, p. 461.

Ulrich, A. and F.J. Hills. Principles and Practices of Plant Analysis, in *Soil Testing and Plant Analysis*, Part 2, Hardy, G.W., Ed. Soil Science Society America Special Publication No 2, 1967, pp. 11–24.

Ward, S.C., G.E. Pickersgill, D.V. Michaelson, and D.T. Bell. Responses to Factorial Combinations of Nitrogen, Phosphorus and Potassium Fertilisers by Saplings of *Eucalyptus saligna* Sm., and the Prediction of Responses by DRIS Indices. *Aust. For. Res.*, 15, pp. 27–32, 1985.

Ward, S.C. and J. Koch. Biomass and Nutrient Distribution in a 15.5 Year Old Forest Growing on a Rehabilitated Bauxite Mine. *Aust. J. Ecol.*, 21, pp. 309–315, 1996.

Westermann, R.L., Ed. *Soil Testing and Plant Analysis*, 3rd ed. Soil Sci. Soc. Amer. Book Series No. 3. Soil Science Soc of America, Madison, WI, 1990, p. 784.

Chapter 17

Land Degradation and Its Control Strategies in China

C.F. Yang

INTRODUCTION

Human-induced land degradation as a central theme of desertification in the world refers to the decline or loss of biological or economic productivity caused by human activities. According to the assessment of the United Nations Environmental Program (UNEP), about 3.6 billion ha of land in over 100 countries have been affected by desertification or land degradation. The direct economic loss is equivalent to 42.3 billion U.S. dollars, and indirect loss is two to three times or even ten times higher. The annual average direct investment planned to be used in combating intensively degraded land is 15.2 to 38 billion U.S. dollars in the world, and the indirect investment is several times higher (UNEP, 1992). The eventual result of land degradation, in addition to poverty, is a possible threat to regional and global security.

Since China is one of the countries subjected to comparatively serious land degradation in relation to its large population and limited land resources available per capita, control strategies suitable to land degradation should be adopted for the purpose of sustainable development.

PRESENT STATUS OF LAND DEGRADATION IN CHINA

Two types of land degradation process induced by human activities are identified. One refers to development of badland which is unfavorable to human economic activities due to topsoil loss and terrain deformation. The second denotes decline or loss of soil fertility caused by changes of chemical properties and loss of nutrient elements from soil.

Wind Erosion

Wind erosion is serious in arid, semiarid and dry subhumid areas in China. Wind erosion has caused 371,000 km^2 of land or 3.86% of the total land area to desertify. About one-third of the total desertified area is distributed in the agropastoral area in Inner Mongolia and along the Great Wall. The most serious desertified area is found in Horqin Desert, Mu Us Desert, and Qahar Grassland (Table 17.1) (Zhu, 1994).

Slightly and moderately desertified land areas predominate in China, accounting for 47.2% and 14.2%, respectively, whereas severely and very severely desertified land areas account for 18.3% and 10.2%, respectively. It should be noted that once the land is

Table 17.1. Percentage of Desertified Area to Total Regional Area in Agropastoral Area of China.

Region	I	II	III	IV	V	VI	VII	VIII
Desertified land (%)	47.8	26.1	25.6	47.4	93.6	43.4	27.3	61.4

Note: I: Jirem League of Inner Mongolia in Horqin Sandy Land; II: Chifeng City (Ju Ud League) in Horqin Sandy Land; III: Bashang of Chengde Prefecture of Zhangjiakou, Hebei; IV: Houshan Area of Ulanqab League of Inner Mongolia; V: Ih Ju League of Inner Mongolia in Mu Us Sandy Land; VI: Nanwuqi of Xilin Gol League of Inner Mongolia; VII: Yanchi (southwest of Erdos Grassland); VIII: Yulin Prefecture of northern Shaanxiland.

desertified, the expansion rate will be increasing rapidly. For instance, the rate of development of desertified land of slight, moderate, and severe degrees in the central section of the agropastoral area is 4.1%, 9.2%, and 12.6%, respectively. It is thus essential to combat potential slight, or even moderate, desertification.

Water Erosion

Water erosion is another important process of land degradation. It is mainly distributed in the Loess Plateau, hilly areas of southern China and rocky mountain areas of northern China. Water-eroded areas have expanded from 1.53 million km^2 in 1953 to 1.796 million km^2 in the 1990s. The cultivated area covers 25.4% of the total water-eroded area, and that in the basin of the Yangtze River and the Yellow River is 31.8% and 19.3% of the total, respectively. Five billion tonnes of topsoil is lost due to water erosion each year throughout the country, of which cultivated land topsoil loss accounted for 66% of the total. Slight and moderate water erosion constitute the important portion, accounting for 51.2% and 27.8% of the total water-eroded area (Table 17.2) (Gong, 1994).

Of the 1.796 million km^2 of water-eroded land area in China, 21.1% suffers an obvious decline or loss in biological or economic productivity.

Salinization

Currently, saline soil covers 818,000 km^2, of which 369,000 km^2 is recently formed, while 449,000 km^2 is considered as relic saline soil, while another 173,300 km^2 is liable to be salinized (Wang, 1993). Saline soil is mainly distributed in patches in the basin of inland rivers and the coastal zone of northern China, while secondary salinization mainly occurs in irrigated areas, particularly in northwestern arid zones and subhumid irrigated areas of plains in northern China (Table 17.3). Because secondary salinization is caused by inappropriate irrigation practices and is mainly distributed in irrigated areas for farming practices, it is also harmful to people.

Loss of Soil Fertility

In the past four decades, soil fertility conditions have been improved to a considerable extent. However, the increased production also means the increased removal of nutrient elements from soil. So if no efforts are made to keep the balance of soil nutrient elements and prevent occurrence of the nutrient deficiency, soil fertility will decline. Some research has already proved that deficiency of certain elements has happened in farmland. For in-

Table 17.2. Extent of Water Eroded Land in China.

Erosion Severity	Area (in 10^4 km^2)	Percentage (%)
Slight	91.9	51.2
Moderate	49.9	27.8
Severe	24.5	13.6
Very severe	13.3	7.4
Total	179.6	100

Table 17. 3. Distribution and Area of Secondary Salinized Land in China.

Region	Area (in 10^3 km^2)
Northeast China	7.6
North China	20.6
Loess Plateau	3.5
Northwestern arid zone	20.8
Qinghai-Xizang Plateau	0.1
Lower and middle reaches of Yangtze River	7.0
South China	1.1
Southwest China	1.9

stance, K deficiency was only found in limited farmland in southern China in the 1960s but nowadays K-deficit land has accounted for 54% of the total farmland in southern China. In the whole country, 52.6% of the farmland is subject to P deficiency, 25.6% to B deficiency, and Mo-, Mn-, Zn-, and Cu-deficient occupy 34.8, 15.8, 38.0, and 5.2% of farmland, respectively (Zhao, 1989).

Soil Acidification and Pollution

With the development of industry, land degradation caused particularly by discharge of pollutants from industrial and mining activities has become increasingly a focus of attention. Statistics revealed recently by the National Environmental Protection Agency showed that the area affected by acid rain (pH < 5.6) increased from 1.75 million km^2 in 1985 to 2.8 million km^2 in 1993 in China. The provinces affected by acid rain (pH<4.5) expanded from partial areas in Chongqing and Guiyang in 1986 to vast areas of Sichuan, Guizhou, Hunan, Hubei, Jiangxi, Guangdong, Fujian, and Zhejiang of southern China in 1993. About 70% of tea plantations in southern China are affected by acid rain. It is reported that grain loss due to pollution totaled 11.65 million t each year (Chen, 1990).

CAUSES OF LAND DEGRADATION IN CHINA

Land degradation is a complex of physical and human factors. Natural factors only provide external conditions for land degradation, whereas human activities such as overgrazing, excessive fuelwood cutting, deforestation, industrial and mining activities, steep slopeland reclamation, even improper forest management, etc., are the major causes of land degradation.

Physical factors present in northern China are: loose sandy sediments on ground surface; coincidence of the dry season and windy season; and in southern China, intense rainfall, thick weathering crust, and loose soil.

The impact of human activities on land degradation is complicated. In northern China, desertification is caused by overgrazing, overreclamation, irrational water resources utilization, and vegetation destruction by industrial and mining activities. In southern China, causes of degraded land are deforestation, steep slope reclamation, improper afforestation, and industrial mining and transportation (Zhu et al., 1996).

Although the proportion of land degraded due to industrial and mining activities is small, the negative effects are serious because it develops fast and with severe effects. The contiguous area of Shanxi-Shaanxi-Inner Mongolia is the largest energy base in China where environmental problems such as water erosion, wind erosion, and soil pollution caused by resource exploitation are becoming more and more serious. The nondegraded land area only accounts for 0.52% of the total land area in the region. According to the investigations carried out in Jiangxi Province (Xiao, 1994), vegetated area destroyed due to mining occupied 582 km^2; the amount of waste disposed from tailings was 612 million m^3; and the amount of topsoil loss was 58 million m^3 in the province in the last ten years. In 1989, land affected area due to capital construction and road construction totaled 140 km^2; total amount of earth, 113 million m^3; disposed earth and stone, 57 m^3; area of piles totaled 2,462 km^2; collapse and slides caused due to road construction occurred at 33,692 sites.

Population growth has accelerated land resources utilization. The total population of China was 580 million in 1953, increasing to 1.21 billion in 1995. The growth of population has increased pressure on land and caused destruction of vegetation resources. For example, expansion of desertification has resulted in Gulibenhua county of Naiman Banner of Horqin steppe, because of reclamation of the fixed sand dunes and interdune pastural land in the mid-1960s to the mid-1970s in Inner Mongolia. In Jiangxi Province, fuelwood consumption exceeded 50% of biomass accumulation, the number of villages suffering from fuelwood deficiency increased from 15 in the 1950s to 35 in the 1980s. As a consequence, the area of bare mountain land has been increased and water erosion has accelerated.

Extensive farming is another critical factor causing land degradation. In many degraded areas, conditions are harsh, economically depressed, and production technologies are backward. In the vast desertified area, population is sparse and extensive cultivation is practiced. The input-output ratio in agricultural production is only 1.08, 60% lower than the national level. In order to meet the needs of increased population, the arable land had to be cultivated to compensate for the decline of land productivity. Consequently, part of the cultivated land became seriously desertified due to wind erosion. For instance, an average expansion of 6.87 to 22 ha of cultivated land is observed in part of the villages in northern Shangdu County of Inner Mongolia where desertification has been developing rapidly. Of this, wind-eroded cultivated land makes up 26% in Xijinzi Township.

REMARKABLE ACHIEVEMENTS IN COMBATING LAND DEGRADATION IN CHINA

Remarkable achievements (Zhang, 1993; Information Office of State Council of People's Republic of China, 1996) have been gained in combating land degradation in the past 47 years since the founding of New China, in spite of the relatively serious land degradation problems existing in the country.

Soil and Water Conservation

In the past years, the focus of soil and water conservation planning has been small catchments. More than 9,000 small catchments covering 400,000 km² dealing with problems of mountains, rivers, farmland, forests, and roads have been tackled in a comprehensive way. In the basin of the Yellow River, 73,300 km² have been brought under control in ten years, an area equal to the total sum of the previous 30 years. With the existing water conservation projects, an annual increase of grain yield has been equivalent to 11 billion kg; fruits, 23.5 billion kg; fodder, 18 billion kg; an annual increase of water-hold capacity of over 18 billion m³, and a reduction of amount of soil eroded of 1.14 billion tons. Up to the 1990s, 550,000 km² have been brought under control in the country. Of them, 93,000 km² is devoted to terraces and warp land; 353,000 km² is forestry land for soil and water conservation, as well as for economic purposes; and 34,000 km² are preserved for forests and grasses. Meanwhile, a great many soil and water conservation projects were carried out which effectively changed the production conditions of degraded land and accelerated the pace for eradicating poverty.

Combating Desertification

Since the founding of New China, about 10% of desertified land has been brought under control. Some one million ha of shelterbelts, trees, and grasses for stabilizing sand dunes were planted; 1.73 million ha of desertified land were exploited; 44 million ha of seriously degraded grassland were rehabilitated and reconstructed; forest belts were built on one million ha of farmland seriously affected by wind and sand; and an increase of grain output by 10 to 30% was realized (Information Office of State Council of People's Republic of China, 1996). Nine experimental stations directed against different types of desertified land were set up by the Chinese Academy of Sciences, and models for combating desertification with Chinese characteristics were gradually developed. For example, a prevention and control system was set up in Shapotou, where annual precipitation under harsh climatic and nonirrigated conditions is less than 200 mm. As a consequence, safe transportation along the Baotou-Lanzhou railway was ensured and substantial ecological, economic, and social benefits were obtained. This experimental station has become a prototype in desertification control in China. The sand prevention technique was listed in the Global 500 roll by UNEP. Another example is Linze experimental field in Gansu. Established in 1975, a new oasis of 3,300 ha protected by shelterbelts was developed on the desertified land at the periphery of the original, seriously affected oasis. This has not only protected the original oasis but also has allowed 125 households of farmers to move in and achieve an annual per capita income reaching 700 yuan. The spread of the achievement in the Hexi Corridor of Gansu resulted in the development of nearly 27,000 ha of new oasis with social and economic benefits.

Afforestation

The Chinese Government has always attached importance to forestry establishment. Notable forestry projects include the Three North (North, Northeast, and Northwest China) Shelterbelt System, the protection system of the middle and upper reaches of Yangtze River, the protection system of coastal zones, the afforestation project in Taihang Mountains, and crisscross forest belts in farmland and forests of agricultural plain areas. At present, the completed area of the Three North Shelterbelt System makes up 1.85 billion ha, forest coverage have increased from 5.1 to 8.2%, 40,000 ha desertified land has become woodland, and more than 1.3 million ha of sandy land became farmland, pasture, and orchards. The country's forest cover has increased to 13.9%, of which forest area, total standing tree storage and forest storage are 9.047 million ha, 1.213 billion m^3, and 996 million m^3 higher, respectively, than the statistics of the 1984 to 1988 period.

Grassland Protection

Periodic achievements in grassland construction have been reported in China. The area of aerial sown grass and improved pastures has reached 11.75 million ha, and pasture fenced on 83.33 million ha. Tremendous achievements were gained on the 49 key pasture-livestock farming complex demonstration projects initiated by the State. Up to the end of 1994, 5.64 million ha of grass were artificially planted, which effectively prevented and controlled land degradation and mitigated the impact of drought on livestock farming.

Prevention and Control of Environmental Pollution

As environmental pollution is an essential factor of land degradation, China has taken measures for prevention and control of environmental pollution in recent years. By the end of 1995, the 170 prevention and control projects assigned by the State in 1990 and a great many projects assigned by local governments have all been finished within the required time. The national industrial wastewater treatment rate increased by 13% in 1995 compared with 1990; smoke and dust elimination from industrial waste gas and synthetic utilization of industrial solid waste all increased by over 14%. During the 8th-Five Year Plan period (1991–1995), the nationally industrial wastewater treatment capacity was increased by 64.68 million t per day, industrial waste gas treatment capacity was increased by 10.7 billion m^3 per hour, and urban sewage treatment capacity by 3.1 million t per day. By the end of 1995, the industrial wastewater treatment rate of above county-level enterprises has reached 76.8%, the industrial waste gas treatment rate has reached 80.9%, and the synthetic utilization rate of industrial solid waste has reached 43%.

Construction of Nature Reserves

Since the establishment of the first national reserve in the Dinghushan mountains of Guangdong province in 1956, remarkable progress has been made in scientific survey, legislation, social education, rare and endangered species protection, international cooperation and exchange, as well as construction and management of nature reserves, especially since the reform and opening up of China in 1978. At present, 799 nature reserves of various types, or 7.2% of the country's total land area, have been set up throughout China. Nature reserves at national, provincial, municipal, and county levels can be found (Table 17.4). Ten nature reserves of Changbaishan, Dinghushan, Wolong, Wuyishan, Fanjingshan, Xilin Gol, Mount Bogda, Shennongjia, Yancheng, and Xishuangbanna have joined the World Biosphere Protected Network. Additionally, 512 scenic spots, including 119 nationally significant ones, were set up and 710 forestry parks were identified.

Table 17.4. Size and Distribution of Nature Reserves at Various Levels of Administration.

	National	Provincial	Municipal	County
Number	99	344	82	274
Percentage (%)	12.4	43.0	10.3	34.3
Area (10,000 ha)	1718	5138	39	290
Percentage (%)	23.9	71.5	0.5	4.0

Land Rehabilitation After Mining

In June 1986, China issued the Land Management Method specifying that the public and private companies section should be responsible for rehabilitating those areas that can be reclaimed after mining. In November 1988, the State Council issued the special Land Reclamation Specification; since then, reclamation of land in mining areas has been strengthened. Statistics revealed that reclaimed cultivated land totaled 3.5 million ha from 1987 to 1995 throughout the country. The reclamation rate has risen from 1% in the 1980s to 4%, of which the reclamation rate of State-owned large and medium sized mining areas accounted for 27%, and that of the coal mining industry, 28%.

FUNDAMENTAL CONTROL STRATEGIES TO COMBAT LAND DEGRADATION IN CHINA

To Pursue the National Objectives of Environmental Protection

The Ninth-Five Year Plan for National Economic and Social Development and Outline of Long-term Target for the year 2010 points out that objectives of environmental protection are to basically control the aggravating trend of environmental pollution and ecological destruction and to improve the environmental qualities in some cities and regions. According to this demand, the State Council specifies that discharge of all pollutants from all sources of industrial pollution in China should reach the national set criteria, and that ecological deterioration must be basically arrested.

To Achieve Sustainable Development Strategy and Adjust Industrial Structure

Since economic structure is considered as the principal reason causing environmental pollution, resources consumption, and ecological destruction, measures that should be adopted in the future in China are: (1) to close or suspend production of 15 types of local and small enterprises that are backward in production technology, causing serious pollution and resource and energy waste, bringing all enterprises to the set standard prior to the year 2000; (2) to strengthen urban infrastructure construction, manage urban environment in a comprehensive way, perfect urban functions, and mitigate urban and rural environmental pollution; (3) to adopt economic measures such as fiscal, financial, taxation, and credit means to encourage development of industries and products of energy saving and low energy consuming; (4) to practice family planning, enhance population quality, and reduce population pressure on land; (5) to forbid steep slopeland reclamation, to abandon farming for forest, grassland, and water bodies, to strengthen farmland capital construction, increase agricultural input, popularize the scientific application of fertilizers and pesticides, and to relieve damage from agricultural production to land.

To Strengthen Legislation and Intensify Supervision Management

The Chinese Government has attached great importance to legislation on environmental protection and has issued a number of laws and regulations concerning land degradation prevention and control, such as Land Management Law, Law of Soil and Water Conservation, Law of Environmental Protection, Law of Marine Environmental Protection, Law of Water Pollution Prevention and Control, Law of Prevention and Control of Atmospheric Environmental Pollution, Law of Solid Waste Prevention and Control, Law of Mineral Resources, Forest Law, Grassland Law, Regulations on Nature Reserves, Regulations on Land Reclamation, Forest Protection and Fire Prevention Regulations, Regulations on Tree Disease and Pests Quarantine, etc. Legislation work will be further enhanced and the existing laws and regulations will be further revised in combination with the development of a socialist market economy.

To Strengthen Rehabilitation of Land Degradation

First, experimental and demonstration studies on comprehensive management of degraded land should be carried out so as to explore critical ways and methods for tackling degraded land. Secondly, engineering concerning the "Three North" forest shelterbelts, shelterbelts in the middle and upper reaches of the Yangtze River, shelterbelts in the coastal zone and afforestation in the plain areas and in the Taihang Mountains, ecoengineering projects for the shelterbelts of the middle reaches of the Yellow River, shelterbelts in the Huaihe and Taihu drainage basins, shelterbelts in the Pearl River drainage basin and the Liaohe drainage basin should be established continuously. Thirdly, investment in soil and water conservation will be strengthened in the next five years, taking small catchment as a unit. Finally, the number of nature reserves can reach 1,000 by the year 2000 in the whole country, with the area of nature reserves accounting for 9% of the total national land area.

To Protect Cultivated Land Resources and Develop Ecofarming

Up to the end of 1995, about 2,100 county-level units have finished fundamental field protection planning work throughout the country, and over 70% of the cultivated land has been effectively protected. Efforts will be made in the future to practice field protection systems; to organize and implement fertile soil projects; to encourage farmers to combine land use with soil fertility increase; to apply chemical fertilizers in combination with organic manure and to increase application of organic manure; to assist in popularizing formulation fertilization and techniques for deep-applying chemical fertilizer; to energetically study and spread high efficiency, economical and safe pesticides; to expand comprehensive prevention and control techniques of disease and insect pests centered around biological prevention and control; to speed up recollection of plastic films for agricultural purposes; and to mitigate farmland pollution and increase land quality. In addition, to pay close attention to the construction of the 50 ecofarms by the State; ecofarming should be popularized in a big way. The number of ecofarming experimental counties will reach 250 in the year 2000 throughout the country.

To Conduct Scientific Research and Monitoring

Science and technology are important in guaranteeing success in land degradation management. Land degradation monitoring is the prerequisite for scientific decision-making by governments at various levels, and the necessary way to clarify damage status and causes. A monitoring system of land degradation will be gradually perfected; scientific site locating will be exercised; and the application of high technology will provide a scientific basis for combating land degradation.

To Mobilize and Promote Public Awareness and People's Participation

It is thus very necessary, in improving population quality, to mobilize fundamental education and environmental education, so as to improve public awareness in conscientiously protecting the environment. For local people's participation in combating land degradation, it is important to mobilize to the utmost the public, and persons of various circles, to conscientiously get involved in the cause of combating land degradation.

To Strengthen International Cooperation and Introduce Necessary Technology and Capital

Land degradation is an environmental problem attracting worldwide attention. China has initiated international cooperation concerning environment and land degradation in many aspects, and gained promising achievements. As advanced technology and capital are the necessary conditions for successful land degradation prevention and control, international cooperation in the field should be actively promoted so as to hasten the process of combating land degradation in China.

REFERENCES

Chen, H. The Status of Pollution of Soil in China, *Adv. Soil Sci.* 1, pp. 53–56, 1990 (in Chinese).
Gong, Z. Problem Soil in China. *The Collection and Analysis of Land Degradation Data*, FAO, 1994, pp. 89–98.
Information Office of State Council of People's Republic of China, Environmental Protection in China, 1996, p. 45.
UNEP. Status of Desertification and Implementation of the UN Plan to Combat Desertification, UNEP, Nairobi, 1992.
Wang, Z. *Saline Soil in China*. Science Press, 1993, p. 525 (in Chinese).
Xiao, G. Water Erosion and Its Control in Jiangxi Province. *Bull. Soil Water Conserv.*, 14(3), pp. 39–43, 1994 (in Chinese).
Yang, R. Water Erosion and Control Measures on Cultivated Land in China. *Bull. Soil Water Conserv.*, 14(2), pp. 32–36, 1994 (in Chinese).
Zhang, W. et al. The Disasters of Desertification in China. *Acta Natural Disaster*, 3(3), pp. 23–30, 1994 (in Chinese).
Zhang, Y. The Status of Water Erosion and Its Control in China. *Bull. Soil Water Conserv.*, 13(1), pp. 17–23, 1993 (in Chinese).
Zhao, Q. Land Resources and Its Regionalization of Utilization. *J. Soil*, 21(3), pp. 113–119, 1989 (in Chinese).
Zhu, Z. Sandy *Desertification in China*. Science Press, 1994, p. 250 (in Chinese).
Zhu, Z. et al. Features of Distribution and Assessment for Control Measures of Desertification in China. *China Environ. Sci.*, 16(5), pp. 328–334, 1996 (in Chinese).

Chapter 18

Restoration of Degraded Lands in Hong Kong

S.L. Chong

INTRODUCTION

The Territory Development Department (TDD) is involved in the planning, coordination, and implementation of new town development programs to house a predicted population of 3.6 million people in the New Territories (Territory Development Department, Hong Kong, 1989, 1993). Since the 1970s, nine new towns have been or are still being developed. At present, about 2.6 million people are already living in the developed areas.

The hilly terrain in Hong Kong has presented both constraints and opportunities in providing land for their development. Hills were identified as a source for extracting fill material for land reclamation from the sea and the low-lying areas. The extraction activities, however, created landscape wounds that have greatly affected the landscape quality of the New Territories. A restoration program of these hillside "borrow areas" began in the early 1980s.

The hill slopes of Hong Kong consist of mainly decomposed granite which is very friable and free-draining. These factors, together with the frequent occurrence of tropical storms and typhoons in summer and the high frequency of hill fires in the dry winter, contribute to the many badly eroded landscapes in the Territory.

The hill slopes within the new town boundary are designated as "green belt" in the Outline Development Plan of New Towns for the purpose of protection and conservation. Many of these slopes, however, are eroding rapidly, resulting in large areas of deep gullies and bare soil devoid of vegetation. The increasing number of people relocating to the new towns has caused additional disturbance to the green belt areas due to more intensive public use of hill slopes for recreation. The erosion control program started in 1983, and has since extended beyond the green belt areas in some new towns.

In 1989, the Metroplan Landscape Strategy, a government planning study, identified around 3000 ha of visually degraded landscape in the urban fringe of the metropolitan area of Hong Kong Island, Kowloon, and Tsuen Wan (Lands and Works Branch, Hong Kong, 1989). These included eroded and bare hill slopes, burned and illegally occupied hill slopes, abandoned squatter areas, quarries, and hilltop installations. A pilot rehabilitation planting program was instituted in 1991 to rehabilitate about 60 ha of badly eroded hill slopes in Kowloon. The program now forms part of the yearly restoration works carried out by the TDD Landscape Service. The objectives of TDD Landscape Service are:

FIGURE 18.1. The restoration of a borrow area in progress; showing slope grading, reconstruction, and hydroseeding.

- to ensure that landscape conservation and development is incorporated into TDD's engineering works program
- to undertake landscape works relating to engineering projects, borrow area restoration, erosion control planting in green belt areas, and the rehabilitation of degraded land under the Metroplan Landscape Strategy.

RESTORATION TECHNIQUES

Borrow Areas

In planning the restoration of borrow areas, TDD landscape staff have to agree to the general layout, final contours, haul route, and the drainage system of the borrow area before the commencement of borrowing activities. The project engineer is responsible for the excavation. The layout design is prepared to minimize environmental and visual impact during and after borrowing activities. The final contours are designed to blend with the existing and surrounding landform. The contours at the edges are designed to run smoothly into existing contours for visual continuity. The drainage system for the recontoured areas is designed to follow the natural pattern as closely as possible.

The restoration process begins with the grading of the slope and forming of new ridgelines or peaks as necessary to replicate the original landform. The earthworks and drainage reconstruction are carried out from the top, working down the site (Figure 18.1). On rock bases where vegetation is difficult to establish, a sufficient depth of topsoil is spread over to provide a suitable medium for grass and tree growth. The aim at this stage is to establish a vegetative cover rapidly on the cut slopes to stabilize the soil. This was achieved by grass hydroseeding (Figure 18.2). The grass mix and the quantity to be used depends upon whether hydroseeding is to be carried out in the 'wet' or 'dry' season. *Cynodon dactylon* (Bermudagrass) and *Paspalum notatum* (Bahiagrass) could be used for hydroseeding in both seasons; *Lolium perenne* (Manhattan ryegrass), in dry season only; and *Chloris gayana* (Rhodes grass), *Eragrostis curvula* (Weeping love grass), and

FIGURE 18.2. A view of a graded and hydroseeded slope ready for seedling planting in the wet season.

Table 18.1. Commonly Used Pioneer and Native Species in Restoring Borrow Areas.

Pioneer Species	Native Species
Acacia confusa	*Castanopsis fissa*
Acacia auriculiformis	*Cinnamomum camphora*
Acacia mangium	*Gordonia axillaris*
Casuarina equisetifolia	*Liquidambar formosana*
Eucalyptus torrelliana	*Litsea glutinosa*
Eucalyptus citriodora	*Machilus breviflora*
Eucalyptus robusta	*Machilus thunbergii*
Melaleuca leucadendron	*Sapium discolor*
Tristania conferta	*Schima superba*
Pinus elliottii	*Tutcheria spectabilis*

Eremochloa ophiuroides (centipede grass) in the wet season only. At the time of spraying, fertilizer, mulch, a soil binding agent, and a dye are added to the hydroseeding mix.

Tree planting is carried out in the wet season following the successful establishment of grass. One- to two-year old tree seedlings are planted in pits at 1 to 1.5 m spacing, depending on the ground conditions and the species used. Approximately 50 g of a slow-release fertilizer is placed in each planting pit. Table 18.1 lists the pioneer and native species commonly used in the borrow area restoration works.

On exposed sites and upper slopes, a planting mix of two or three pioneer species such as *Acacia* and *Pinus* species are used and randomly planted. At the middle slope, a planting mix of *Acacia* and *Eucalyptus* species with about 30% native species is commonly used. On lower slopes, where the site conditions are more favorable, many different combinations of pioneer and native species may be used.

During the first year of establishment, a further 50 g of fertilizer is added and the grass cut to reduce competition. If the overall survival rate is less than 80%, replacement planting is carried out. In the third year, low pruning to the *Acacia* species is carried out to encourage height growth. The plantation is thinned as necessary to assist the regeneration and the growth of native species.

Eroded Land

Invariably, the highly visible eroded hill slopes with bare patches along the ridges occur in the higher areas of the green belt. Aerial photographs are initially used to identify the extent of the problem before carrying out the ground survey. The survey also determines whether repairs to the ground are required before planting can commence. Such remedial measures include the repair of eroded gullies by blocking them with sandbags to encourage silting and vegetation growth; the upgrading of certain heavily used pedestrian pathways and steps; and the provision of firebreaks. These minor engineering works are usually carried out in the dry months preceding seedling planting.

For erosion control planting, seedling trees of pioneer species are planted as close as possible, about one meter from center to center. Planting is carried out at the beginning of the wet season around April to ensure good establishment before the onset of the dry season around October. In some years, adjustment of the planting program has been necessary in order to take into account the early arrival of the rainy or dry season. Planting at the end of the planting season carries a greater risk of failure due to the more unpredictable weather conditions.

Tree seedlings are collected from the nursery of the Agriculture and Fisheries Department, transported and delivered by foot to the site from the nearest point accessible by vehicle. The seedlings are planted in pits by the TDD landscape term contractor on the same day. In recent years, however, helicopters have been used for the transportation of planting materials to the remote sites. For planting in Lantau and Lamma Islands, for example, seedlings were barged to the nearest pier before being uplifted by helicopter to the planting sites. To increase the survival rate and more rapidly establish the newly-planted seedlings, a water-absorbing agent and a long-lasting slow-release fertilizer in tablet form was added to each planting pit.

Degraded Land in the Urban Fringe

The Metroplan Landscape Strategy proposed a tree planting and woodland extension program within the urban fringe to control soil erosion and to restore the degraded landscape caused by various authorized and unauthorized developments and to enhance areas for potential recreational uses.

The first phase of the pilot rehabilitation planting scheme was carried out at Jordan Valley on about 20 ha of badly denuded land in 1991. The planting was completed in 1993. The second phase of planting began in 1994 and was completed in 1996 covering a further 40 ha of degraded land at Lam Tin and Lei Yue Mun.

The restoration techniques used were similar to those used on the eroded areas described above. Native species were included, however, in the planting mix to encourage the development of a wildlife habitat. In addition to the extensive planting of seedling trees, TDD undertook the experimental planting of a tropical grass, *Vetiveria zizanoides*, to assist in the problem of runoff in the eroded areas. The grass was planted in rows along the contours of the eroded slopes at 150 mm spacing between the seedling planted areas. The vetiver grass has provided an effective living barrier against the washdown of loose material, maintaining a stable growing environment for the seedling trees. The grass is noninvasive and provides a useful aid to the reestablishment of woodland in eroded areas. Once the trees have grown to the point of overshadowing the grass, it will die out, having served its purpose as a pioneer species in the landscape restoration process. The Jordan Valley project was the first erosion control project to use this grass in the Territory.

A REVIEW OF RESTORATION RESULTS

The level of success or failure of the restoration works carried out by TDD in the past 15 years for the various types of degraded land is best assessed against the announced objectives, which are :

- For the borrow areas, the immediate objective is to provide a vegetation cover to the borrow areas in the shortest possible time, and the ultimate objective is to produce a new landscape which is physically stable, ecologically sustainable, and visually acceptable.
- For the eroded land in the green belt, the objective is to control soil erosion and to reestablish vegetation on the bare slopes.
- For the degraded land in the urban fringe, the objective is to mitigate the visual impact of these urban scars and to enhance the potential recreation areas for the enjoyment of the public.

Borrow Areas

The development of hydroseeding techniques in the early 1980s has provided a quick means to achieve a temporary vegetation cover to the borrow areas. Under the right conditions, grass seeds can germinate within a month and the graded slopes can be covered with a luxuriant growth of grass in about two months. The grass cover reduces the likelihood of surface erosion from direct raindrop impact, improving the site conditions for tree seedling growth, and helps to reduce the adverse visual impact of the site and to blend the site back into the surrounding landscape.

Planting trials were undertaken in the early 1980s to find an optimum planting mix of a well-tried group of introduced pioneer species which could thrive on borrow areas and provide a more suitable environment for the natural regeneration of native species. The trials also evaluated the viability of interplanting native species among the pioneer species.

The results so far indicate that single species planting; e.g., *Acacia* or *Eucalyptus* or *Pinus* species, would not achieve an ecologically stable landscape within a time frame of 15 years. *Acacia* develops a closed canopy within three years of planting. The grass dies off and even shade-tolerant species cannot grow beneath. Extensive silvicultural works such as thinning and pruning operations are required to improve the ground conditions for ferns or native vegetation to emerge. The planting of a single *Eucalyptus* species on borrow areas also does not encourage the regeneration of native species except on sheltered lower slopes and on better soil. The slender nature of *Eucalyptus* trees and their slower growth rate on exposed sites provide little shade for a number of years after planting. The ground remains dry, which encourages wild grasses and *Dicranopteris linearis* (a fern which colonizes first on dry or burned areas) to thrive. The regeneration of native species will not commence until the ground conditions have improved. The worst result was noted when a single *Pinus* species was planted. The dropped needles were noted to completely sterilize the ground, stopping the colonization of other species.

FIGURE 18.3. A fully restored borrow area eight years later, richly afforested and integrated into the surrounding landscape.

The planting of mixed pioneers provides a visually more attractive woodland due to their different growth habit and form (Figure 18.3). *Eucalyptus* and *Casuarina* species form the upper canopy with *Acacia* and *Pinus* species as the subdominants. The vertical gap between the dominants and the subdominants and the slender crown of *Eucalyptus* and *Casuarina* species permit enough light to penetrate the canopy, but not enough for the wild grasses to thrive. More shade-tolerant ferns can be seen to colonize the understorey in three to four years, and native shrubs and the seedlings emerge in about 5 to 6 years. In areas where mixed planting was more than 10 years old, naturally regenerated native species formed the entire understorey which was impenetrable (Figure 18.4). The common native species found to be present include various *Machilus* species, *Sterculia lanceolata* and *Schefflera octophylla*.

Where only native species had been planted on borrow areas, the survival rate was found to be low. For those that survived the planting shock, the growth was mostly stunted. The site remains dry and exposed, making it difficult for the transplanted natives to thrive. The survival and growth rate of native species interplanted with pioneer species seemed to be better. The faster-growing pioneers would provide shelter for them within a year or two. The best result was obtained through the planting of *Acacia, Eucalyptus,* and various native species in the proportion of 50%, 25%, and 25%. The microclimate improved and more shade-tolerant ferns began to colonize the understorey in three to four years, with the formation of a humus layer soon after. In about 6–8 years, the naturally regenerated natives would outperform the transplanted natives, however.

Eroded Land

For erosion control planting, only a well-tried group of pioneer species, which included *Acacia* and *Eucalyptus* species, was selected. The former is a nitrogen-fixer and the latter has lignotubers which assist survival and growth under adverse environmental conditions such as fire and drought. There has been a marked improvement in the green belt areas. Soil erosion has been reduced and unsightly scars have been removed from the landscape. Woodland has been well established in some early restoration sites (Figure 18.5).

FIGURE 18.4. Luxuriant growth of naturally regenerated native vegetation under mixed pioneers ten years later. Thinning operation in stages would assist the growth of native species.

FIGURE 18.5. A well-established woodland on formerly eroded hill slopes.

Degraded Land

In the short time of the urban fringe rehabilitation program in the Kowloon metropolitan area, there has been a drastic visual improvement in the area. The size of the eroded and bare areas has been reduced with trees growing vigorously in their place. Soil erosion has been steadily arrested as the tree canopy closes. Wildlife is starting to return, with an increase in bird and insect life in the revegetated areas. The area is also becoming popular with morning walkers and hikers keen to enjoy the spectacular views across Kowloon and Hong Kong. Walking trails are extensively used and lookout points are popular meeting areas for morning walkers.

The success of the pilot rehabilitation program in the above locations is beyond expectation (Figure 18.6). The rehabilitation program has now been extended to cover the de-

FIGURE 18.6. A rehabilitated site in the urban fringe after two years.

graded land on Hong Kong and Lantau Island. In Tsuen Wan, a program to restore 90 hectares of degraded landscape at a cost of $11.5 million is planned. Planting will commence in the planting season of 1997.

MANAGEMENT IMPLICATIONS AND CONCLUSION

The immediate objective for the visual restoration of degraded land has been achieved successfully using the techniques adopted. Generally, the visual impact of the degraded landscape can be substantially minimized in about 3 to 4 years. In some cases, the restored land has fully reintegrated with the surrounding landscape to the extent that it is largely invisible to the eyes of the general public. The established woodland comprising pioneer species and native understorey now stands where there were once degraded grassy hill slopes, bare and eroded, forming unsightly scars on the landscape.

The role played by the pioneer species has been of critical importance to the success in restoring the degraded landscape even though the resulting woodland might not be as diverse in species or as attractive to wildlife as natural woodland in the first 10 years after restoration. The process of vegetation succession could, however, be speeded up by the use of forestry management practices to achieve a greater species diversity which would, in turn, assist the development of a stable, visually attractive and ecologically sustainable landscape.

From the systematic observation of the growth and changes of the restored landscape and the results of limited silvicultural operations being carried out to the woodland at the borrow areas in the past 15 years, one can conclude that a silvicultural regime would need to be set up in order to accelerate the development of an indigenous woodland.

The close planting spacing adopted is necessary to provide mutual protection for the newly planted tree seedlings, especially on upland eroded areas and very exposed borrow sites. It is therefore essential to low-prune the pioneers, particularly *Acacia* species, to provide room for the understorey to form, and to carry out phased thinning operations to open up the closed canopy to assist the regeneration and growth of the more light-demanding native species. The silvicultural regime should be site- and species-specific, due to the different growth rates and tree form of the pioneer species growing under different site conditions. In the first two years of the establishment of the trees, weeding and fertilizing should be prescribed. On the lower slopes, pruning to not more than one-third of the tree height should be carried out in the third year for the *Acacia* species. On upland and exposed areas, pruning operations should be carried out on an as-needed basis. The pruning of lower branches of *Acacia* species will increase the vertical spacing between the ground and the crown, thereby reducing the risk of ground fire in the dry months and assisting tree growth and the development of the understorey. Thinning operations should be scheduled in phases whenever the tree canopy has closed. During each operation, not more than 20% of the trees shall be taken out to avoid typhoon damage to the remaining trees. The opening up of the canopy will provide the necessary light for the steady and healthy growth of native understorey and will assist the regeneration and growth of more light-demanding species. These thinning operations will help to increase species diversity and the rate of growth of the natives.

For potential recreational areas in the urban fringe of Kowloon and Hong Kong Island, additional land management measures should be included to reduce the risk of fire and other biotic factors. Only minimal facilities should be provided such as rain shelters and morning exercise areas. Proper footpaths or walking trails should be constructed in sensitive areas. Strategic viewpoints should also be formed for visitors to enjoy the spectacular views across Kowloon and Hong Kong from a vantage point rich in species diversity.

ACKNOWLEDGMENTS

The author would like to thank Mr. Shing-see Lee, Director of Territory Development, for the permission to present this paper. Useful comments and encouragement from Mr. C.E. Fair and Mrs. N. Hemey-Yiu are gratefully acknowledged.

REFERENCES

Chong, S.L. Healing the Wounds—Restoring Hong Kong's Borrow Area—A TDD Challenge, in *Urban Explosion in Asia—A Review: Proceedings of the 8th IFLA Eastern Regional Conference 1996*, Hong Kong, 1996, pp. 120–125.

Lands and Works Branch, Hong Kong. *Metroplan Landscape Strategy for the Urban Fringe and Coastal Areas.* Government Printer, Hong Kong, 1989.

Territory Development Department, Hong Kong. *The Development Challenge.* Government Printer, Hong Kong, 1989.

Territory Development Department, Hong Kong. *20 Years of New Town Development.* Government Printer, Hong Kong, 1993.

Chapter 19

Reforestation in the Countryside of Hong Kong

S.P. Lau and C.H. Fung

INTRODUCTION

Hong Kong has a total land area of about 1,092 km². It has a hilly topography with flat land in the lower valleys or at head of sea inlets. About three-quarters of the territory is still countryside. In this region, the original climax vegetation cover was believed to be ever-green broad-leaved forest. However, due to centuries of human impact, particularly the 'slash and burn' farming practiced in tropical and subtropical regions, large areas of forest have turned into grassland or secondary woodland. In addition, farmers also harvested woodland for timber and firewood uses. As a result, most of the hillsides were barren, or covered with coarse grass and low shrub, at the beginning of the last century.

The reforestation policy has been changing with the social and economic development of the local society. In the early days, forests were for amenity around the urban development on Hong Kong Island. Later, reforestation was for soil and water conservation. After the second World War, a replanting scheme was introduced to cater to the firewood demand of the local market and to reinstate the barren landscape of the Hong Kong countryside. In the last two decades, a multiple-use forestry policy has been adopted for soil conservation, nature conservation, countryside education, and recreation.

FORESTRY POLICY

1841–1949

The main objective of reforestation in this period was to protect catchment areas from soil erosion and to minimize silting up of the reservoir so that the water supply could be maintained. Large-scale reforestation was started with the construction of new reservoirs in the New Territories such as the Kowloon Reservoir area in 1902, the catchment area for Tai Po, known as Tai Po Forest Reserve, in 1925 and the Shing Mun Reservoir in 1936. Trees were also planted to cover bare hillsides. From 1873 to 1877, an average of over 15,000 trees were planted per annum. From 1880 to 1883, about 810,700 trees were planted per year. Urban planting was also commenced to improve the amenities in and near the settlement of Hong Kong (Daley, 1970).

1949–1964

During the second World War, the importation of firewood and timber was impossible, and many trees were felled. This tendency continued immediately after the war, because there was an influx of refugees into Hong Kong. Thus, the principal aims of forestry in this period were to encourage sound forestry operation by villagers in the New Territories, to replace the forest cover removed during the years 1939–1946, to protect vegetation on the hillsides, especially those on the catchment areas, and to carry out scenic and roadside tree-planting.

Government forest reserve and plantations were established and villagers were encouraged to produce timber and fuel for local use in this period. By 1963, there were about 4,047 ha of government plantations and about 809 ha of government-assisted village plantations. From 1953 to 1964, the government planted 5,317 ha of forest in the countryside (Table 19.1).

1964–1970

As more and better alternative fuels became available, the market for firewood virtually disappeared. Different aspects of land uses had also gained in prominence, and new objectives for reforestation were, therefore, recommended. Revision of forestry policy to suit the needs of conservation, recreation and education was introduced. On the other hand, government plantation and natural regenerated woodland suffered from the occurrence of many hill fires. In such circumstances, replacement planting of trees damaged by fire was also carried out. During this period, about 990 ha of trees were planted by the government (Table 19.1).

1970 Onward

Recently, the policy was adopted by the Agriculture & Fisheries Department (AFD) to conserve local flora, fauna, and natural habitats. This has been achieved through law enforcement, conservation advice on development proposals, planning strategies, environmental impact assessments, and identification of Sites of Special Scientific Interest (SSSI). The government has also promoted public awareness and participation in nature conservation through publicity and educational activities.

With the enactment of the Country Parks Ordinance and the establishment of the Country Parks Authority and Country Parks Board (now renamed as Country and Marine Parks Authority and Country and Marine Parks Board) in 1976, the first Special Areas and Country Parks were formally designated in 1977. Twenty-two Country Parks have now been established for nature conservation, recreation, tourism, and countryside education. The AFD has provided and maintained a green and clean environment in these country parks, special areas, and around new towns through fire prevention, seedling production, tree planting, and management of woodland (Table 19.2). The Forest and Countryside Ordinance (Cap. 96) is also enforced to protect forests, plantation, and natural vegetation including rare species against removal from the field.

GOVERNMENT WOODLANDS OUTSIDE COUNTRY PARKS

The government is carrying out a program of reforestation in areas between the country park boundaries and the urban areas for the purposes of environmental improvement and erosion control. Reforestation is also being implemented to rehabilitate disturbed landscapes such as landfills or borrow areas. Since 1981, the AFD has assumed the responsibility for management of woodlands on unallocated government land.

Table 19.1. Statistics of Trees Planted in Hong Kong, 1949–1977.[a]

Reforestation

Years	Trees (thousands)	Areas (ha)	Years	Trees (thousands)	Areas (ha)
1949–50	165	N/A	1963–64	N/A	304
1950–51	293	N/A	1964–65	N/A	184
1951–52	331	N/A	1965–66	N/A	138
1952–53	900	N/A	1966–67	N/A	110
1953–54	N/A	195	1967–68	N/A	188
1954–55	N/A	198	1968–69	N/A	195
1955–56	N/A	344	1969–70	N/A	175
1956–57	N/A	431	1970–71	N/A	119
1957–58	N/A	700	1971–72	N/A	89
1958–59	N/A	861	1972–73	N/A	109
1959–60	N/A	1,042	1973–74	N/A	89
1960–61	N/A	674	1974–75	N/A	117
1961–62	N/A	438	1975–76	N/A	69
1962–63	N/A	130	1976–77	229	N/A

[a] Data source: Agriculture and Fisheries Department, Hong Kong Government.

Table 19.2. Statistics of Trees Planted in Country Parks, 1977–1995.[a]

Years	Trees (thousands)	Years	Trees (thousands)
1977–78	254	1986–87	545
1978–79	233	1987–88	360
1979–80	261	1988–89	303
1980–81	328	1989–90	445
1981–82	432	1990–91	349
1982–83	250	1991–92	354
1983–84	312	1992–93	320
1984–85	232	1993–94	323
1985–86	436	1994–95	346

[a] Data source: Agriculture and Fisheries Department, Hong Kong Government.

The locations of government planting are mainly identified by the Territory Development Department (TDD). It includes rehabilitation of quarries, borrow areas, and landfill sites. In addition, government also carries out urban-fringe reforestation and subsequent maintenance. With the provision of funds and successful establishment of seedlings, these woodlands are handed over to the AFD for management. Routine maintenance works like grass cutting, fertilizing, tree pruning, and fire belt formation are then carried out by private contractors under the supervision of the AFD. The extent of managed woodland is increasing, and during the year 1995–1996, there were 275.09 ha of woodland in 28 sites with about 1,480,500 trees (Table 19.3).

The development projects under the Ports and Airport Development Strategy (PADS) have considerable effects on the natural environment. The construction of the airport and

Table 19.3. Statistics of Government Woodlands Managed by the AFD (Outside Country Parks).[a]

Years	Trees (no.)	Trees (cumulative no.)	Area (ha)	Area (cumulative ha)
1988–89	36,700	36,700	10.78	10.78
1989–90	242,435	279,135	41.28	52.06
1990–91	378,527	657,662	53.08	105.14
1991–92	190,290	847,952	45.66	150.80
1992–93	181,933	1,029,885	43.13	193.93
1993–94	326,689	1,356,574	51.73	245.66
1994–95	19,700	1,376,274	4.43	250.09
1995–96	104,300	1,480,574	25.00	275.09
Total	1,480,574	1,480,574	275.09	275.09

[a] Data source: Agriculture and Fisheries Department, Hong Kong Government.

its support highway systems have changed the surrounding landscape such as the northern coast of Lantau Island and Tung Chung. With the implementation of the project, the AFD has taken up the responsibility of coordinating the implementation of ecological measures related to the construction of the PADS projects.

The major objective of these ecological measures, particularly the reforestation aspects, is to recreate 60 ha of woodland to compensate for the loss of 20 ha of woodland habitats on Chek Lap Kok and North Lantau. In this regard, about 80,000 seedlings of 50 different native species have been planted. In addition, a total of 260,000 seedlings will be planted on the hill slopes near Tung Chung in three phases.

TREE SPECIES

Pinus massoniana (Chinese Red Pine) was the major tree species used in reforestation before the war. It was an indigenous species well adapted to local acidic and infertile soils. There was also a plentiful supply of seed from China, while seed of local indigenous species was difficult to obtain at that time. Most importantly, it was traditionally a major source of timber, poles, and fuel. When the territory was leased in 1898, the government acknowledged the importance of the pine trees to the villagers by granting them Forestry Licenses to grow these trees on Crown land for these purposes.

Pine trees were planted widely in 1882. The estimated number of *P. massoniana* planted by the villagers over 10 to 15 years was about 25,000,000. The government also used a lot of pine in its plantations. For instance, 50,000 sites between Cheung Sha Wan and Tai Po Road were planted with pine trees in 1921. Immediately after the second World War, the use of pine was continued and extended to cover the hill lands and the catchment areas. In 1951–1953, *P. massoniana* made up about 63% of the total plantings in the countryside.

From 1960s onward, *P. massoniana* was found to be unsuitable for reforestation because it was affected by pine wilt nematode, which caused "sudden death" of the infected trees. It was also susceptible to damage by fire. It was then gradually replaced by more broad-leaved species and woody legumes like *Acacia, Tristania conferta, Schima superba, Castanopsis fissa,* and *Liquidambar formosana,* etc. The use of pine and broad-leaved mixtures promised better growth and greater resistance to fire damage. Moreover, it promotes a higher de-

Table 19.4. Species of Trees Planted in
Country Parks, 1995.[a]

Tree Species

Acacia auriculiformis[b]
A. confusa[b]
Antidesma bunius
Bridelia monoica
Castanopsis fissa
Casuarina equisetifolia[b]
Choerospondias axillaris
Cinnamomum camphora
Cunninghamia lanceolata
Eucalyptus torelliana[b]
Gordonia axillaris
Keteleeria fortunei
Liquidambar formosana
Litsea glutinosa
Machilus breviflora
M. chinensis
M. ichangensis
M. oreophila
Myrica rubra
Pinus elliottii[b]
Quercus edithae
Sapium discolor
Sapium sebiferum
Schima superba
Sterculia lanceolata
Syzygium jambos[b]
Tristania conferta[b]
Tutcheria spectabilis

[a] Data source: Agriculture and Fisheries
Department, Hong Kong Government.
[b] Exotic Species.

gree of biodiversity, accelerates ecological succession, and provides good habitat for wild-life conservation. The percentage of broad-leaved tree species used in reforestation has increased from 62% in 1973 to 80% in 1995 (Tables 19.4 and 19. 5).

THE FUTURE

The main objective of reforestation in Hong Kong is to provide a quick soil cover at low establishment and maintenance cost. The secondary objectives are to improve the landscape, to enhance biodiversity, and to provide recreational and educational opportunities. The percentage of native species used will continue to increase biodiversity. More collaboration between scientists and managers is necessary to meet the challenges of choosing the right species and managing the woodlands more effectively to achieve good results.

Table 19.5. Species of Trees Planted in Government Woodland Managed by the AFD (Outside Country Parks).[a]

Tree Species	Tree Species
Acacia auriculiformis[b]	*Glyptostrobus pensilis*
A. confusa[b]	*Gordonia axillaris*
A. farnesiana[b]	*Hibiscus tiliaecus*
A. mangium[b]	*Jacaranda acutifolia*[b]
Acanthus ilicifolius	*Ketelerria fortunei*
Albizia lebbek	*Leucaena glauca*
Aleurites montana[b]	*L. leucacephala*
Alnus formosana[b]	*Liquidambar formosana*
Aporusa chinensis	*Litsea cubeba*
Aquilaria sinensis	*L. glutinosa*
Bauhinia spp.	*Macaranga tanarius*
Bombax malabricum	*Machilus breviflora*
Callitist glauca	*M. orephalia*
Castanopsis fissa	*Melaleuca leucadendron*
Casuarina equisetifolia[b]	*Melastoma sanguieumn*
C. stricta[b]	*Melia azaderach*[b]
Celtis sinensis	*Peltophorum pterocarpum*
Cerbera manghas	*Pinus elliottiib*
Cinnamomum camphora	*Rhaphioelpis indica*
Delonix regia[b]	*Sapium discolor*
Eucalyptus citriodora[b]	*S. sebiferum*
E. robusta[b]	*Schefflera octophylla*
E. torelliana[b]	*Tristania conferta*[b]
Ficus microcarpa	

[a] Data Source: Agriculture & Fisheries Department, Hong Kong Government.
[b] Exotic Species.

REFERENCES

Agriculture and Fisheries Department. *Annual Departmental Report 1949/50 to 1959/60.* Hong Kong Government, Hong Kong.
Agriculture and Fisheries Department. *Annual Departmental Report 1960/61 to 1963/64.* Hong Kong Government, Hong Kong.
Agriculture and Fisheries Department. *Annual Departmental Report 1964/65 to 1995/96.* Hong Kong Government, Hong Kong.
Daley, P.A. Man's Influence on the Vegetation of Hong Kong, in *The Vegetation of Hong Kong: Its Structure and Change*, Thrower, L.B., Ed. Royal Asiatic Society, Hong Kong Branch, Hong Kong, 1970, pp. 44–56.

Chapter 20

The Role of Plantations in Restoring Degraded Lands in Hong Kong

X.Y. Zhuang and M.L. Yau

INTRODUCTION

Hong Kong lies in the northern margin of the Asian tropics. The climax vegetation is subtropical evergreen broad-leaved monsoon forest (Chang et al., 1989). However, due to long-term human disturbances, all the original forest was eradicated before 1900.

Large-scale tree planting was started on degraded areas in Hong Kong in 1880, but most of the early plantations were destroyed during the Japanese occupation between 1942 and 1945. The whole Territory of Hong Kong was very barren and soil erosion was severe after the war. Some active measures were taken by the government to improve rehabilitation of the forest in Hong Kong. First, large areas of eroded barren lands were planted with trees. In addition to *Pinus massoniana*, other species such as *Lophostemon confertus* (previously *Tristania conferta*), *Pinus elliottii*, *Melaleuca quinquenervia* (previously misidentified as *M. leucadendron*), *Acacia confusa*, *Schima superba*, and *Castanopsis fissa* were also used in plantations after 1945. Next, the Hong Kong government formally proposed to shift the purpose of forestry from production to conservation in 1953 (Robertson, 1953). All forests were by now protected from cutting. During the late 1970s, about 40% of the Territory was protected as Country Parks, Special Areas, and Sites of Special Scientific Interest (SSSI). Nowadays, forest coverage in Hong Kong is 14%, two-thirds of which is natural secondary forest and one-third plantations (Dudgeon and Corlett 1994).

A study of secondary forest showed that native woodland habitat can be restored in Hong Kong through natural succession in the absence of disturbance (Zhuang, 1993). Despite the fact that one-third of the total forest cover is plantation, no studies have been performed on the role of plantations in ecological succession and conservation in Hong Kong. Plantations using fast-growing tree species can speed up the formation of forest cover and improve physical conditions on degraded lands. However, information on invasion and establishment of native species in the understory and overstory, species composition and diversity in the plantations, and factors affecting the successional process in the plantations are wanting. This chapter aims to provide quantitative baseline information on the development of native flora in the major plantations of Hong Kong. Factors affecting species composition, abundance, and diversity under the plantations are discussed.

METHODS

Study Site

Lying on the transitional belt between the subtropics and tropics and in a coastal area, Hong Kong has a maritime, tropical monsoon climate. The summer is long, hot, and humid whereas the winter is short, cool, and dry, a set of conditions that is favorable for the growth of most plants and animals. The mean annual temperature is 23°C, while mean annual rainfall is 2214 mm. Eighty percent of rainfall occurs in summer, from May to September (Anon., 1992).

Field Surveys

Field surveys were conducted between October 1995 and January 1996 in four types of plantations: *Lophostemon confertus*, *Acacia confusa*, and *Schima superba* and mixed plantations of the latter with *Melaleuca quinquenervia* or *Cunninghamia lanceolata*. A total of eight 400 m² plots of plantations of similar age (30 to 50 years old) and at similar elevations were surveyed (Table 20.1, Figure 20.1) so as to eliminate the age and elevation effect as far as possible. Within each plot all the trees and shrubs with a diameter at breast height (dbh) greater than 2 cm were measured, and species in the understory were recorded. Basal area, tree density (no./ha) and Shannon's Index were computed for each plot. Results obtained were also compared with other studies of natural secondary forest and younger plantations. Only nonplanted species recorded with a dbh greater than 2 cm were included in quantitative analyses.

Due to limited time and resources, only a composite surface soil (1–5 cm) sample was collected within each plot. pH, organic matter, total nitrogen, potassium and phosphorus, and available potassium and phosphorus were analyzed.

RESULTS

Floristic Composition in Plantations

A total of 139 nonplanted, mostly native pioneer species, belonging to 60 families and 108 genera, was found in these plantations. Approximately 50% were tree taxa, representing 20% of the total tree flora in Hong Kong. Nineteen native species occurred in at least three studied plots (Table 20.2). The most common native species in plantations were bird-dispersed species which are also common in natural secondary forests (Table 20.2). Some rare species, such as *Cryptocarya chinensis*, *Castanopsis concinna*, and *Ania hongkongensis*, were also recorded in plantations, but they were only found in the understory of the plantations near relict woodlands such as those at Shing Mun Country Park and on Tai Mo Shan.

Species Diversity in Plantations

Species richness and diversity indices of the surveyed plots varied from 1.98 to 2.27 (Table 20.3). In general, numbers of invading species were highest in *Lophostemon confertus*. Shannon's Index (H') was the highest in the plantations of *Acacia confusa*, but was still lower than that of natural secondary forests (H' = 2.53, Table 20.3). No obvious differences were shown in numbers of recorded species with dbh greater than 2 cm.

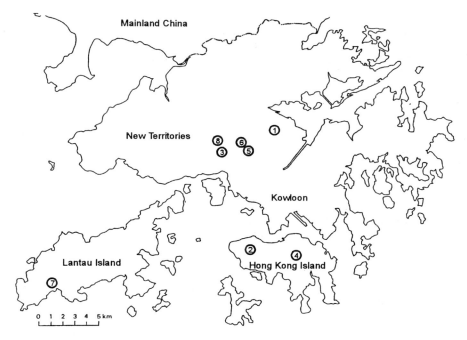

FIGURE 20.1. Locations of the eight study sites in the Territory of Hong Kong. Refer to Table 20.1 for site description.

Table 20.1. Description of the Study Sites.

Site No.	Planted Species	Location	Elevation (m)	Age (yr)[a]
1	*Lophostemon confertus*	Tai Po Kau	180	40
2	*Lophostemon confertus*	Hatton Road	200	?[b]
3	*Lophostemon confertus*	Tai Mo Shan	280	?
4	*Schima superba*	Tai Tam	220	35
5	*Schima superba* + *Cunninghamia lanceolata*	Shing Mun	250	45–50
6	*Schima superba* + *Melaleuca quinquenervia*	Shing Mun	250	45–50
7	*Acacia confusa*	South Lantau	150	35
8	*Acacia confusa*	Tai Mo Shan	400	35

[a] Information on ages of the surveyed plantations were obtained from the Agriculture and Fisheries Department of the Hong Kong Government.
[b] ? = unknown.

Comparisons were made between plantation plots of this study and other *Acacia* plantations of younger ages (Table 20.4). Shannon's Indices of 15 year-old *Acacia* plantation plots at the Clearwater Bay Golf and Country Club, Hong Kong was zero; no nonplanted species with dbh over 2 cm were recorded, although nonwoody native plants and tree seedlings were observed. Shannon's Indices of the 25 year-old plantations at Shek Kwu Chau ranged between 1.24 and 1.64, higher than those at the Clearwater Bay plots but lower than those recorded in this study. Species diversity, therefore, appears to increase

Table 20.2. Native Species Occurring in at Least 3 Plots of the Plantations Surveyed.

Species	Life Form	Plots Occurred (total plots=8)	Fruit Type
Acronychia pedunculata[a]	tree	4	drupe
Alangium chinense	tree	3	drupe
Aporusa dioica[a]	tree	8	capsule
Ardisia quinquegona	shrub	3	drupe
Artocarpus hypargyrea	tree	3	multiple
Daphniphyllum calycinum[a]	tree	4	drupe
Diospyros eriantha	tree	3	berry
Diospyros morrisiana	tree	3	berry
Evodia lepta	tree	3	capsule
Ilex asprella[a]	shrub	6	drupe
Litsea rotundifolia var. *oblongifolia*[a]	shrub	5	drupe
Machilus breviflora	tree	3	drupe
Machilus longipedunculata[a]	tree	5	drupe
Machilus sp.	tree	3	drupe
Machilus velutina	tree	3	drupe
Psychotria rubra[a]	shrub	8	berry
Schefflera octophylla[a]	tree	5	berry
Schima superba	tree	4	capsule
Sterculia lanceolata[a]	tree	4	follicle

[a] Fruits dispersed by bird.

Table 20.3. Diversity in the Three Types of Plantations Surveyed.

Type	Total No. Species[a]	No. of Species (>2 cm dbh)	Basal Area (m^2 ha^{-1})	Tree Density (no. ha^{-1})	Shannon's Index (H')
Lophostemon confertus	49 (4)[b]	19 (8)	27.78 (9.24)	3408 (1052)	1.98 (0.59)
Schima superba and its mix	43 (17)	19 (6)	45.72 (20.63)	3888 (1149)	2.03 (0.62)
Acacia confusa	36 (16)	19 (8)	32.34 (4.39)	3225 (451)	2.27 (0.40)
Natural secondary forest[c]	—	—	—	—	2.53 (0.32)

[a] Understorey species plus species with dbh > 2 cm.
[b] Numbers in brackets are standard deviation.
[c] Of age 30–50 years old, N = 25 (Zhuang and Corlett, 1996).

Table 20.4. Diversity in Plantation of *Acacia confusa* of Different Ages.

Site	Age (yr.)	No. Plots[c]	No. Species	Tree Density (no. ha^{-1})	Basal Area (m^2 ha^{-1})	Shannon Index
Shek Kwu Chau[a]	25	3	17–20	1400–2300	16.35–30.58	1.24–1.64
Clear Water Bay[b]	15	2	5–6	1200–5300	11.33–15.58	0

[a] Zhuang and Chen (unpublished data).
[b] Zhuang and Yau (unpublished data).
[c] Plot size = 100 m^2.

with age of the plantations. However, human activities and management practices may also affect species diversity in plantation plots. For example, plantations at the Clearwater Bay Golf and Country Club were weeded frequently after planting. Native species naturally established in the understory were probably eradicated with the weeds. Some seedlings and saplings of *Schefflera octophylla* were observed after the weeding was discontinued for a few years.

Comparison of Soil Characteristics

Results of chemical analyses showed that soil fertility varied with different types of plantations. Organic matter, total nitrogen, available phosphorus, and available potassium were highest in the plantations of *Acacia confusa* (Table 20.5). Compared to the *Schima superba* plantation, soil in the plantation of *Lophostemon confertus* had a higher content of organic matter, total nitrogen, and available phosphorus, but a lower content of potassium.

DISCUSSION

Plantation as an Important Approach to Restore Forest Cover in Degraded Lands

The study of natural forest succession showed that grassland on degraded lands of Hong Kong can develop into closed-canopy secondary forest through natural succession in the absence of disturbance, over about 30–40 years (Zhuang, 1993). However, the time span required is longer on more degraded sites. Planting of fast-growing species can facilitate forest restoration on degraded lands. Therefore, plantation is an active and important restoration approach, especially in severely degraded areas. However, most of the commonly planted species used in plantation are exotic species, such as *Pinus*, *Eucalyptus*, and *Acacia* species. All these species produce dry fruits which cannot be used by local animals. Planting indigenous fast-growing, bird-dispersed species can promote the rehabilitation of native forest and improve the ecological function of plantations.

Plantation as an Agent to Facilitate Natural Succession

Pinus massoniana used to be a major native plantation species in Hong Kong, especially before 1942, but its importance has greatly declined recently in Hong Kong due to the infestation of pine wood nematode (*Bursaphelenchus xylophilus*) (Dudgeon and Corlett, 1994). However, it is commonly observed that many natural forests have developed from previous pine woodlands. A study on natural succession under pine plantations indicated that pine plantation could provide a nurse role for the establishment of native species on hillsides characterized by harsh conditions (Chan and Thrower, 1986).

The present study showed that plantations can facilitate natural succession in the absence of human disturbance. The major function of plantation is to create less harsh habitats on degraded lands by increasing soil humidity and reducing competition of weeds by shading, thereby enhancing successful invasion and establishment of native species. It should be noted that species establishing in plantations are different from those invading open sites. Most common species invading plantations can germinate under and are slightly tolerant of shading; e.g., *Schefflera octophylla*. The highly shade-intolerant species, such as *Pinus massoniana*, *Trema orientalis*, and *Sapium discolor*, seldom occur in plantations. On the

Table 20.5. Comparison of Soil Characteristics of Three Plantation Types Surveyed.

Type[a]	Organic Matter (mg kg^{-1})	Nitrogen (mg kg^{-1})	Phosphorus (mg kg^{-1})	Potassium (mg kg^{-1})	Available Phosphorus (mg kg^{-1})	Available Potassium (mg kg^{-1})	pH
1	7.62(1.22)[b]	2.32(0.33)	0.21(0.03)	11.21(8.57)	3.08(0.32)	87(31)	4.8(0.9)
2	5.28(0.32)	1.91(0.23)	0.24(0.09)	14.98(8.51)	2.64(0.43)	106(12)	5.2(0.4)
3	11.23(2.64)	4.51(1.38)	0.37(0.11)	19.63(19.8)	7.62(5.43)	131(6)	3.9(0.1)

[a] 1=*Lophostemon confertus* plantation; 2=*Schima superba* and its mixed plantations; 3=*Acacia confusa* plantation.

[b] Numbers in brackets are standard deviation.

other hand, the development of native species in plantations also depends on management activity and other disturbance. Periodic weeding can inhibit the invasion and development of light-demanding pioneer species in plantations.

Plantations can also promote natural succession by attracting the activities of wildlife by providing roosting sites and foods. Wildlife including birds and mammals can help disperse seeds of native species by consuming fruits. As shown in this study, the common species invading plantations are usually bird-dispersed species with fleshy fruits. Another study has shown that use of native bird-dispersed species in plantations can promote natural succession in plantations (Robinson and Handel, 1993).

Many species commonly found in our study plots are potential candidates for revegetating barren sites. For instance, *Acronychia pedunculata*, *Archidendron lucidum*, *Daphniphyllum calycinum*, *Diospyros morrisiana*, *Machilus longipedunculata*, and *Schima superba* are pioneer species and colonize grassland and shrubland. Other species such as *Ardisia quinquegona*, *Psychotria rubra*, and *Schefflera octophylla* are more or less shade-tolerant. They can be mix-planted with other fast-growing species. The major limitations of using native species in plantations in Hong Kong and South China are a lack of knowledge of ecophysiological characteristics of most of the native flora, and a shortage of a propagule supply of native species. For example, among the 19 species listed in Table 20.2, only 4 (*Psychotria rubra*, *Sheffflera octophylla*, *Schima superba*, and *Sterculia lanceolata*) are commercially available in Hong Kong, often in limited supply. Research on ecophysiological characteristics of and silvicultural techniques for native species can contribute immensely to the utilization of native flora.

Results of soil fertility analyses showed that the plantations of *Acacia confusa* had higher soil fertility than plantations of other species. It has also been reported that this species can adapt well to severely degraded soil (Chan et al., 1991). The roots of this species can form a symbiosis with nitrogen-fixing bacteria, which may contribute to the higher soil fertility in *Acacia* plantations. Therefore, it is a suitable species for use on severely impoverished lands.

Our study showed evidence that existing plantations of *Acacia confusa*, *Lophostemon confertus*, and *Schima superba* mainly contribute to soil conservation, speeding up of forest formation, and amenity purpose. Their ecological functions, including facilitation of natural succession, conservation of biodiversity, and usage by wildlife, are generally low, but can be improved by using more native species in plantations. Therefore, more research on propagation and planting techniques of native species is required.

ACKNOWLEDGMENT

This study was supported by Ecosystems Ltd. Information on the age of the plantations surveyed was obtained from the Agriculture and Fisheries Department of the Hong Kong government. Soil analyses were kindly performed by Mr. Qiming Lu, South China Agricultural University. The manuscript was commented on and edited by Ms. Mary Felley, Mr. Thomas Dahmer, and two anonymous reviewers, for which we express the deepest thanks.

REFERENCES

Anon. *Surface Observations in Hong Kong 1992*. Government Printer, Hong Kong, 1992.
Chan, G.Y.S., M.H. Wong, and B.A. Whitton. Effects of Landfill Gas on Subtropical Woody Plants. *Environ. Manage.*, 15 (3), pp. 411–431, 1991

Chan, Y.K. and L.B. Thrower. Succession Taking Place under *Pinus massoniana*. Memoirs of the Hong Kong Natural History Society, 17, pp. 59–66, 1986.

Chang, H., B. Wang, Y. Hu, P. But, Y. Chung, Y. Lu, and S. Yu. Hong Kong Vegetation. Supplement to *Acta Scientiarum Naturalium Universitatis Sunyatseni*, 8(2), 1989 (in Chinese).

Dudgeon, D. and R.T. Corlett. *Hills and Streams. An Ecology of Hong Kong*. University Press, Hong Kong, 1994, p. 234.

Robertson, A.F. *A Review of Forestry in Hong Kong with Policy Recommendations*. Hong Kong Government Printer, Hong Kong, 1953.

Robinson, G.R. and S.N. Handel. Forest Restoration on a Closed Landfill: Rapid Addition of New Species by Bird Dispersal. *Conserv. Biol.*, 7(2), pp. 271–278, 1993.

Zhuang, X.Y. Forest Succession in Hong Kong. Ph.D. Thesis, Department of Botany, University of Hong Kong, Hong Kong, 1993, p. 273.

Zhuang, X.Y. and R.T. Corlett. The Conservation Status of Hong Kong's Tree Flora. *Chinese Biodiversity*, 4 (supplement), pp. 36–43, 1996.

Chapter 21

Trees and Greening of Hong Kong's Urban Landscape: Subaerial and Soil Constraints

C.Y. Jim

INTRODUCTION

The history of city growth in Hong Kong is epitomized by a relentless quest for developable land. The rather rugged topography has some 80% of the terrain as steep hillslopes not easily convertible to urban usage. The location of the harbor around which flat land is particularly limited has reinforced the problem of land shortage. The consequent urban form has been dictated by this rather immutable geographical reality. In its 150 years or so of history, land for buildings, roads, and the plethora of infrastructural needs has had to be acquired at considerable cost via two laborious means (Tregear and Berry, 1959). On the one hand, platforms have had to be carved out of the hills in an upslope sprawl; on the other, land has had to be reclaimed from the sea. The valuable land parcels thus obtained have been by necessity put to high-intensity uses. The urban morphology of Hong Kong, being of extremely high-density and high-rise from the city core up to its periphery, including some of its suburban outliers, is therefore a faithful reflection of the inherent physical constraints.

Under the circumstances, the possibility for open spaces and greenery can only be very limited. With all sort of activities and needs juxtaposed and cramped together, few habitats are left for vegetation. Any plants that manage fortunately or fortuitously to find occasional niches in the tightly-packed environment often have to persevere in an inordinately stressful environment. Vegetation in formal green spaces such as the urban parks and gardens, with limited total acreage, can fare better due to the generally genial milieu. Along the thoroughfares, however, the conditions for plant survival are particularly trying. Both above- and belowground tree growth is highly confined, and existing growth spaces are frequently intruded upon or degraded due to the unremitting construction and redevelopment programs. Whereas many other cities can take pride in generous provision of open green spaces, for some decades Hong Kong has to make do with a gross undersupply of nature in the built-up areas.

Starting from the 1970s, Hong Kong has begun to embark on a massive program of new town development. Some two-thirds of the current population of 6.3 million has been siphoned off into the hitherto-rural New Territories in a large-scale internal shift of people. The pressure to squeeze more population into the old urban core around the harbor has to a large extent been relieved. Meanwhile, the continual urban renewal and new reclaimed land has furnished many opportunities to mold the city so that it can be ushered away from the previous rather bleak if not harsh existence (Jim, 1989 and 1993). With an official policy of thinning out the population while at the same time improving the quality of life in the older neighborhoods, more and better habitats should be earmarked for greenery. The rising expectation for a more livable city could to a large extent be satisfied by a more liberal penetration of greenery.

Trees in urban areas interact intensively with their environment which is often stressful (Bernatzky, 1978; Harris, 1992), and this is particularly so in densely-packed urban Hong Kong. This chapter attempts to assess the status of tree planting in the core urban area of Hong Kong, with emphasis on the subaerial and subterranean conditions for tree growth. The multiple physical and physiological obstacles toward a greener city are surveyed, followed by a specific study of the soil environment which is often neglected in urban tree-planting endeavors and a major cause of planting failures (Bullock and Gregory, 1991; Craul, 1992; Watson and Neely, 1994; Bradshaw et al., 1995).

STUDY AREA AND METHODS

The study area covers the old urban core of Hong Kong centered around the Victoria Harbour area, including Hong Kong Island, Kowloon, New Kowloon, and Tsuen Wan. The rather coterminous urban sprawl originated from the incipient City of Victoria situated in the present Central District and founded in early 1840s. Officially known as the Metroplan area (Strategic Planning Unit, 1990), its 211 km^2 of land accommodates some 3.95 million people; that is about 62% of the total population. With an average density of 18,720 persons km^{-2} and a maximum spot density of 66,000 persons km^{-2} (Census and Statistic Department, 1995), it has the highest concentration of population in the world. It is physically separated by a range of hills from the more recent generation of new towns established in the New Territories. Compared to the new towns, it has inherited a host of urban blight problems associated with an excessively high-density mode of development and a legacy of inadequate attention to environmental planning. It will take a good deal of effort and determination to bring the quality of the environment in the old city up to the higher standards adopted in the new towns.

The information on urban trees in the study area has been obtained through two comprehensive surveys, in 1985 and 1994 (Jim, 1986 and 1994a), and results of urban tree research in the last 15 years (Jim, 1990). Detailed computer databases have been set up with the help of Microsoft Excel and SPSSPC for Windows software on both occasions to record in coded form a wide range of attributes related to individual trees and their immediate environs (Jim, 1996). The studies have been designed to assess, in quantitative or at least semiquantitative terms, the many impediments to proper tree growth, including a series of on-site measurements of tree and growth-space dimensions.

Specific information on soil conditions for tree growth in the cramped urban habitats has been obtained by detailed study of soil properties both in the field and in the laboratory. Initially, soil pits of about 1m × 1m × 1m dimension where trees were to be planted were dug. Profile and other site and morphological characteristics were then evaluated in situ (Hodgson, 1974). This step was followed by taking composite soil samples, both disturbed and undisturbed (obtained by driving a stainless cylinder into the cut face), from each of the identified soil layers. Adopting standard laboratory procedures (Page et al., 1982; Klute, 1986), the soil samples were air-dried, ground, sieved, and analyzed for a range of physical and chemical properties. A total of 50 soil pits have been evaluated, situated in various urban habitats, including pavement, tree strip, and various roadside locations, urban parks, and other incidental planting sites. This chapter reports mainly on the conditions for tree growth at roadsides.

ABOVEGROUND CONSTRAINTS

Some 20,000 trees are found at roadside habitats in the study area. Most trees are small (<15 cm trunk diameter, <5 m height and < 5 m crown spread), and exceptionally large specimens are rarely found (Jim, 1994b). Although a total of 149 species are represented, the trees are dominated by a small number of popular species, with top 14 species taking up two-thirds of the stock. There is also a clear dominance by exotic species and by the broad-leaved evergreen growth form. About half of the trees are found in grassed planting strips situated either adjacent to pavements or in central divider locations, where the space for tree expansion is relatively more generous. About one-third of the trees are planted in tree pits on pavements where the limitations to growth are often quite acute. Most of the larger trees are trapped in incongruous pavement sites, whereas the more spacious tree strips accommodate most of the smaller trees.

Trees growing along roadsides are prone to a wide range of constraints and problems, both actual and potential. The most severe limitation is the grave shortage of growth space in relation to the final size of the trees. Most cities encounter this keen contest for spaces among trees, roads, and buildings only within the core area (Grey and Deneke, 1986 ; Miller, 1988). In Hong Kong, this phenomenon is pervasive almost throughout the city, extending from the center to the periphery. The narrow growing sites are often beset by a collection of physical obstacles to both vertical and lateral growth. The common occurrence of building awnings above a part of, or the whole, pavement is a serious and quite immutable obstacle to tree growth (Figure 21.1). Haphazardly installed advertisement signs pose an additional competitor for the valuable plantable space. The close proximity of trees to the carriageway, the associated need to provide safe sightlines, and the numerous traffic signs also take away many otherwise usable planting sites.

As the most limiting physical factor to tree growth is related to the room for crown expansion, a statistical summary of the situation will be helpful. Figure 21.2 indicates that tree strip is the predominant type of growth space which accommodates most of the small trees. Tree pit is the second-rank growth space; those recently installed and properly equipped with grills hold most of the newly planted small trees in this category. Many currently small trees, however, will reach a much larger final size and in due course outgrow the capacity of the confined growth space. The trees in pits with no grill, being the older generation of planting mode, are already showing this undesirable trend. This problem will become more acute in both types of tree pits which are mainly found at narrow pavements. For the tree strip, the wider plantable space and less adjacent hindrance will be more able to contain the larger trees.

The crowns of large trees can overhang above the carriageway if they manage to reach above the necessary headroom clearance for vehicles. As such, they can have a chance to develop rather freely, albeit asymmetrically, toward the road. Many roadside species, however, will not be able to attain such a stature and hence will be boxed in by adjacent structures and traffic safety requirements. A study of the lateral restriction to crown expansion of individual trees has been summarized in Figure 21.3. Many trees have their crowns hindered on the property side by buildings, with their complement of awnings and advertisement signs. The almost ubiquitous local practice of building up to the lot bound-

FIGURE 21.1. The common occurrence of building awnings above narrow pavements imposes a serious constraint on tree growth in urban Hong Kong.

ary without any setback has imposed an uncompromising if not insuperable barrier to tree growth. Many trees are impeded on the road side due to vehicular-traffic needs. Many trees, especially the older ones situated at pavement sites (pits with no grill, irregular hole, and paved to trunk), are impeded on both property and road sides. Tree strip has the highest percentage of trees that can prosper without crown restrictions, many of which are found in traffic-island type of situations associated with high-capacity roads and tunnel portals.

The roadside growth space for many trees in effect becomes a narrow and unyielding corridor, severely restricting the potential to develop crowns of natural proportion. The same encumbrances will have deleterious effects on environmental conditions and hence tree physiology. The heavy shading, gaseous and particulate air pollution, vehicular abrasion, and the generally dry and hot environment certainly do not help. Such trees with reduced health are more prone to disease and pest infestation. Under the circumstances, most trees will not have a chance to mature even into mediocre specimens in terms of both tree form and vigor. The intense conflict among buildings, roads, and trees also necessi-

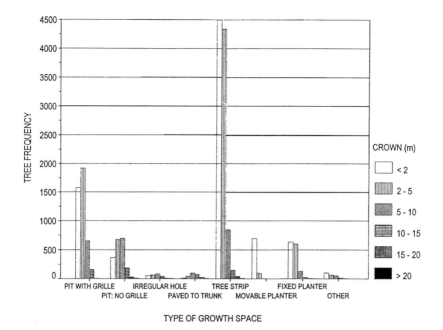

FIGURE 21.2. Roadside tree frequency in urban Hong Kong in relation to eight types of growth space and crown diameter (N=19,154 trees, Cramer's V=0.20, p<0.001).

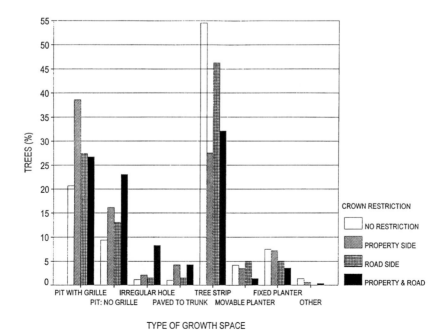

FIGURE 21.3. The percentage of roadside trees in urban Hong Kong in relation to eight types of growth space and crown restriction (N=19,154 trees, Cramer's V=0.12, p<0.001).

tates a frequent and laborious regime of arboricultural inputs. Young trees may not be affected by these rampant obstacles, but as they grow up the problems will appear.

For the confined roadside habitats common in the city, the choice of species with reference to their final attainable size can only be very limited (Jim, 1990). Unfortunately, too many trees with a potential to reach sizeable dimensions have been erroneously inserted due to shortsighted and overzealous planting. The truism that there is a need to match the geometry of planting sites with that of fully-grown trees has been too liberally violated. The management burden will gradually accentuate as these inappropriately-chosen species grow up. Some problems of tree-habitat incongruity related to currently large specimens, however, are attributed to shifts in lot boundaries and road alignment. In a city that is perhaps too frequently rebuilt to an ever-increasing density, the already meager growth space is often squeezed in the process. While trees are assiduously planted in new development areas and along new roads, the existing large trees in the older neighborhoods are often damaged or felled due to infilling and intensification of land use.

GENERAL SOIL CONSTRAINTS

City growth in Hong Kong involves massive destruction of the natural environment to an extent more extreme than other places. In sprawling up the hillslopes, any existing vegetation has been quite thoroughly eradicated. The soil has suffered a similar fate by widespread removal, truncation, burial, mixing, or contamination with foreign materials. In the land reclaimed from the sea, the other mode of acquiring developable land in Hong Kong, the initial state has have been a similarly desolate surface of earth fill that includes building debris. The spoiled landscape has then received buildings, roads, and other artificial accoutrements to create the city fabric. In the past, roadside trees have been inserted only after the various land uses have taken their share; that is, in the remnant or residual spaces often in odd locations. In recent years, there has been a conscious effort to earmark amenity strips along roads to cater to the greening aspiration of the community.

The provision of plantable spaces does not necessarily follow that they will be habitable to trees. A good-quality subaerial environment only tells half of the story. The subterranean conditions for tree survival have been routinely taken for granted, if not altogether neglected (Perry, 1982; Proudfoot, 1992). The only obstacles to root growth are, unfortunately, generally construed to be underground cables, pipes, and large chunks of concrete or rock situated at a shallow depth. The on-site "soil," so long as it has a consistence that can be dug by normal implements and generally free from the above impediments, is considered as a suitable medium to support tree growth.

The underground space at roadsides in Hong Kong, unfortunately, is equally as stressful as the aboveground situation. Soils at designated tree strips and tree pits are often no more than the poorest quality building rubble and waste of all descriptions. Little attempt is made to provide a proper soil specification other than a cursory and doubtful one on "topsoil" (Bradshaw, 1989; Bradshaw et al., 1995). The subsoil, in particular, is regularly ignored as though it will always perform its functions regardless of its composition and property. Soils at planting sites are rarely checked in the field and tested in the laboratory for their suitability to tree growth. Ameliorative measures such as the use of amendments, the removal of harmful substances, or replacement with good-quality materials, are rarely administered.

The width of the exposed soil measured on the ground surface can serve as an indicator of its ability to support tree growth. Sealed surface around a tree will deprive the soil of

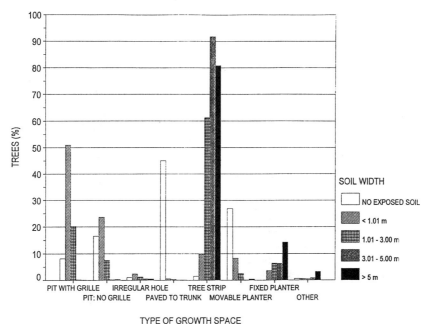

FIGURE 21.4. The percentage of roadside trees in urban Hong Kong in relation to eight types of growth space and soil width (N=19,154 trees, Cramer's V=0.48, p<0.001).

moisture replenishment by infiltration and aeration; that is, entry of oxygen to sustain respiration and venting of carbon dioxide released from biological activities. The city's roadside habitats are often sealed with concrete paving or asphalt road covering. Recent attempts to replace concrete pavement with unit pavers may provide a limited amount of relief to this constraint.

Figure 21.4 shows that a sizeable proportion of pavement sites, particularly paved to the trunk and pit with no grill, suffers from this problem. For the same sites, where there is exposed soil, they tend to be quite narrow; that is, in the mainly <1 m and secondarily 1–3 m categories. For the tree strips, the exposed soil is much wider, with many in the 3–5 m and >5 m classes, but a significant proportion still lies in the somewhat narrow <3 m groups. The fixed planters to a limited extent furnish a relatively better habitat for trees with their better provision of free passage for air and water. They can additionally overcome the common restriction of shallow underground utilities which would otherwise preclude tree planting. Fixed planters, however, can have serious problems in restricting root volume and water supply in dry periods (Lindsey and Bassuk, 1992).

A good planting site should have as much of its surface remaining unsealed as possible. A study of plantable width in comparison with soil width for the existing 20,000-odd road-side trees is summarized in Figure 21.5. The diagonal line indicates the tree strip sites with plantable width equal to soil width that is the best possible configuration. Sites falling below this line represent that the soil width falls short of plantable width, and imply soil problems association with the sealed surface. The majority of the points fall into the latter category, and many cluster close to the plantable-width axis and the origin, showing that most trees are trapped in highly confined sites with even more restricted soil exposure.

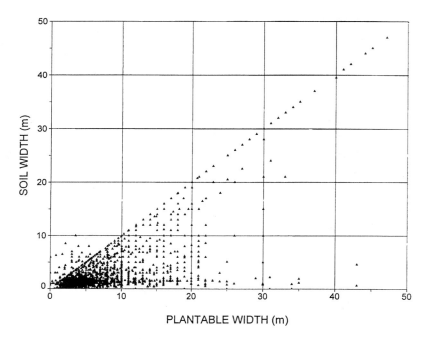

FIGURE 21.5. The relationship between plantable width and soil width for roadside trees in urban Hong Kong (N=19,154 trees, r=0.74, R^2=0.55, p<0.001).

SPECIFIC PHYSICAL SOIL CONSTRAINTS

Physical soil properties have important bearing on tree growth (Glinski and Lipiec, 1990; Mullins, 1991). The study of soils at various roadside locations in Hong Kong provides detailed information and insights on the wide range of profile, physical and chemical properties for a humid-tropical city. The results of the field and laboratory analysis of 50 sites have been condensed and interpreted in Tables 21.1 and 21.2.

In terms of soil depth, some sites do not satisfy the notional minimum of 1 m of solum necessary for normal tree-root growth (Table 21.1). The major limitations are rocky substrate, excessively stony subsurface layer, building rubbles and remnants of old building foundations, disused utility installations, and the masses of current buried utilities (Figures 21.6 and 21.7). The soils are mainly some kind of decomposed granite which is the most common parent material in Hong Kong. The materials in roadside pits often exist in sharply-defined layers of different composition, texture, and structure (Figure 21.8). They have been laid down artificially during the roadwork rather than being the horizons derived from pedogenic processes. Such a layered organization can impede the free movement of water and create a perched water table in an upper layer. Water will not move down unless the upper layer has been saturated, and this can cause moisture deficiency in the lower layer (Baver et al., 1972).

The physical state of the soils is characterized particularly by an inordinately high content of stones (> 2 mm diameter) which in some samples exceed 80% by weight. Besides materials inherited from the weathered granite, a lot of artificial materials, mainly building debris, are present. The high stone content that tends to increase with depth may pose a serious mechanical impedance to root elongation and spread. The sand content (0.05 to

Table 21.1. Profile and Physical Properties of Roadside Soils in Urban Hong Kong.

Attribute	Common State	Assessment
Profile and color:		
Depth (cm)	65 to 100+ cm usable solum	sometimes too shallow for tree growth
Horizon or layer	multiple layers	marked layering of materials
		heterogeneous materials
	gradual to clear boundaries	perched water table
		delayed moisture percolation
General description	compacted and very firm	signs of heavy compaction
Color	moist: brown to dark brown	mainly weathered granite material
	dry: dull brown to yellow-orange	
Stone and particle size (USDA):		
Stone content (%)	22 to 82, usually increases with depth	can be excessively stony
		often contain construction debris
		impediment to root growth
Sand (%)	72 to 87	usually very high sand content
		excessive drainage
Silt (%)	7 to 18	usually low to medium content
Clay (%)	4 to 13	often inadequate for nutrient retention
Textural class	loamy sand to sandy loam	predominantly coarse texture
Structure and consistence:		
Structure type	low-grade small to medium angular blocky, to massive (single-grain for sand texture)	damaged or poorly developed structure
Aggregate stability (%)	16 to 58% remain after wet sieving	weak to somewhat stable aggregates
Aggregate slaking	none to complete	some aggregates easily destroyed
		by sudden wetting (air explosion)
Ped (or clod) strength	firm to strong when dry	peds can withstand some mechanical
		disturbance without yielding
Failure characteristics	semideformable to brittle	mode of deformation nonplastic
Maximum stickiness	slightly- to nonsticky	consistence echoes low clay content
Maximum plasticity	nonplastic	consistence echoes low clay content
Bulk density and porosity:		
Bulk density (Mg m^3)	1.5 to 2.2	often higher than normal soil
		inadequate available-moisture retention
		inadequate for aeration
		inadequate for infiltration
		inadequate for percolation
Total porosity (% v/v)	23 to 38	can be inadequate for proper aeration, drainage and root growth

Table 21.2. Chemical Properties of Roadside Soils in Urban Hong Kong.

Attribute	Common State	Norm for Humid-Tropical Soils[a]
"pH, carbon, nitrogen, and phosphorus:"		
pH	7.5 to 8.3	5.5 to 7.0
Loss on ignition (%)	3.22 to 6.73	(n.a.)
Organic carbon (%)	0.11 to 2.1	4 to 10
Organic matter (%)	0.22 to 4.20	8 to 20
Total nitrogen (%)	0.00 to 0.42	0.2 to 0.5
Carbon/nitrogen ratio	0.00 to 5.01	7 to 10
Extractable phosphorus ($\mu g\ g^{-1}$)	2.12 to 3.37	5 to 15
Exchangeable nutrient:		
Cation exchange capacity ($cmol\ kg^{-1}$)	4.65 to 12.61	15 to 25
Exchangeable sodium ($cmol\ kg^{-1}$)	0.20 to 3.56	0.5 to 1.0
Exchangeable potassium ($cmol\ kg^{-1}$)	0.29 to 0.68	0.2 to 0.6
Exchangeable calcium ($cmol\ kg^{-1}$)	0.40 to 2.06	4 to 10
Exchangeable magnesium ($cmol\ kg^{-1}$)	0.15 to 0.78	0.5 to 1.0
Base saturation (%)	22.4 to 56.1	20 to 60
Salinity:		
Exchangeable sodium percentage (%)	4.3 to 28.2	<15
Sodium adsorption ratio	0.27 to 2.11	<13
Conductivity ($dS\ m^{-1}$)	0.05 to 0.82	2 to 4
Chloride ($\mu g\ g^{-1}$)	12.2 to 42.5	200 to 350
Heavy metal:		
Total copper ($\mu g\ g^{-1}$)	11.2 to 85.1	30
Total zinc ($\mu g\ g^{-1}$)	37 to 504	90
Total lead ($\mu g\ g^{-1}$)	32 to 420	35
Total cadmium ($\mu g\ g^{-1}$)	0.30 to 2.34	0.35
Total nickel ($\mu g\ g^{-1}$)	2.0 to 27.5	50

[a] Refers to the medium range or average value of soil chemical attributes normally found in humid tropical soils (Landon, 1991).

2.0 mm diameter) is similarly consistently high, in the 72% to 87% range, whereas the silt (0.002 to 0.05 mm diameter) and clay (<0.002 mm diameter) contents are much lower. The texture is therefore predominantly very coarse, mostly in the classes of loamy sand or sandy loam. Such a skewed particle-size distribution provides a good deal of large pores for an excessive drainage and aeration regime, but an inadequate supply of medium pores for the retention of available moisture (Rawls et al., 1991). The shortage of clay-sized particles limits the cation exchange capacity; thus nutrient retention can only be restricted.

The structure of the highly disturbed soil materials is mainly poorly developed, and in extreme cases it is either of massive or single-grained categories. The soil structure has been very much damaged during the mechanical handling and subsequent compaction treatments. Some of the small aggregates, however, remain stable after the wet sieving procedure, indicating that the high content of iron and aluminium sesquioxides common in the granitic materials has been able to bind the elementary particles into relatively stable compound units. The consistence, being slightly- to nonsticky and nonplastic, and fails in the semideformable to brittle modes, reflect the low clay-colloid and high sand mineral composition.

FIGURE 21.6. The underground environment below the pavement and the adjacent road is often densely occupied by buried utility lines in urban Hong Kong.

FIGURE 21.7. The cut face illustrates the shallow penetration of tree roots in a recently felled large Chinese banyan (*Ficus microcarpa*), and the common competition for underground space between tree roots and utility lines in urban Hong Kong.

As is the case for urban soils in other cities (Patterson, 1977; Patterson and Mader, 1982), the roadside soils in Hong Kong are beset by poor structure associated with soil compaction. The bulk density of the soils, a common indicator of compaction, covers the 1.5 to 2.2 Mg m³ range. Whereas the lower end is a normal value for local natural soils, the

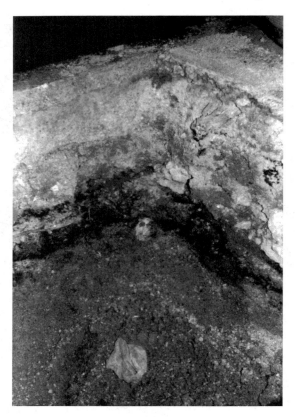

FIGURE 21.8. The soil in this tree pit at a pavement site in urban Hong Kong shows the common problem of sharp layering due to differences in composition, texture, and structure, and the compaction resulting in poor structure.

higher end denotes compaction to a severe degree. More importantly, the compaction tends to extend to the subsoil rather than being limited to the topsoil. The entire solum of one meter or so can be affected by compaction. The total porosity results of 23% to 38% echo the magnitude of structural degradation with pore collapse and the accompanying problems for root growth and functions (Zisa et al., 1980; Cochran and Brock, 1985; Gilman et al., 1987). The infiltration and transmission of water, the storage of available moisture, and aeration can be seriously hampered at a bulk density above 1.6 Mg m^{-3}. Both water and nutrient absorption can suffer (Greacen and Sands, 1980; Cannell, 1977; Raghavan et al., 1990).

SPECIFIC CHEMICAL SOIL CONSTRAINTS

The chemical properties of the roadside soils (Table 21.2) indicate additional constraints to tree growth. The results suggest a common occurrence of poor-quality planting media at the urban habitats. The slightly alkaline pH is somewhat aberrant in comparison with predominantly acidic reaction of local natural soils. The influence of lime originated from construction materials, especially cement and concrete, is obvious and perhaps somewhat unexpectedly widespread. Some nutrient elements which are less readily soluble in an alkaline edaphic environment, such as phosphorus and the suite of micronutrients (Fe, Mn, Zn, Cu, Co, and B) could become inadequate (Ware, 1990). The common tropical tree species used in landscape planting may not be able to attain optimal growth in an unaccustomed chemical environment.

The amount of organic carbon for most samples, at merely 0.11% to 2.1%, is consistently very much below the norm expected of a humid-tropical soil. The lack of inherited organic matter in the original soil materials and the absence of subsequent amendment cannot possibly provide even an average level of this essential substance. This general shortage of organic matter in turn severely limits the supply of nitrogen and phosphorus. The supply of these two important nutrient elements, for some samples going below the detection threshold, is extremely insufficient and could therefore be regarded as the limiting factor to tree growth.

With a common shortage of both organic and inorganic colloids, the ability to retain nutrient cations in readily-available forms, reflected by the cation exchange capacity (CEC), can only be very restricted. All samples are rated as having low CEC using normal humid-tropical soils as the yardstick. The concentration ranges of the exchangeable bases (Na, K, Ca, and Mg) for most soils fall below the expected norms. Due to the inherently low CEC, the base saturation values are not particularly subdued.

A small number of sites are situated close to the shoreline and occupy land reclaimed from the sea. The influence of sea-water penetration into the soil profile either by direct inundation of a high groundwater table, or by capillary rise, could cause a chemical problem due the presence of too much soluble salt and chloride ions. This phenomenon is fortunately much less demonstrated than expected. The heavy downpour in the summer months could have flushed any inherited or accumulated salts away from the normal rooting depth.

The heavy metal analysis provides a hint to the extent of pollution influence on the roadside soils. The lack of baseline study on the level of heavy metals in local soils precludes the use of a more reliable benchmark to evaluate the urban soils. Compared to the notional average concentration in tropical soils (Landon, 1991), however, some elements, particularly Zn and Pb, show markedly elevated concentrations. These two metals are commonly released from vehicular traffic (Thornton, 1991) due to the use of leaded gasoline and the zinc-plating of vehicle bodies. The cases for Cu, Ni, and to a lesser extent Cd, demonstrate a weaker sign of unnatural accumulation. The mobility of some metals in the soil due to the heavy leaching regime may have dampened the magnitude of pollution effects.

CONCLUSION

The urban morphology of Hong Kong is very much dictated by the inescapable reality of a grave shortage of developable land at the right location. It is a city that has had to develop on a gradually accreted area of often massively damaged landscape. The extent of land degradation necessitates extensive restorative measures to render it suitable for amenity vegetation. In the past, landscape plants have too often been inserted in the wrong habitats with little regard to proper site selection and improvement. Consequently, many extant trees are found struggling in incongruous environs with below-par performance. The nature of such low-caliber site conditions has to be studied in detail so that the common problems of tree establishment and survival can be identified, and the relevant ameliorative inputs applied correctly and in good time.

This study endeavors to furnish the fundamental scientific data to secure an understanding of the site-specific adverse conditions facing tree growth in the cramped city environment of Hong Kong. With a greater emphasis on and more resources diverted toward a greener city, it is all the more important to ascertain and achieve better site conditions for better quality trees. In the old city areas, the latitude for significant improvement in site conditions for trees can only be very limited. Comprehensive urban renewal may provide promises to earmark dedicated space for greenery. For new development areas, the need to reserve high-grade sites at the land-use planning stage cannot be more emphatically stressed. This study has highlighted the type and magnitude of pertinent obstacles in both subaerial and soil environments, and they can be used in a landscaping plan to specify the ingredients of individual planting sites. There is a need to guard against future intrusion of nonconforming land uses and impediments into the designated amenity space.

The next logical stage of research can focus on the possible ameliorative measures to bring relief to selected large and outstanding specimen trees (Jim, 1994b). The need to establish a detailed site and soil specification for use by the landscape profession, to be adhered to in future planting schemes, is necessary to upgrade the quality of the greening program. There should be an attempt to create elite sites to nurture elite trees which are so gravely lacking in the city. The ingrained attitude of neglecting the soil conditions for trees should be overhauled. It will be worthwhile to advocate the complete substitution of poor site soil with a properly-constituted mixture to enhance tree growth (Couenberg, 1994; Jason and Bassuk, 1995; Watson et al., 1996). The applicability of innovative site preparation procedures (Kopinga, 1985a and 1985b; Kuhns et al., 1985a and 1985b) could be field tested in the local context to further improve the conditions for tree survival in an inordinately difficult milieu for plants.

ACKNOWLEDGMENTS

I would like to convey my sincere gratitude to the research grant supports provided by the Croucher Foundation and the Hong Kong Urban Council. Assistance kindly furnished by Lawrence Cheung, L.C. Choi, Grace Kwong, W.H. Leung, Jeannette Liu, and Samson Yip is gratefully acknowledged. The comments given by the two anonymous referees in improving the manuscript are deeply appreciated.

REFERENCES

Baver, L.D., W.H. Gardner, and W.R. Gardner. *Soil Physics*, 4th ed. John Wiley & Sons, New York, 1972, p. 498.

Bernatzky, A. *Tree Ecology and Preservation*. Elsevier, Amsterdam, 1978, p. 357.

Bradshaw, A.D. The Quality of Topsoil. *Soil Use Manage.*, 5, pp. 101–108, 1989.

Bradshaw, A.D., B. Hunt, and T. Walmsley. *Trees in the Urban Landscape*. Spon, London, 1995, p. 272.

Bullock, P. and P.J. Gregory, Eds. *Soils in the Urban Environment*. Blackwell, Oxford, 1991, p. 174.

Cannell, R.Q. Soil Aeration and Compaction in Relation to Root Growth and Soil Management. *Appl. Biol.*, 2, pp. 1–86, 1977.

Census and Statistics Department *Hong Kong Annual Digest of Statistics 1995 Edition*. Hong Kong Government, Hong Kong, 1995, p. 307.

Cochran, P.H. and T. Brock. *Soil Compaction and Initial Height Growth of Planted Ponderosa Pine*. U.S. Department of Agriculture Forest Service, Pacific Northwest Forest and Range Experiment Station, Research Note PNW-434, Portland, Oregon, 1985, p. 4.

Couenberg, E.A.M. Amsterdam Tree Soil. In *The Landscape Below Ground*. International Society of Arboriculture, Savoy, IL, 1994, pp. 24–33.

Craul, P.J. *Urban Soils in Landscape Design*. John Wiley & Sons, New York, 1992, p. 396.

Gilman, E.F., L.A. Leone, and F.B. Flower. Effect of Soil Compaction and Oxygen Content on Vertical and Horizontal Root Distribution. *J. Environ. Horticulture*, 5(1), pp. 33–36, 1987.

Glinski, J. and J. Lipiec. *Soil Physical Conditions and Roots*. CRC Press, Boca Raton, FL, 1990, p. 250.

Greacen, E.L. and R. Sands. Compaction of Forest Soils—Review. *Australian J. Soil Res.*, 18(2), pp. 163–189, 1980.

Grey, G.W. and F.J. Deneke. *Urban Forestry*, 2nd ed. John Wiley & Sons, New York, 1986, p. 299.

Harris, R.W. *Arboriculture: Care of Trees, Shrubs, and Vines in the Landscape*, 2nd ed. Prentice-Hall, Englewood Cliffs, NJ, 1992, p. 674.

Hodgson, J.M., Ed. *Soil Survey Field Handbook: Describing and Sampling Soil Profiles*. Rothamsted Experimental Station, Soil Survey England and Wales, Harpenden, Herts, 1974.

Jason, G. and N. Bassuk. A New Urban Tree Soil to Safely Increase Rooting Volumes Under Sidewalks. *J. Arboriculture*, 21(4), pp. 187–201, 1995.

Jim, C.Y. *Urban Tree Survey 1985: Pavement Trees Managed by the Urban Council*. Urban Council, Hong Kong, 1986, p. 84.

Jim, C.Y. Tree Canopy Cover, Land Use and Planning Implications in Urban Hong Kong. *Geoforum*. 20(1), pp. 57–68, 1989.

Jim, C.Y. *Trees in Hong Kong: Species for Landscape Planting*. Hong Kong University Press, Hong Kong, 1990, p. 434.

Jim, C.Y. *Urban Tree Survey 1994 Roadside Trees Managed by the Urban Council*. Urban Council, Hong Kong, 1994a, p. 470.

Jim, C.Y. *Champion Trees in Urban Hong*. Urban Council, Hong Kong, 1994b, p. 294.

Jim, C.Y. Roadside Trees in Urban Hong Kong: Part I Census Methodology. *Arboricultural J.*, 20, pp. 221–237, 1996.

Klute, A., Ed. *Methods of Soil Analysis, Part 1: Physical and Mineralogical Methods*, 2nd ed. American Society of Agronomy, Madison, WI, 1986, p. 1188.

Kopinga, J. Site Preparation Practices in the Netherlands, in *METRIA 5: Selecting and Preparing Sites for Urban Trees*, Kuhns, L.J. and J.C. Patterson, Eds., Pennsylvania State University, University Park, PA, 1985a, pp. 62–71.

Kopinga, J. Research on Street Tree Planting Practices in the Netherlands, in *METRIA 5: Selecting and Preparing Sites for Urban Trees*, Kuhns, L.J. and J.C. Patterson, Eds., Pennsylvania State University, University Park, PA, 1985b, pp. 72–84.

Kuhns, L.J., P.W. Meyer, and J.C. Patterson. Creative Site Preparation, in *METRIA 5: Selecting and Preparing Sites for Urban Trees*, Kuhns, L.J. and J.C. Patterson, Eds., Pennsylvania State University, University Park, PA, 1985a, pp. 92–100.

Kuhns, L.J., P.W. Meyer, and J. Patterson. Creative Site Preparation. *Agora* (The Landscape Architecture Foundation Journal), 5(1), pp. 7–10, 1985b.

Landon, J.R., Ed. *Booker Tropical Soil Manual*. Longman, Burnt Mill, 1991, p. 474.

Lindsey, P. and N. Bassuk. Redesigning the Urban Forest from the Ground Below: A New Approach to Specifying Adequate Soil Volumes for Street Trees. *Arboricultural J.*, 16, pp. 25–39, 1992.

Miller, R.W. *Urban Forestry: Planning and Managing Urban Greenspaces*. Prentice Hall, Englewood Cliffs, NJ, 1988, p. 404.

Mullins, C.E. Physical Properties of Soils in Urban Areas, in *Soils in the Urban Environment*, Bullock P. and P.J. Gregory, Ed. Blackwell, Oxford, 1991, pp. 87–118.

Page, A.L., R.H. Miller, and D.R. Keeney, Eds. *Methods of Soil Analysis, Part 2: Chemical and Microbiological Properties*, 2nd ed. American Society of Agronomy, Madison, WI, 1982, p. 1159.

Patterson, J.C. Soil Compaction—Effects on Urban Vegetation. *J. Arboriculture*. 3, pp. 161–167, 1977.

Patterson, J.C. and D.L. Mader. Soil Compaction: Causes and Control, in *Urban Forest Soils: A Reference Workbook*, Craul, P.J., Ed. College of Environmental Science and Forestry, State University of New York, Syracuse, New York, 1982, pp. 3.1 to 3.15.

Perry, T.O. The Ecology of Tree Roots and the Practical Significance Thereof. *J. Arboriculture*, 8(8), pp. 197–211, 1982.

Proudfoot, D. Compacted Soil—The Silent Killer. *Tree Care Industry*, April 1992, pp. 22–24.

Raghavan, G.S.V., P. Alvo, and E. McKyes. Soil Compaction in Agriculture: A Review Toward Managing the Problem, in *Soil Degradation, Advances in Soil Science*, Vol. 11, Lal, R. and B.A. Stewart, Eds. Springer-Verlag, New York, 1990, pp. 1–36.

Rawls, W.J., T.J. Gish, and D.L. Brakensiek. Estimating Soil Water Retention from Soil Physical Properties and Characteristics, in *Advances in Soil Science*, Vol. 16, Stewart, B.A., Ed. Springer-Verlag, New York, 1991, pp. 213–234.

Strategic Planning Unit. *Metroplan: The Foundations and Framework*. Planning, Environment and Lands Branch, Hong Kong Government, Hong Kong, 1990, p. 51.

Thornton, I. Metal Contamination of Soils in Urban Areas, in *Soils in the Urban Environment*, Bullock, P. and P.J. Gregory, Eds. Blackwell, Oxford, 1991, pp. 47–75.

Tregear, T.R. and L. Berry. *The Development of Hong Kong and Kowloon as Told in Maps*. Hong Kong University Press, Hong Kong, 1959, p. 31.

Ware, G.H. Constraints to Tree Growth Imposed by Urban Soil Alkalinity. *J. Arboriculture*, 16(2), pp. 35–38, 1990.

Watson, G.W. and D. Neely, Eds. *The Landscape Below Ground*. International Society of Arboriculture, Savoy, IL, 1994, p. 222.

Watson, G.W., P. Kelsey, and K. Woodtli. Replacing Soil in the Root Zone of Mature Trees for Better Growth. *J. Arboriculture*, 22(4), pp. 167–173, 1996.

Zisa, R.P., H.G. Halverson, and B.B. Stout. *Establishment and Early Growth of Conifers on Compact Soils in Urban Areas*. U.S. Department of Agriculture Forest Service, Northeastern Experimental Station Research Paper NE-451, Bromall, PA, 1980, p. 8.

Chapter 22

Field Evaluation of Tree Species for Afforestation of Barren Hill Slopes in Hong Kong

K.W. Cheung

INTRODUCTION

The climax vegetation cover of Hong Kong, South China (latitude of 22°N), should be a subtropical evergreen broad-leaved forest (Hou, 1994). Due to centuries of settlement, most of the original forests have disappeared. As a result of frequent hill fires and poor soil conditions, the most common present vegetation cover is grassland. In some areas, serious soil erosion occurs and vegetation cover is lacking. Reestablishment of trees on these barren hillsides by natural succession is rather slow and requires favorable conditions such as protection from fires and shielding from strong winds. In heavily eroded areas, regeneration through natural processes is impossible. Afforestation is one of the limited means to rehabilitate these barren areas.

Extensive planting of trees on the hillsides of Hong Kong commenced in the 1870s. The main species used was the native pine, *Pinus massoniana*. However, systematic afforestation in the territory was only begun after the appointment of the first trained forestry officer in 1937. The 70 years of forestry work was destroyed during the second World War and nearly all forests had been felled by 1945 (Agriculture and Fisheries Department, undated).

To replace the forests lost and to conserve the water catchment areas, large-scale afforestation works commenced in 1949 by the then Agriculture and Fisheries Department (AFD). *P. massoniana* was still the major species planted. Since then, however, more species, especially broad-leaved species, have been used. These included some fast-growing introduced species such as *Acacia confusa*, *Casuarina* species, *Eucalyptus* species, *Melaleuca leucadendron*, and *Tristania conferta*. Native species such as *Castanopsis fissa*, *Liquidamber formosana*, and *Schima superba* were also planted.

With the increase in diversity of the afforestation species, field trials were continuously carried out to test the suitability of different species for planting in the territory. However, most of the results were translated into routine practice but were not published. To obtain some quantitative data of the afforestation species, small-scale trial plots were established in the 1980s.

The primary objective of these small-scale trial plots was to evaluate the growth performance of different species used in afforestation. Commonly used species as well as less well-known species were tested. These included some rare species such as *Ailanthus fordii*, *Camellia granthamiana* (IUCN, 1978), and *Keteleeria fortunei* (Fu and Jin, 1992) and newly introduced species like *Acacia auriculiformis*.

MATERIALS AND METHODS

Some of the successful trial plots set up in various country parks since 1983 are shown in Table 22.1. One-year-old seedlings were used and the numbers of trees planted in each trial plot ranged from 50 to 500. The tree seedlings were normally planted in a rectangular plot. Since 1989, two to three samples of the surface 20 to 30 cm of soil were collected randomly at each trial plot and sent to the Soil Laboratory, Tai Lung Experimental Station of the AFD for analysis. The extraction methods basically followed that of Metson (1956) for pH, Jackson (1962) for organic matter and available phosphorus, and Association of Official Analytical Chemists (1970) for total nitrogen (Kjeldahl nitrogen) and exchangeable potassium.

One year after planting, weeding and addition of fertilizer (N:P:K = 15:15:15) were carried out. Fertilizer was applied at a rate of 100 g per seedling for *Acacia confusa* and 50 g per seedling for other species (AFD, 1994). Normally, tending of these plots continued for two more years. Subsequent management of some of the plots such as Plots 1/83, 5/83, and 7/93 was continued by the nearby Country Park Management Centres. Tree height was measured annually. Basal diameter (for young trees) or diameter at breast height (for older trees) was also measured after 1990. The general condition of the plants was also noted.

Besides evaluating growth performance, some trial plots also had other objectives. For example, Plot 3/93 was established to test the effect of care/maintenance works on the tree seedlings' survival and growth. The seedlings were divided into two groups. One group received care including weeding, addition of fertilizers, pruning, and watering in prolonged dry weather, while the other group was left unattended.

RESULTS AND DISCUSSION

The general soil conditions of the hill slopes in the territory are rather poor. Topsoil is normally very thin or absent. The results of the soil analysis in Table 22.2 indicate the poor soil conditions present. All the soil samples were strongly acidic, with average pH values lower than 5. The soil texture is basically sandy loam to sandy clay loam with a high sand content. Levels of organic matter and macronutrients, viz., nitrogen, phosphorus, and potassium, were generally poor, especially phosphorus (less than 0.1% Kjeldahl nitrogen, 70 mg kg^{-1} available P_2O_5 and 140 mg kg^{-1} exchangeable K_2O). Levels of micronutrients such as copper, iron, manganese, and zinc (results not shown) were also deficient. Therefore, these soils provided a hostile environment for tree establishment and growth.

Comparing the soil characteristics of various periods after afforestation, Plots 5/83, 43/85, 3/93, and 1/94 afforestation did not change soil acidity and phosphorous level significantly. However, there were significant increases in organic matter, total nitrogen, and exchangeable potassium, except Plot 3/93, which did not exhibit the increase in organic matter and total nitrogen. These results thus indicated that afforestation can improve some soil conditions. The addition of fertilizer may have some contribution to the improvements for the younger Plot 1/94. However, this was unlikely for the older Plots 5/83 and 43/85 as the soil analyses were carried out quite some time after the applications of fertilizer.

Results in Tables 22.3 and 22.4 showed two general trends: (1) introduced species had faster growth rates than native species in terms of increase in tree height and (2) growth rates for native species were higher in older plots. The fastest growth rate of 71.1 cm yr^{-1}

Table 22.1. AFD Small-Scale Trial Plots in the Country Parks (CP) of Hong Kong

Plot No.	Month/Year Established	Location	Species	Altitude (m)	Aspect	Planting Spacing (m)	Planting Method	Site Conditions Before Planting
1/83	4/83	Hok Tau (Pat Sin Leng CP)	Cunninghamia lanceolata*	140	W	1.8 x 1.8	pit	grassland affected by hill fires
2/83	3/83	Shing Mun CP	Ailanthus fordii*	200	SW	1.8 x 1.8	notch	grassland with patches of shrubs
3/83	4/83	Shing Mun CP	Keteleeria fortunei*	340	ENE	2 x 2	pit	exposed grassland
4/83	4/83	Shing Mun CP	Keteleeria fortunei*	300	SSE	2 x 2	pit	sheltered grassland
5/83	4/83	Yuen Tun (Tai Lam CP)	Castanopsis fissa* Schima superba*	220	SE NW	1.2 x 1.2	notch	newly burnt scrubland
1/85	5/85	Tai Tong (Tai Lam CP)	Acacia confusa, A. auriculiformis, A. mangium, Casuarina littoralis, Ficus retusa*, Keteleeria fortunei*, Pinus elliotii and Tristania conferta	100	NW	1 x 1	pit	heavily eroded, no vegetation cover
43/85	10/85	Sham Tseng (Tai Lam CP)	Casuarina cunninghamia and C. littoralis	dna	dna	1.5 x 1.5	dna	grassland with patches of shrubs
3/93	7/93	Tai Tam Sapling Nursery (Tai Tam CP)	Aquilaria sinensis*, Camellia granthamiana* and Endospermum chinense*	260	dna	1 x 1	pit	bare ground of the Tai Tam sapling nursery
1/94	4/94	Lin Fa Shan (Tai Lam CP)	Acacia confusa, A. auriculiformis, Adenanthera pavonina*, Cinnamomum camphora*, Quercus edithae* and Q. myrsinaefolia*	500	E	1.5 x 1.5	notch	exposed grassland
2/94	6/94	Wong Lai Tun (Tai Lam CP)	Acacia confusa, A. auriculiformis, Adenanthera pavonina*, Cinnamomum camphora*, Quercus edithae* and Q. myrsinaefolia*	220	SW	1.5 x 1.5	notch	heavily eroded, no vegetation cover

* Native species; dna = data not available.

Table 22.2. Results of Soil Analyses of Some Trial Plots.

Plot No.	Species	Sampling Date	pH	% Organic Matter	% Total Nitrogen	Available P (mg 100g⁻¹)	Exchangeable K (mg 100g⁻¹)	% Sand	% Silt	% Clay
2/83	Ailanthus fordii	3/89	5.04	2.1	0.1	< 0.5	6.6	75	7	18
3/83	Keteleeria fortunei	3/89	4.69	9.0	0.17	< 0.5	11.6	69	16	22
4/83	Keteleeria fortunei	3/89	4.85	4.6	0.12	< 0.5	10.9	65	17	21
5/83	Schima superba and Castanopsis fissa	9/86	4.40	1.4	0.06	trace	3.5	80	4	16
		3/96	4.25	1.9	0.14	< 0.5	23.0	72	7	21
43/85	Casuarina cunninghamia and C. littoralis	3/89	4.49	1.0	0.01	< 0.5	4.8	dna	dna	dna
		3/96	4.04	1.8	0.17	< 0.5	20.3	70	9	21
3/93	Aquilaria sinensis,	6/93	4.4	2.5	0.13	< 2.5	6.8	71	5	24
	Endospermum chinense and Camellia granthamiana	2/96	4.2	1.4	0.12	< 0.5	19.7	75	6	18
1/94	Acacia, Adenanthera, Cinnamomum and Quercus spp.	6/94	4.38	0.76	0.05	1.8	1.6	68	4	28
		3/96	4.29	3.20	0.28	<0.5	7.7	69	10	21
2/94	Acacia, Adenanthera, Cinnamomum and Quercus spp.	6/94	4.6	5.52	0.2	< 0.5	9.1	75	11	14

dna = data not available.

Table 22.3. Results of the AFD Small-Scale Trial Plots

Species	Plot No.	Age of Plant (yr)	No. of Plants# (Survival Rate)	Average Growth Rate (cm yr^{-1}) Height	Basal Diameter	Diameter at Breast Height
Acacia auriculiformis	1/85	10	225 (45%)	56.7	dna	0.5
	1/94	3	170 (85%)	51.4	0.8	dna
	2/94	3	100 (100%)	51.8	1.1	dna
Acacia confusa	1/85	10	32 (6%)	26.1	dna	0.2
	1/94	3	183 (92%)	42.1	0.8	dna
	2/94	3	91 (91%)	30.1	0.7	dna
Acacia mangium	1/85	10	90 (18%)	47.8	dna	0.8
Adenanthera pavonina*	1/94	3	15 (15%)	10.0	dna	dna
	2/94	3	92 (92%)	15.1	0.4	dna
Ailanthus fordii*	2/83	13	57 (48%)	34.6	dna	0.4
Castanopsis fissa*	5/83	14	195 (98%)	41.1	dna	0.5
Casuarina cunninghamia	43/85	11	45 (dna)	41.8	dna	1.0
Casuarina littoralis	1/85	10	242 (48%)	71.1	dna	0.9
	43/85	11	28 (dna)	27.6	dna	0.4
Cinnamomum camphora*	1/94	3	44 (44%)	20.1	dna	dna
	2/94	3	82 (82%)	17.9	0.4	dna
Cunninghamia lanceolata*	1/83	12	200 (dna)	54.7	dna	1.0
Ficus retusa*	1/85	10	9 (30%)	6.1	dna	dna
Keteleeria fortunei*	3/83	13	268 (54%)	24.5	dna	0.6
	4/83	13	246 (49%)	45.3	dna	0.7
	1/85	10	285 (57%)	21.4	dna	0.3
Pinus elliotti	1/85	10	173 (35%)	44.0	dna	0.5
Quercus edithae*	1/94	3	62 (62%)	20.0	0.3	dna
	2/94	3	59 (59%)	18.9	0.4	dna
Quercus myrsinaefolia*	1/94	3	52 (52%)	13.5	0.2	dna
	2/94	3	89 (89%)	10.7	0.3	dna
Schima superba*	5/83	14	195 (58%)	28.0	dna	0.5
Tristania conferta	1/85	10	251 (50%)	50.5	dna	0.4

* Native species, # number of surviving plants at the last measurement.
dna = data not available.

Table 22.4. Growth Rates of Native Species of Trees on Trial Plot No. 3/93.

Species	Tending[a]	Age (yr)	Average Height (cm)	Average Growth Rate (cm yr^{-1})	Number of Plants[b]
Aquilaria	with	4.5	113.6	25.2	20
sinensis	without	4.5	98.7	21.9	21
Camellia	with	4.5	80.8	18.0	24
granthamiana	without	4.5	62.6	13.9	23
Endospermum	with	4[c]	154.8	38.7	21
chinense	without	4.5	89.1	19.8	18

[a] Tending involved weeding, addition of fertilizers, pruning, and watering during prolonged dry weather.
[b] Number of surviving plants at the last measurement.
[c] The *Endospermum* saplings were transplanted out of the sapling nursery at year 4.

was observed in *Casuarina littoralis* in Plot 1/85. However, it grew much slower in Plot 43/85. *Acacia auriculiformis* had the most consistent growth rate, with growth rates averaging over 50 cm yr^{-1}. Some native species also showed comparable growth rates of over 40 cm yr^{-1} but only in older trees. These included *Castanopsis fissa*, *Cunninghamia lanceolata*, and *Keteleeria fortunei*. Satisfactory growth rates of over 30 cm yr^{-1} were observed in *Ailanthus fordii* and *Endospermum chinense*.

Introduced species also showed higher increase in basal diameter in young trees. However, there was no such difference among species in older plots in terms of increase in diameter at breast height. Higher rates were observed in *Acacia mangium*, *Casuarina* species, and *Cunninghamia lanceolata*. Other species had similar low rates.

As shown in Figures 22.1 to 22.4, the five native species all had a slower initial growth rate (increase in height) for the first 5 to 10 years. However, growth accelerated after this slow growth period. Similar growth patterns were observed in *Cunninghamia lanceolata*, *Cinnamomum camphora*, and *Quercus* species by Tian (1984). This pattern explained the generally low growth rates of less than 20 cm yr^{-1} observed for native species in younger trees. For introduced species, no such slow-growing initial stage was observed. Rather, steady growth rates were recorded in both young and old plantings.

The growth rates for introduced species, though faster than natives, were lower than those observed in south China. This is also true for native species. Growth rates of some of the species reported in south China are shown in Table 22.5. These high grow rates, however, were usually measured under favorable conditions such as suitable climate, good soil, and good management of the plantations. Poor soil conditions would be one of the major reasons for the generally lower growth rates found in the present trials. Other factors such as climate, altitude, etc., might also be important. For example, Huang and Shen (1993) reported that strong illumination, high temperature, and dry winter have been shown to be responsible for poor growth of *Cunninghamia lanceolata* in southern areas of Guangdong Province, where Hong Kong is located. The results of Plots 3/83 and 4/83 showed that *Keteleeria fortunei* performed better in more sheltered conditions (Figure 22.3).

The results of Plot 3/93 as shown in Table 22.5 indicate the importance of care/maintenance. Plants receiving attention had higher growth rates in terms of increase in tree height. The effect was more prominent for *Endospermum*, which had doubled growth rates for plants receiving care. The success of tree seedlings to establish in a site depends on, among other things, the development of a healthy root system. The establishment of the

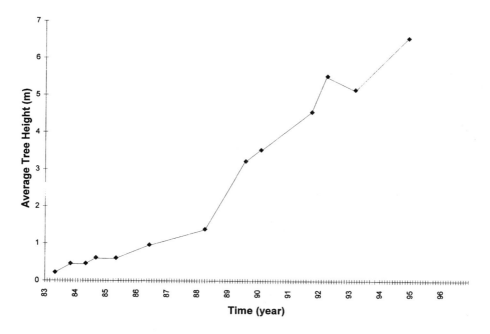

FIGURE 22.1. Growth curve of *Cunninghamia lanceolata* in Plot 1/83.

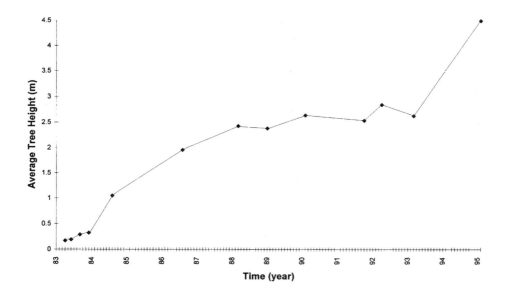

FIGURE 22.2. Growth curve of *Ailanthus fordii* in Plot 2/83.

root system needs to take place as early as possible after planting. Suitable maintenance of the seedlings during the first one or two years will also be crucial, especially for species demanding good conditions such as *Endospermum*. Failing to give good care will result in weaker plants or slower growth rates. General observations of larger-scale plantations of

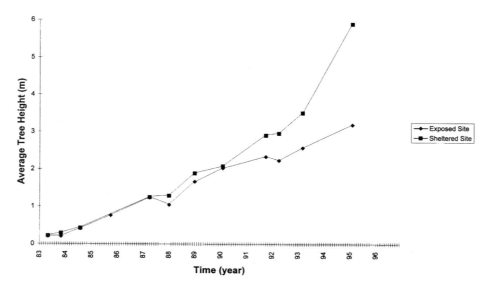

FIGURE 22.3. Growth curves of *Keteleeria fortunei* at different site conditions in Plots 3/83 and 4/84.

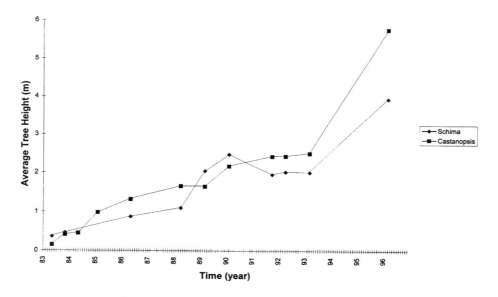

FIGURE 22.4. Growth curves of *Castanopsis fissa* and *Schima superba* in Plot 5/83.

similar species in the country parks also support the above finding. Properly managed plantations usually have better performance in terms of tree height, girth, and form.

Regarding the formation of canopy, general observations of the plantations found that introduced species formed a closed canopy faster than native species. For *Acacia* species, a preliminary canopy could be formed within the first five years with few understorey plants. Hence, these species are suitable for erosion control and suppression of grasses. The latter is rather important in controlling hill fires in the countryside by reducing fuel accumulation.

Table 22.5. Growth Rates of Some Tree Species in South China.

Species	Age (year)	Average Increase in Height (cm yr^{-1})	Average Increase in Diameter[a] (cm yr^{-1})	Source
Acacia auriculiformis	6[c]	66.2	0.86[b]	Hu et al.,
	6[d]	80.1	0.88[b]	1994
	5.3	157	1.5	Pon and You,
	10	150	2.0	1994
Acacia confusa	6[p]	29.6	0.62[b]	Hu et al.,
	6[m]	33.3	0.58[b]	1994
	5.3	80	0.67	Pon and You, 1994
Acacia mangium	5	216	4.7	Pon and You,
	5.3	170	2.0	1994
	10	148	2.8	
Casuarina species	dna	100	1.5	Pon and You, 1994
Cinnamomum camphora	5	100	2.4	Tian, 1984
Cunninghamia lanceolata	18	80	1.1	Huang and Shen, 1993
Pinus elliotti	4	83	2.7[b]	Kong, 1994
Schima superba	4	81	1.7[b]	Kong, 1994

[a] diameter at breast height unless stated otherwise.
[b] basal diameter.
[c] pure stand.
[d] mixed stand.
dna = data not available.

More open canopies were seen in plots of native species. This led to more understorey growth which competed with the trees. Competition with understorey plants, especially tall grasses, can hinder the growth of tree seedlings and this might contribute to the lower initial growth rates of the native species. On the other hand, however, the diversity of the site increased from recruitment of some native trees and shrubs. Provided that the growth of trees was not seriously affected, this might not be undesirable from a biological diversity standpoint.

CONCLUSION

The trial plot results indicate that both introduced and native species could be used for afforestation in Hong Kong successfully. Introduced species usually have faster initial growth rates and are suitable for planting at sites requiring fast establishment of vegetation cover. They are more suitable for erosion control and planting in sites with poor conditions. Local species tend to have slower initial growth rates and are more suitable for enrichment of existing vegetation cover. However, longer time, say over 10 years, will be needed to establish a preliminary woodland. Moreover, suitable maintenance of the plants will be necessary, especially during the first two years after planting.

ACKNOWLEDGMENT

I would like to express my sincere gratitude to the present and past staff of the Field Investigation Unit (formerly the Field Experimental Unit) of the AFD who carried out the field works. I am also indebted to Mr. C.C. Lay and Mr. Anthony Leung, the former Ecologists of the Department, who planned most of the trial plots. I would also like to thank Mr. Patrick Lai, the present Ecologist, who allowed me to use the data from the trial plots. Special thanks are given to Mr. W.M. Lam and the staff of the Tai Tam Sapling Nursery for the planning and management of Plot 3/93, and Dr. M.L. Kwong, Mr. K.N. Tam, and staff of the Tai Lung Experimental Station for carrying out the soil analyses and providing useful information.

REFERENCES

Agriculture and Fisheries Department. (Undated). *History of Forestry in Hong Kong.* Information sheet published by the Department.

Agriculture and Fisheries Department. *Country Parks Division Technical Circular No. 1/94.* Agriculture and Fisheries Department, Hong Kong, 1994.

Association of Official Analytical Chemist. *Official Methods of Analysis of the Association of Official Analytical Chemist,* 11th ed. Association of Official Analytical Chemist, Washington, DC, 1970, p. 1015.

Fu, L. and J. Jin. *China Plant Red Data Book—Rare and Endangered Plants.* Vol. 1. Science Press, Beijing, 1992, p. 741.

Hou, H. Vegetation of China with Reference to Its Geographical Distribution, in *Researches on Vegetation Ecology—A Commemoration for Famous Ecologist Professor Hou Xueyu,* Jiang, S. and C. Chen, Eds., Science Press, Beijing, 1994, pp. 53–76.

Hu, S., X. Li, T. Lue, and K. Chen. Cultivation Experiment of Fast Growing Fuel Forests in Terms of Ecology, in *Researches on Vegetation Ecology—A Commemoration for Famous Ecologist Professor Hou Xueyu,* Jiang, S. and C. Chen, Eds., Science Press, Beijing, 1994, pp. 352–363.

Huang, M. and G. Shen. *Silvicultural Technology of China.* Chinese Forestry Press, Beijing, 1993, p. 635.

IUCN. *The IUCN Plant Red Data Book.* IUCN, Morges, 1978, p. 540.

Jackson, M.L. *Soil Chemical Analysis.* Constable & Co. Ltd., London, 1962, p. 497.

Kong, D. A Preliminary Study on Restoration of the Degeneration Vegetation in Subtropical Red Soil Hilly Area in Jianxi Province, in *Researches on Vegetation Ecology—A Commemoration for Famous Ecologist Professor Hou Xueyu,* Jiang, S. and C. Chen, Eds., Science Press, Beijing, 1994, pp. 375–387.

Metson, A.J. *Method of Chemical Analysis for Soil Survey Samples.* New Zealand Department of Scientific and Industrial Research, Wellington, 1956, p. 207.

Pon, Z. and Y. You. *Growing Exotic Trees in China.* Beijing Science and Technology Press, Beijing, 1994, p. 758.

Tian, Y., Ed. *Afforestation,* Vol. 2. China Forestry Press, Beijing, 1984, p. 354.

Chapter 23

Relationship Between Biofuel Harvesting and Upland Degradation in Hong Kong and Guangdong

R.J. Chen, R.T. Corlett, and R.D. Hill

INTRODUCTION

More than half the world's population depends primarily on biomass fuel (biofuel) for domestic purposes (United Nations, 1988). As long as there is no degradation of the vegetation, the biofuel system makes no net contribution to change in atmospheric CO_2 and little damage to the upland environment. As a result, ecologists are increasingly interested in successful experiences in biofuel harvesting as a sustainable development and with their impacts on the upland environment, although there have been very few studies in this field.

In Hong Kong, villagers depended mainly on biofuel for cooking until the 1950s (Hayes, 1983), but only a few still used fuelwood after 1980.

China had a rural population of 894 million in 1990 (PRC Ministry of Agriculture, 1991). Rural energy consumption makes up 38% of its total, and biomass which is harvested mainly from uplands is estimated to be 90% of the rural energy consumption (Xie, 1989). Like many other provinces of China, Guangdong suffers from a scarcity of biofuel. This scarcity is decreasing wildlife and accelerating land degradation as a result of overharvesting of vegetation (Xie, 1989). The degradation of uplands starts with the degeneration of vegetation, continues with the loss of biota, and ends with soil erosion. China has a vast area of degraded uplands. Its successes and failures in the improvement and sustainable utilization of uplands have influences on the uplands of the world. Alternative fossil fuels are not available or are too expensive for most villagers in Guangdong. Villagers cannot cook food without biofuels, so it is impractical that uplands are entirely closed for conservation. Thus, it is important to develop a sustainable system of biofuel production which guarantees both that they have fuel enough to cook and that upland resources regenerate well. In South China, the regressive succession of vegetation results mainly from overharvesting. However, the fact that a decrease in soil nutrients, or even soil erosion also results from heavy harvesting has not been proven by published work in South China. The importance of biofuels, and the serious consequences of their shortage and the role of heavy biomass harvesting in upland degradation, make it necessary to study the change of nutrients in uplands under harvest.

This study was initiated to compare the differences in soil nutrient contents among Hong Kong, Shenzhen, and Heshan uplands, and to observe the effects of three harvesting patterns on soil nutrients at the three study sites.

METHODS

Study Sites

The Hong Kong study site was an upland grassland (22°27'26"N, 114°05'35"E) with many shrubs in the Lam Tsuen Country Park. This site has been protected from harvesting since the 1950s. The soils on this study site were Acrorthoxes forming on black and coarse volcanic rocks of middle and lower Jurassic age (160–195 million years) (Geotechnical Control Office, 1989), weakly acidic (pH 5.0), about 27 m deep, and clay. The grassland at this site carried 17 herbaceous species and 15 woody species, and was dominated by the grass species, *Ischaemum* spp., *Arundinella setosa*, and *Cymbopogon tortilis*; the fern, *Dicranopteris linearis*; and the shrub species, *Baeckea frutescens* and *Helicteres angustifolia*.

The Shenzhen study site was located on uplands (22°35'50"N, 114°28'30"E) of Dapeng, Shenzhen. This site was established on an uncut hillside and another often-cut hillside. The often-cut hillside, only 0.5 km away from the rural households, is harvested for domestic biomass fuel once a year; and the uncut one, 2 km away from the rural households, has been free from harvesting since 1979. The two hillsides originally were quite similar in vegetation and soil, according to the general information provided by the Dapeng Office of Agriculture. The soils at this site were Orthoxes occurring on granitic rocks, acidic (pH 4.9), 26 m deep, and clay. This site carried 16 herbaceous species and 17 woody species, and was dominated by the grass species, *Ischaemum* spp., *Arundinella setosa*, and *Cymbopogon tortilis*; the fern, *Dicranopteris linearis*; and the shrub species, *Rhodomyrtus tomentosa* and *Rhaphiolepis indica*.

The Heshan study site was situated on uplands (22°43'35"N, 112°55'00"E) of Taoyuan, Heshan. This site was also established on an uncut hillside inside the Heshan Institute of Forestry and another often-cut hillside nearby. The often-cut hillside is harvested for domestic biomass fuel twice a year, and the uncut one, which is fenced in with barbed wire, has been protected from harvesting since 1975. The two hillsides were similar in vegetation and soil before 1975, according to the general information provided by the Institute and the Heshan Bureau of Forestry. The soils at this site were Orthoxes formed on sedimentary rocks, acidic (pH 4.8), about 24 m deep, and clay. This site carried 19 herbaceous species and 16 woody species, and was dominated by *Ischaemum* spp., *Dicranopteris linearis*, *Eriachne pallescens*, *Rhodomyrtus tomentosa*, *Baeckea frutescens*, and *Melastoma dodecandrum*.

Although just within the geographical boundaries of the tropics, the study areas experience a strongly seasonal climate. More than 82% of the total annual rainfall (1994 mm) falls between April and September, which is also the hot season (unpublished data provided by the Hong Kong Royal Observatory, Heshan and Shenzhen Bureaux of Meteorology). Winter is dry and cool, with short-lived cold surges from the north bringing extreme temperature minima below 5°C. The mean air temperatures vary from 14.3°C in January to 28.5°C in July with an annual mean of 22.2°C.

Soil Sampling

Soil samples were collected from June to August 1990 on the hillside of the Hong Kong study site, and the uncut and often-cut hillsides of the Shenzhen and Heshan study sites. Five sampling locations were randomly selected on each hillside as five replicates. Since very few roots were found below 40 cm, two soil samples were drawn with an auger from the upper soil layer (0–20 cm) and lower soil layer (20–40 cm) at each sampling location.

More than 200 g of each sample was transferred to the laboratory for the determination of total and available nutrients.

Soil Sample Analysis

Soil samples were air-dried in the laboratory and reduced to roughly to 100 g by the "cone-and-quarter" method. The air-dried samples were rolled with a wooden rolling pin. One part was passed through a 1-mm sieve for the determination of available nutrients, while the other was further ground in an agate mortar and passed through a 0.15-mm sieve for the determination of total nutrients. Three grams of the ground soil sample was dried in a forced air oven at 105–110°C to constant weight. This oven-dried weight was used as the dry basis for soil analysis (Page et al., 1982).

The soil samples were analyzed for N, P, and K contents by the standard methods (Page et al., 1982) using a Kjeltec system 1026 distilling unit (Tecator AB, 1989), a UV-VIS recording spectrophotometer (Simadzu Corporation, 1990) and an AA/AE spectrophotometer (Allied Corporation, 1985). The samples were digested for total N and P determination by 93–98% H_2SO_4 and 60% $HClO_4$, whereas they were digested for total K determination by 48% HF and 60% $HClO_4$. Available N, P, and K in soil were extracted by 10 N NaOH, 0.025 N HCl-0.03 N NH_4F, and 1 N NH_4OAc, respectively.

Analysis of Data

Analysis of variance (ANOVA) was performed in the Minitab software package (Minitab Inc., 1989). The multiple comparisons by the Tukey method were performed using relevant procedures contained in the SAS/STAT software package (SAS Institute Inc., 1988).

RESULTS

Decrease in Soil N Content

The total N contents at the three study sites ranged from 280 mg kg^{-1} to 690 mg kg^{-1} (Table 23.1). The upper soil layer (0–20 cm) on the often-cut hillside at the Heshan study site had the lowest total N content, while the upper soil layer on the uncut hillside at the Hong Kong study site had the highest total N content. ANOVA showed that there was a significant difference in total N content among sampling locations (df = 9,40; F = 36.62; p < 0.001). The multiple comparison test by the Tukey method indicated that, at the Heshan study site, the uncut hillside had a significantly higher total N content of soil than the often-cut hillside. The same was true of the Shenzhen study site. There was no significant difference in total N content of the upper soil layer on the uncut hillsides among the three sites, although the total N contents were in the order of Hong Kong study site > Shenzhen study site > Heshan study site. The lower soil layer (20–40 cm) on the uncut hillsides showed the same trends in total N content as the upper soil layer. The often-cut hillside at the Shenzhen study site did not have a significantly higher total N content than the often-cut one at the Heshan study site. The uplands in Shenzhen are harvested for domestic biomass fuel once a year, while those in Heshan are harvested twice a year. The insignificant difference between the Shenzhen and Heshan often-cut uplands is probably due to the insufficient time for the existing harvesting frequencies.

The available N contents varied between 41.5–97.5 mg kg^{-1}. ANOVA indicated that there was a significant difference in available N content among sampling locations (df = 9,40; F = 74.36; p < 0.001). The results of Tukey's multiple comparison test on available N contents were different from those on total N contents. At the Heshan study site, the upper soil layer gave a significantly higher available N content on the uncut hillside than on the often-cut hillside, and the lower soil layer did not. The same was true of the Shenzhen study site. Therefore, biomass harvesting affects the upper soil layer first. The uncut hillsides gave the highest available N content at the Hong Kong study site, amounting to 97.5 mg kg^{-1} for the upper soil layer and 70.3 mg kg^{-1}. That may be owing to the better protection of uplands in Hong Kong than in Guangdong.

Decrease in Soil P Content

The total P contents ranged between 80 mg kg^{-1} and 400 mg kg^{-1} (Table 23.1). There was a significant difference in total P content of soil among sampling locations (ANOVA: df = 9,40; F = 234.81; p < 0.001). At the Heshan study site, the upper soil layer had a significantly higher total P content on the uncut hillside than on the often-cut hillside, but in the lower soil layer, the uncut hillside had the same total P content as the often-cut hillside. The Shenzhen study site showed the same trends. Therefore, harvesting has not yet influenced total P content in the lower soil layer. The lower soil layers gave a significantly higher total P content than the upper soil layers on the often-cut hillsides, and they did not on the uncut hillsides. It is probably because phosphorus comes mainly from soil parent-material. The Shenzhen and Heshan uncut hillsides had a similar total P content of soil, and either of them had a significantly higher total P content of soil than the Hong Kong uncut hillside. On the often-cut hillsides, the upper soil layer gave a significantly higher total P content at the Shenzhen study site than at the Heshan study site, but the two lower soil layers gave a similar total P content.

The available P contents differed between 0.9 mg kg^{-1} (the upper soil layer of the Hong Kong uncut hillside) and 3.1 mg kg^{-1} (the lower soil layer of the Shenzhen uncut hillside). There was a significant difference in available P content among sampling locations (ANOVA: df = 9,40; F = 61.32; p < 0.001). At the Heshan study site, the uncut hillside gave a significantly higher available P content of soil than the often-cut hillside. The Shenzhen study site showed the same trends as the Heshan study site. There was no significant difference in available P content of soil between the Shenzhen and Heshan uncut hillsides, but both of them had a significantly higher available P content of soil than the Hong Kong uncut hillside. The Shenzhen often-cut hillside gave a higher available P content of soil than the Heshan often-cut hillside, although it was not significant.

Decrease in Soil K Content

The upper soil layer on the Heshan often-cut hillside gave the lowest total K content, 20.84 mg g^{-1}; and the lower soil layer on the Shenzhen uncut hillside gave the highest total K content, 29.03 mg g^{-1} (Table 23.1). The results of ANOVA showed a significant difference in total K content of soil among sampling locations (df = 9,40; F = 15.71; p < 0.001). At the Heshan and Shenzhen study sites, the uncut hillsides gave a significantly higher total K content of soil than the often-cut hillsides. There was no significant difference in total K content of the upper soil layer among the three uncut hillsides. The

Table 23.1. Nutrient Content of the Soils at the Hong Kong, Shenzhen, and Heshan Study Sites (means of five replicates ± standard error). Uncut = uncut hill, Often-cut = often-cut hill, upper = 0–20 cm layer of soil, lower = 20–40 cm layer of soil. The superscript letters from A to F indicate significance of difference in total N content according to Tukey's multiple comparisons (a = 0.05), and those from G to K, from L to O, from P to S, from T to W and from X to Z are for available N, total P, available P, total K and available K respectively. Means with the same superscript letter are not significantly different.

Sampling Location	N Mean	SE	P Mean	SE	K Mean	SE
Total nutrient (mg kg^{-1})						
Hong Kong study site						
Uncut upper	^{A}0.69	0.04	^{O}0.08	0.01	^{TUV}25.87	1.41
Uncut lower	^{BC}0.57	0.04	^{O}0.11	0.01	^{TU}26.81	1.46
Shenzhen study site						
Uncut upper	^{AB}0.61	0.02	^{L}0.40	0.01	^{TU}27.05	0.29
Uncut lower	^{BCD}0.52	0.01	^{L}0.40	0.01	^{T}29.03	0.31
Often-cut upper	^{EF}0.35	0.02	^{M}0.25	0.01	^{W}21.40	0.27
Often-cut lower	^{DF}0.43	0.01	^{L}0.39	0.01	^{VW}23.28	0.36
Heshan study site						
Uncut upper	^{ABC}0.59	0.01	^{L}0.38	0.01	^{UV}25.18	0.30
Uncut lower	^{CD}0.50	0.02	^{L}0.39	0.01	^{UV}25.38	0.38
Often-cut upper	^{F}0.28	0.01	^{N}0.18	0.01	^{W}20.84	0.18
Often-cut lower	^{EF}0.34	0.02	^{L}0.39	0.01	^{W}21.64	0.17
Available nutrient (x10^{-2} mg kg^{-1})						
Hong Kong study site						
Uncut upper	^{G}9.57	0.25	^{S}0.09	0.01	^{Y}3.41	0.29
Uncut lower	^{H}7.03	0.38	^{S}0.11	0.01	^{Y}3.18	0.26
Shenzhen study site						
Uncut upper	^{H}6.46	0.13	^{PQ}0.28	0.01	^{X}5.59	0.04
Uncut lower	^{I}5.13	0.07	^{Q}0.31	0.01	^{X}5.32	0.10
Often-cut upper	^{JK}4.16	0.06	^{R}0.19	0.00	^{Y}3.09	0.08
Often-cut lower	^{IJ}5.08	0.06	^{R}0.19	0.02	^{Y}3.12	0.05
Heshan study site						
Uncut upper	^{H}6.78	0.21	^{Q}0.25	0.01	^{Y}3.48	0.07
Uncut lower	^{I}5.32	0.14	^{Q}0.26	0.01	^{Y}3.65	0.04
Often-cut upper	^{K}4.15	0.14	^{R}0.17	0.01	^{Z}2.13	0.06
Often-cut lower	^{I}5.20	0.20	^{R}0.18	0.01	^{Z}2.27	0.09

upper and lower soil layers on the Shenzhen often-cut hillside gave a higher total K content of soil than those on the Heshan often-cut hillside, respectively, although it was not significant.

The available K contents varied from 2.13×10^{-2} mg g^{-1} (the upper soil layer on the Heshan often-cut hillside) to 5.59×10^{-2} mg g^{-1} (the upper soil layer on the Shenzhen uncut hillside). The results of ANOVA showed a significant difference in available K content of soil among sampling locations (df = 9,40; F = 67.79; p < 0.001). At the Heshan study site, the uncut hillside gave a significantly higher available K content of soil than the often-cut hillside. The Shenzhen uncut hillside had the highest available K content of soil, while the Heshan and Hong Kong uncut hillsides had a similar available K content of soil. The Shenzhen often-cut hillside had a significantly higher available K content of soil than the

Heshan often-cut one. Accordingly, loss of nutrients increases with frequency of biomass harvesting.

DISCUSSION

The change in soil nutrients in uplands which is considered very complex may be affected by many factors including soil parent materials, climate, soil weathering, soil erosion, and land use. The results of the present study showed that biomass harvesting reduces nutrient contents of upland soil at the same study site. This indicates that the loss of nutrients in the form of harvested biomass is more than the natural gain of nutrients such as excreta of wildlife and biological N-fixation. In South China's Guangdong Province, soil nutrients losses with years and the loss of soil nutrients causes upland degradation. All these result mainly from heavy or even overharvesting of biomass for domestic fuel, which is in line with Huang (1988). In Heshan, the area of uplands is not large enough to meet all the requirements of biomass fuel so that the hillsides are harvested twice or even more times a year. It is considered natural in Heshan that the uplands have a "shaven" appearance in winter. Villagers in Shenzhen substitute liquefied petroleum gas (LPG) for some biomass fuel, reducing the consumption of biomass so that the uplands are harvested only once a year. Increased loss of soil nutrients from Heshan uplands is probably due to frequent harvesting of biomass fuel there. Chen (1993) found that a larger amount of nutrients in harvested biomass were annually removed from Heshan uplands than Shenzhen uplands. At the three sites of the present study, the lower soil layers did not show significantly more loss of nutrients on the often-cut hillsides than on the uncut hillsides, but the upper soil layers did so. It is probably because the effects of the existing harvesting patterns in Guangdong are not great enough to go deep into the lower layer of soil. It is worth studying whether or not the longer practice of the existing harvesting patterns would have an effect on the lower soil layer.

The plant gives the highest calorific value and lowest nutrient content at its maturity (Bell, 1984; Rodriguez-Barrueco et al., 1984). Most plants in Guangdong are mature from September to November of the year. One cut a year takes place from September to November in Shenzhen, while two cuts a year take place from April to May and from September to November in Heshan. The plant grows vigorously from April to May in Guangdong, and so young parts of it have a high nutrient content and low calorific value. Therefore, biomass harvesting from April to May removes more nutrients from uplands.

If well managed, uplands are renewable; if poorly managed, uplands can be degraded. From a sustainable point of view, reduced loss of nutrients in harvested biomass from uplands is preferable. In order to minimize this removal of nutrients from uplands, vegetation must be harvested at the optimum frequency during the optimum season. One cut a year after August is the optimum harvesting pattern which guarantees both that villagers have biomass fuel enough to cook and that upland resources keep in a sustainable state (Chen, 1993). Hassink (1992) thought that harvesting reduced yield of grassland, and fertilizer application ameliorated this detrimental effect. But fertilization of the vast uncultivated uplands used for biomass fuel in Guangdong would be impossibly expensive.

ACKNOWLEDGMENTS

This research was supported by a Hui Oi Chow Scholarship and a research grant from the Hsin Chong—K.N. Godfrey Yeh Education Fund. The laboratory work was done at the Kadoorie Agricultural Research Centre, University of Hong Kong. The authors express gratitude to D.K.O. Chan, C.J. Grant, and C.T. Wong for their advice in conducting field studies. The authors also thank Peiquan Gu, Runfang He, C.M.F. Mak, and D.W.H. Yu for their help in many ways.

REFERENCES

Allied Corporation. *Operator's Manual for AA/AE Spectrophotometer*, Waltham, MA, 1985.

Bell, D.T. Seasonal Changes in Foliar Macro-Nutrients (N, P, K, Ca and Mg) in *Eucalyptus saligna* Sm. and *E. wandoo* Blakely Growing in Rehabilitated Bauxite Mine Soils of the Darling Range. *Plant Soil*, 81, pp. 377–388, 1984.

Chen, R.J. Utilization of Upland Phytomass for Fuel. Ph.D. thesis, University of Hong Kong, Hong Kong, 1993.

Geotechnical Control Office, Civil Engineering Services Department, Hong Kong. *Territory of Hong Kong*. Government Printer, Hong Kong, 1989.

Hassink, J. Effect of Grassland Management on N Mineralization Potential, Microbial Biomass and N Yield in the Following Year. *Netherlands J. Agric. Sci.*, 40, pp. 173–185, 1992.

Hayes, J. The Use of Hill Land, in *The Rural Communities of Hong Kong: Studies and Themes*, Hayes J., Compiler, Oxford University Press, Hong Kong, 1983, pp. 179–182.

Huang, B.W. Utilization and Improvement of Uplands in South China: Significance and Feasibility. *Soil Water Conserv. China*, 4, pp. 2–9, 1988.

Minitab Inc., *Minitab Reference Manual* (Release 7), PA, 1989.

Page, A.L., et al. *Methods of Soil Analysis* (Part 2). American Society of Agronomy, Madison, WI, 1982.

PRC Ministry of Agriculture. Rural Energy. *The China Official Yearbook of Agriculture*. Agriculture Press, Beijing, 1991.

Rodriguez-Barrueco, C., C. Miguel, and P. Subramaniam. Seasonal Fluctuations of the Mineral Concentration of Alder [*Alnus glutinosa* (L.) Gaertn.] from the Field. *Plant Soil*. 78, pp. 201–208, 1984.

SAS Institute Inc. *SAS/STAT User's Guide*, Release 6.03 Edition. SAS Institute Inc., Cary, NC, 1988.

Shimadzu Corporation. *Instruction Manual for UV-VIS Recording Spectrophotometer*. Shimadzu Corporation, Kyoto, 1990.

Tecator AB. *Manual for Kjeltec System 1026 Distilling Unit*. Tecator AB, Hoganas, 1989.

United Nations. *Energy Balances and Electricity Profiles 1986*. New York, 1988.

Xie, Y.H. Rural Energy Problems in Guangdong Province. *Tropical Geography*, Guangdong, 9(1), pp. 1–7, 1989.

SECTION THREE

SOIL CONTAMINATION AND REMEDIATION

A Decision Support System for Assessing Remedial Technologies for Metal-Contaminated Sites

P.T. Chiueh, S.L. Lo, and C.D. Lee

INTRODUCTION

The process of determining site remediation alternatives is usually comprehensive and most often time-consuming as well as expensive. Several computer-aided systems have been used to prompt users to determine applicable treatment technologies by the characteristics of site and contaminant (Penmetsa and Grenney, 1993). However, most of these computer systems have been developed for all types of wastes, and access the site and contaminant characteristics by complicated queries and models. For metal-contaminated sites, the computer decision-making process of remedial technology selection needs further study, since metal-contaminated soils are intractable and spatial-dependent (Benker, 1995). In general, detailed descriptions rather than principle classification of remedial techniques for metal-contaminated soil are necessary, and spatial attributes should also be taken into account.

Spatial decision support system (SDSS) combines the technologies of geographical information system (GIS) and decision support system (DSS) to aid decision makers with problems that have spatial dimension (Walsh, 1993). A typical DSS consists of a user interface, model base and database (Richard et al., 1991; Turban, 1990). When linked with GIS, DSS can be converted into SDSS by using the spatial analysis and display capability of GIS (Cressie, 1993).

GIS provides users with the spatial components of information and allows users to organize the information in a real-world manner (Aronoff, 1989; Burrough, 1992). This system has become very common in environmental applications with the increasing availability of digital data in recent years (Reichhardt, 1996). Although there has been some effort to use the overlapping function of GIS to assist with site remediation jobs, GIS is often used with existing environmental modeling tools to achieve management goals (Kilborn et al., 1992; Watkins et al., 1996). On the other hand, DSS is applied in site remediation problems to ameliorate the decision-making process (Penmetsa and Grenney, 1993). The output from these applications has helped some decision makers but still keeps users from viewing the problem spatially. If we link GIS with DSS into a tightly coupled form, GIS can interact with DSS directly and add a spatial dimension that makes decision-making reflect real decision-making processes (Moreno, 1990).

A prototype SDSS, RAS (Remediation Assessment System) was developed for this study with the intention of helping users screen proven technologies and evaluate innovative remedial alternatives. RAS consists of models and the Taiwan geographical database to identify contaminated areas and to provide attribute information needed in the remediation process. Several assessment rules linked with the spatial distribution analysis and mobility prediction of contaminants were also constructed into RAS. These rules were based on published literature and reviewed by a remedial technology workshop at the National Taiwan University. A case study was conducted to illustrate RAS application.

METHODOLOGY

Assessment by RAS consists of site identification, contaminant distribution analysis, contaminant mobility prediction, and technology screening. The assessment process is illustrated in Figure 24.1. By integrating geographical information and the results of analysis and prediction with the assessment rules of each technology, RAS recommends the best remedial alternative for metal-contaminated sites.

Site Identification

Site identification is normally the first step in assessing pollution problems. Users can locate and view contaminated sites after searching by name or by coordinate. Maps can be viewed at different scales, and show different environmental attributes, such as watersheds, soil textures, or depth of groundwater. Furthermore, attributes that are required for sequential analysis are extracted from the database automatically.

Spatial Distribution of Contaminants

A knowledge of the spatial distribution of a pollutant at a contaminated site is often necessary for successful risk analysis (Flatman et al., 1988; Murray and Baker, 1992). A nonparametric geostatistical technique, indicator kriging, was used in RAS to obtain an average estimate value called "E-type estimate" in each unsampled area (Chiueh et al., 1996). The E-type estimate, being the expected value of the probability distribution for the least-squares criterion, is equal to the mean of the conditional cumulative distribution function (CCDF) (Deutsch and Journel, 1992; Journel, 1988):

$$[z(x_0)]_E^* = \int_{-\infty}^{+\infty} zc \, dF(x_0; zc \backslash (n))$$

$$\cong \sum_{k=1}^{K+1} zc_k'[F(x_0; zc_k \backslash (n)) - F(x_0; zc_{k-1} \backslash (n))] \qquad (24.1)$$

where

$$
\begin{aligned}
x_0 &= \text{unsampled point} \\
z(x_0) &= \text{concentration of } x_0 \\
F(x_0; zc \,|\, (n)) &= \text{CCDF of } x_0 \text{ on cutoff } zc \text{ by } n \text{ available data} \\
zc_k, \text{k} &= 1,\dots,\text{K} = \text{K cutoffs retained,}
\end{aligned}
$$

and $zc_0 = zc_{min}$, $zc_{K+1} = zc_{max}$ are the minimum and maximum of the zc-range to be entered as input parameters, respectively. The conditional mean value zc_k' within each class, (zc_{k-1}, zc_k), is obtained by the interpolation procedure specified as input to RAS.

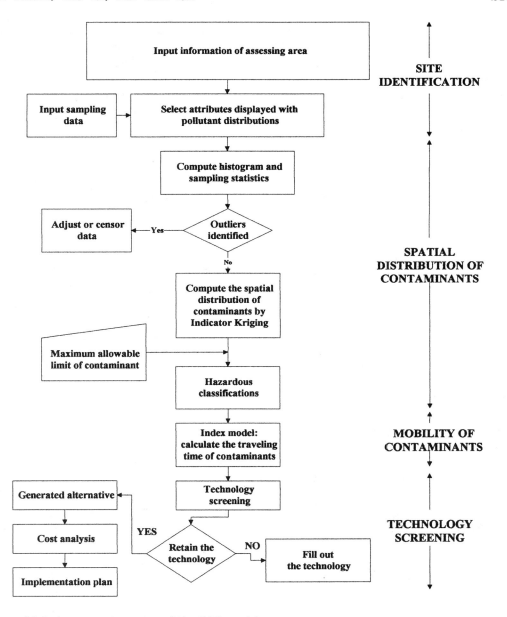

FIGURE 24.1. Assessment process of the RAS model.

When setting the maximum allowable limit as a hazardous threshold, we can obtain the classification of contaminant distribution by E-type estimation and then screen suitable remedial technologies.

Mobility of Contaminants

Knowledge of contaminant mobility through the unsaturated zone is critical for the evaluation of applicable treatment technologies (Addicott and Wagenet, 1985). Although the mobility of heavy metals is slower than that of other contaminants (Alloway, 1995), the mobility index model can still be used to investigate the potential of heavy metals moving down a soil profile and contaminating underlying groundwater. The mobility index model is a simple transport model in which the soil column is taken as an ideal plug flow reactor and all transport mechanisms are described by the retardation factor (*RF*). Thus, the traveling time of pollutants, needed for the screening process, is easily obtained (Liu, 1988; Liu et al., 1993).

$$RF = 1 + \rho(KSD) / \theta \qquad (24.2)$$

$$traveling\ time = depth\ /\ retarded\ velocity \qquad (24.3)$$

where
$$\rho = \text{soil bulk density (g m}^{-3})$$
$$KSD = \text{adsorption coefficient (m}^3\ \text{g}^{-1})$$
$$\theta = \text{soil water content, and}$$
$$retarded\ velocity = percolating\ velocity\ /\ RF,\ \text{(m day}^{-1})$$

Technology Screening

The treatment technologies considered by the current version of this study comprise three categories: on-site processing of contaminated soils, in situ treatment of contaminants, and macroencapsulation of contaminated areas (Smith, 1985). Two major on-site processing technologies applied in this study are extraction and on-site stabilization. In situ treatment technologies under consideration consist of mixing-and-respreading, vegetational uptake, and in situ stabilization. Macroencapsulation involves capping contaminated areas. They are all proven technologies.

Users first need to determine the potential land use of the remedial site and then use RAS to screen appropriate technologies for the site under its specific conditions. Each technology is suited to specific site parameters. Information on these parameters for each of the forgoing technologies is contained in the main knowledge-base file of RAS. RAS queries the user and extracts site-specific information on each of the parameters from the database. Next, the system evaluates each of the technologies to be screened by classifying the site-specific parameteric values using defined values contained in the knowledge-base file and following the rules built with each class of parameters. Table 24.1 gives the evaluation results for seriously contaminated areas of agricultural land use and Table 24.2 lists the parameter classes.

The output of the preliminary screening is a recommendation regarding applicable technologies for the specific site considered by RAS. If the user wants to retain the technology for later consideration, RAS will finish generating remedial alternatives including geographical information and descriptions of technology, and will analyze costs to recommend the best alternative.

MODEL DESCRIPTION

RAS is built in the Arc/View GIS environment, which was developed by ESRI of Redlands, California. The major components are shown in Figure 24.2. Being a SDSS, RAS has the features of a DSS consisting of database, model base, and user interface. A brief description and summary of RAS is provided as follows:

Table 24.1. Evaluation Results for Seriously Contaminated Areas of Agricultural Use.

Projected Land Use[a]	Concentration of Metals	Soil Texture	Ground Water Depth	Contaminant Mobility	Limitation of Technology of Remediation	Remediation Technology
Agriculture	High	Sand	High	High	Replace to stab.	Nonedible crops
Agriculture	High	Sand	High	High	Replace to extract	Nonedible crops
Agriculture	High	Sand	High	Low	Replace to stab.	Nonedible crops
Agriculture	High	Sand	High	Low	Replace to extract	Nonedible crops
Agriculture	High	Sand	Low	High	Replace to stab.	Nonedible crops
Agriculture	High	Sand	Low	High	Replace to extract	Nonedible crops
Agriculture	High	Sand	Low	Low	Replace to stab.	Nonedible crops
Agriculture	High	Sand	Low	Low	Replace to extract	Nonedible crops
Agriculture	High	Clay	High	High	Replace to stab.	Ordinary agr. use[c]
Agriculture	High	Clay	High	High	Replace to extract	Ordinary agr. use
Agriculture	High	Clay	High	Low	Replace to stab.	Ordinary agr. use
Agriculture	High	Clay	High	Low	Replace to extract	Ordinary agr. use
Agriculture	High	Clay	Low	High	Replace to stab.	Ordinary agr. use
Agriculture	High	Clay	Low	High	Replace to extract	Ordinary agr. use
Agriculture	High	Clay	Low	Low	Replace to stab.	Ordinary agr. use
Agriculture	High	Clay	Low	Low	Replace to extract	Ordinary agr. use

[a] Projected use after remediation.
[b] Replace contaminated soils and on site stabilization.
[c] Ordinary agricultural use.

Table 24.2. Parameter Classification.

Parameter	Class 1	Class 2	Class 3	Source
Land use	Agriculture	Industry/ commerce	Recreation	User input
Concentration	High > 10 mg kg^{-1}	Medium 1 mg kg^{-1}– 10 mg kg^{-1}	Low < 1 mg kg^{-1}	Extract from RAS
Soil texture	Sand $> 10^{-6}$ cm day^{-1}	Clay $< 10^{-6}$ cm day^{-1}		Extract from RAS
Groundwater depth	High < 3 m	Low > 3 m		Extract from RAS
Mobility	High < 20 years	Low > 20 years		Extract from RAS

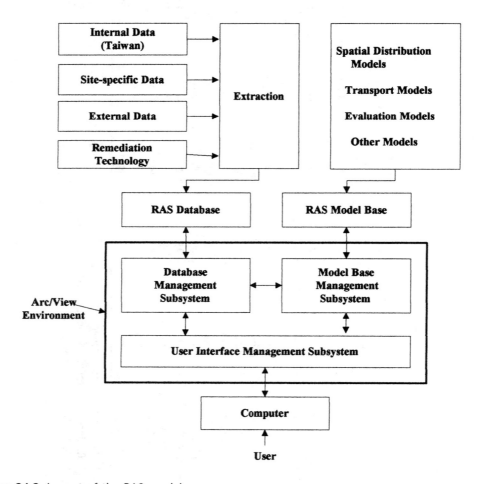

FIGURE 24.2. Layout of the RAS model.

The Database Management Subsystem

The database management subsystem is composed of a database and its management system. The management system should provide users with an environment to manage and integrate data from different sources. Arc/View GIS, the database management system used by RAS, is a client-server system that can be used to access all available data resources from database servers, such as ARCINFO. The data format can be coverages, map libraries, grids, images, and event data. In the RAS database management subsystem, we can display, query, summarize, and organize the database geographically.

RAS pulls out administration, soil texture, traffic route, watershed, and river coverages from the Taiwan database, developed by the Department of Geography at National Taiwan University, to support users in designing implementation plans. Moreover, site-specific sampling data that are necessary for screening suitable technologies can be integrated by tab- or comma-delimited text files and mapped out on the study area.

The Model Base Management Subsystem

The model base management subsystem is capable of interrelating models, such as geostatistical, transport, and technology evaluation models, with the appropriate linkages through the database. The geostatistical model used in RAS is IK technology from the Geostatistical Software Library developed by Deutsch and Journel (1992) at Stanford University.

The User Interface Subsystem

The RAS user interface includes many kinds of dialogue styles. Dialogue styles determine how RAS is directed and what RAS requires as an input and provides as an output. Since RAS is a SDSS, the user interface needs to handle spatial data as well as other action languages. With a completely graphic user interface, Arc/View GIS, this system can carry out all processing requirements of graphic data well.

MODEL APPLICATION

RAS was applied to a case study in Tao-Yuan County, Taiwan. The site was covered by paddy soils and contaminated with cadmium (Cd) from wastewater discharged by a chemical plant that produces plastic stabilizers. Extensive soil sampling was performed on a 6-hectare site out of the entire 84-hectare contaminated area (Chen, 1994). Sampling locations were based on a grid system dividing the 6-hectare study site into 126 blocks of 25 m × 25 m. A postplot of Cd concentration at each sampled location is shown in Figure 24.3. The histogram of the 78 data points retained shows a mean of 9.76 mg Cd kg^{-1} dry wt. of soil, a strong positive skewness, a large coefficient of variation ($\sigma/m = 1.3$), and a long tail; 9% of the data range from 20 mg kg^{-1} to a maximum of 93 mg kg^{-1}. The 9% of the data can be called "outliers" since they tend to distort the general shape of the distribution curve. Outlier values are difficult to analyze and evaluate in a contaminated site, and they can be the result of testing errors or actual site conditions. In IK, all data are transformed into indicator variables prior to variogram analysis and take values of either 0 or 1 to substitute the raw data. Thus, the removal of outlier data prior to variogram analysis is unnecessary.

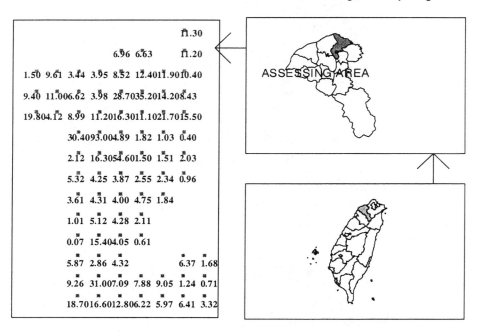

FIGURE 24.3. Posting of the data available over the assessing site (mg Cd/Kg dry wt. of soil).

IK analysis of the study site provided each point with a CCDF model for the uncertainty $z(x)$. However, because potential remedial technologies are usually considered by bulldozer size, we transferred the point data to block data to assess contaminant distribution. The corresponding E-type estimates for setting 10 mg kg^{-1} as the maximum allowable limit (MAL) of cadmium, are given in Figure 24.4. From the figure, the estimates reveal a hazardous area centered at the north and south site, which is relatively consistent with the trend shown by the sampling data. 76 blocks (60% of whole study site) exceeded MAL and were considered to be seriously contaminated areas. The other 50 blocks were considered middle-contaminated areas, since the concentrations of Cd were between 1–10 mg kg^{-1} dry soil.

For prediction of contaminant mobility, the retardation factor calculated from the index transport model was approximately 182, and the traveling time of cadmium was about 110 years. The parameters for the prediction include groundwater depth and several site characteristics, such as soil bulk density, adsorption coefficient of cadmium, soil water content, and percolating velocity. The groundwater depth at the study site was 9 m extracted from the database automatically, and other site parameters are input by users according to specific site conditions.

In the technology screening process, as projected agricultural land use, on-site processing, such as extraction or on-site stabilization techniques, is applicable for all blocks, and in situ mixing-and-respreading is appropriate for middle-contaminated blocks. Thus, there were four remedial alternatives feasible for the site: applying on-site extraction to seriously contaminated areas and in situ mixing-and-respreading to middle-contaminated areas; applying on-site stabilization to seriously contaminated areas and in situ mixing-and-respreading to middle-contaminated areas; applying on-site extraction to whole site; and applying on-site stabilization to whole site. These four remedial alternatives determined

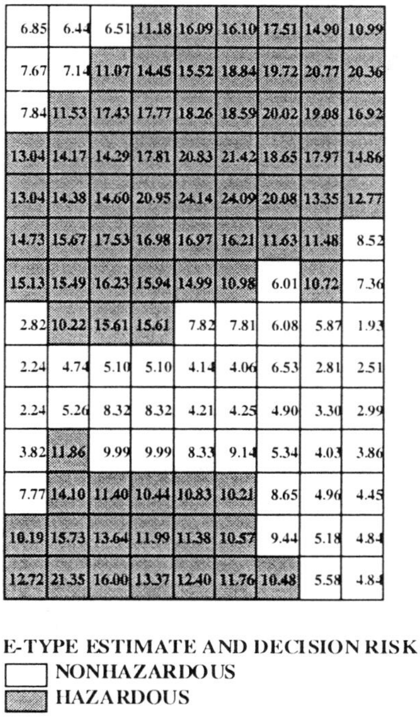

6.85	6.44	6.51	11.18	16.09	16.10	17.51	14.90	10.99
7.67	7.14	11.07	14.45	15.52	18.84	19.72	20.77	20.36
7.84	11.53	17.43	17.77	18.26	18.59	20.02	19.08	16.92
13.04	14.17	14.29	17.81	20.83	21.42	18.65	17.97	14.86
13.04	14.38	14.60	20.95	24.14	24.09	20.08	13.35	12.77
14.73	15.67	17.53	16.98	16.97	16.21	11.63	11.48	8.52
15.13	15.49	16.23	15.94	14.99	10.98	6.01	10.72	7.36
2.82	10.22	15.61	15.61	7.82	7.81	6.08	5.87	1.93
2.24	4.74	5.10	5.10	4.14	4.06	6.53	2.81	2.51
2.24	5.26	8.32	8.32	4.21	4.25	4.90	3.30	2.99
3.82	11.86	9.99	9.99	8.33	9.14	5.34	4.03	3.86
7.77	14.10	11.40	10.44	10.83	10.21	8.65	4.96	4.45
10.19	15.73	13.64	11.99	11.38	10.57	9.44	5.18	4.84
12.72	21.35	16.00	13.37	12.40	11.76	10.48	5.58	4.84

E-TYPE ESTIMATE AND DECISION RISK
☐ NONHAZARDOUS
▨ HAZARDOUS

FIGURE 24.4. Block map of E-type estimates for Cd concentration.

by RAS were subjected to cost analysis. The study recommended alternative 1 (Figure 24.5) as the best choice because of its economic cost.

CONCLUSION

The primary advantage of RAS is its rapid screening of proven technologies for a specific metal-contaminated site. The best remedial alternative determined by RAS includes existing geographical attributes and descriptions of recommended technology. In RAS, the IK model provides spatial distribution information that is essential for designing remedial alternatives, and the index model provides the most convenient analytical tool for assessment possible groundwater contamination. RAS was designed for easy expansion of its assessment rules and databases. If developed to its potential, the expanded system could be applied to other soil remediation problems.

ACKNOWLEDGMENTS

The authors express their gratitude to the National Science Council, Taiwan, R.O.C. for its financial support (Contract No: NSC 85-2621-P-002-001).

REFERENCES

Addicott, T.W. and R.J. Wagenet. Concepts of Solute Leaching in Soil: A Review of Modeling Approaches. *J. Soil Sci.*, 36, pp. 411–424, 1985.

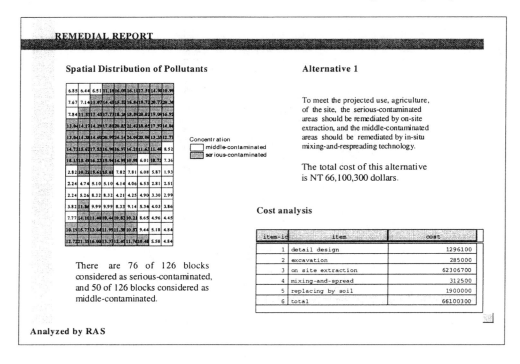

FIGURE 24.5. Alternative recommended by RAS for the study site.

Alloway, B.J. *Heavy Metals in Soils*, 2nd ed., Black Academic & Professional, an Imprint of Chapman & Hall, UK, 1995, p. 420.

Aronoff, S. *Geographic Information Systems: A Management Perspective*. WDL Publications, Ottawa, Canada, 1989, p. 294.

Benker, W.K. Removing Metals from Soil. *Civil Eng.*, October, pp. 69–71, 1995.

Burrough, P.A. Soil Information System, in *Geographical Information Systems—Principles and Applications*, Maguire, D.J., M.F. Goodchild, and D.W. Rhind, Eds., Longman Group UK Ltd., 1992, pp. 153–169.

Chen, Z.S. Summary Analysis and Assessment of Cadmium-Contaminated Agricultural Soils in Taoyuan, Taiwan. Final Rep., National Taiwan University, Taiwan, R.O.C., 1994, p. 153.

Chiueh, P.T., S.L. Lo, and C.D. Lee. Prototype Spatial Decision Support System for Using Probability Spatial Analysis in Soil Contamination Problems. *J. Environ. Eng.* ASCE, 123(5), pp. 514–519, 1996.

Cressie, N. Geostatistics: A Tool for Environmental Modelers, in *Environmental Modeling with GIS*, Goodchild, M.F., B.O. Parks, and L.T. Steyart, Eds., Oxford University Press, New York, 1993, pp. 414–421.

Deutsch, C.V. and A.G. Journel. *GSLIB: Geostatisical Software Library and User's Guide*. Oxford University Press, New York, 1992, p. 340.

Flatman, G.T., E.J. Englund, and A.A. Yfantis. Geostatistical Approaches to the Design of Sampling Regimes, in *Principles of Environmental Sampling*, Keith, L., Ed., American Chemical Society, Inc., Washington, DC, 1988, pp. 73–84.

Journel, A.G. Nonparametric Geostatistics for Risk and Additional Sampling Assessment, in *Principles of Environmental Sampling*, Keith, L., Ed., American Chemical Society, Washington, DC, 1988, pp. 45–72.

Kilborn, K., S.R. Hanadi, and P.B. Bedient. Connecting Groundwater Models and GIS. *Geo Info System*, 2(2), pp. 26–31, 1992.

Liu, C.K. Solute Transport Modeling in Heterogeneous Soils: Conjunctive Application of Physically Based and System Approaches. *J. Contam. Hydrol.*, 3, pp. 97–111, 1988.

Liu, C.K., J.S. Tsai, and L.W. Chiang. Assessing Groundwater Contamination Potential in the Kaohsiung Area, Southern Taiwan, in *Proceedings of Chinese Academic and Professional Convention*, Chicago, IL, pp. 7.65 to 7.68, 1993.

Moreno, D.D. Advanced GIS Modeling Techniques in Environmental Impact Assessment, in *Proceedings of GIS/LIS '90*, pp. 345–356, 1990.

Murray, M.R. and D.E. Baker. Using Probability Spatial Analysis in Assessing Soil Cadmium Contamination, in *Engineering Aspects of Metal-Waste Management*, Iskandar, I.K. and H.M. Selim, Eds., Lewis Publishers, Boca Raton, FL, 1992, pp. 25–47.

Penmetsa, R.K. and W.J. Grenney. STEP: Model for Technology Screening for Hazardous-Waste-Site Cleanup. *J. Environ. Eng.* ASCE, 119(2), pp. 231–247, 1993.

Reichhardt, T. Environmental GIS: The World in a Computer. *J. Environ. Sci. Technol.*, 30(8), pp. 340–343, 1996.

Richard, D.J., P.M. Nanninga, J. Biggins, and P. Laut. Prototype Decision Support System for Analyzing Impact of Catchment Policies. *J. Water Resour. Planning Manage.* ASCE, 117(4), pp. 399–414, 1991.

Smith, A.M. *Contaminated Land*. Plenum Press, New York and London, 1985, p. 433.

Turban, E. *Decision Support and Expert Systems*. Macmillan Publishing Company, New York, 1990, p. 846.

Walsh, M.R. Toward Spatial Decision Support System in Water Resources. *J. Water Resour. Planning Manage.* ASCE, 119(2), pp. 158–169, 1993.

Watkins, D.W., D.C. McKinney, D.R. Maidment, and M.D. Lin. Use of Geographic Information Systems in Ground-Water Flow Modeling. *J. Water Resour. Planning Manage.* ASCE, 122(2), pp. 88–96, 1996.

Chapter 25

The Single and Interactive Effects of Aluminum, Low pH, and Ca/Al Ratios on the Growth of Red Pine Seedlings in Solution Culture

Y. Shan, T. Izuta, and T. Totsuka

INTRODUCTION

Soil acidification caused by acid mine drainage, metalliferous and monmetalliferous ores, and acid deposition near coal combustion power plants and other industry facilities has induced many degraded lands and forest declines. Soil solution aluminum (Al) concentrations are identified to increase exponentially with increases in hydrogen ion concentration below 5.5 (Magistad, 1925). It is well established that high Al concentrations in soil solutions can damage sensitive plants. Al toxicity was considered as a cause of forest decline (Ulrich et al., 1980). Effects of Al on tree growth have given rise to concern about forest decline in Europe, North America, and Eastern Asia (Cronan and Shonfield, 1979; Miyake et al., 1991). Soil acidification reduced the growth of red pine seedlings and soil analysis showed that the Al concentrations were elevated with soil pH reduction (Shan et al., in press). Trees are different in sensitivity to Al or low pH. Therefore, it was asked whether toxicity of Al and low pH was greater to red pine seedlings in the acidified soils. It is necessary to study the interactive effects of low pH, Al toxicity, and calcium (Ca) on plants for the remediation and revegetation in these damaged lands. Toxicity due to low pH or Al per se is difficult to show because of complex soil chemistry and lack of a good understanding of ion uptake by roots in any case. In solution culture, because the concentrations of Al and other elements can be controlled, the single and combined effects of Al, low pH, or Ca/Al ratio can be examined. The present study was therefore undertaken to determine single and interactive effects of Al, low pH, or Ca/Al ratio on growth of red pine (*Pinus densiflora* Sieb. and Zucc.) seedlings through a solution culture experiment.

MATERIALS AND METHODS

Al and pH Treatment

Nutrient Solution Culture

Seeds of red pine were obtained from a seedling nursery in Tokyo. Seeds were sown in a pot (40 cm diameter) filled with red-yellow soil from the surface forest soil. When seedlings were about 6 cm high, they were transplanted into plastic containers (15.8 cm diameter, 19.0 cm height, and 3 L size) filled with nutrient solution. Nutrients were supplied to the

Table 25.1. Salts and Element Concentrations in Nutrient Solution.

Salts	Concentrations (mg L^{-1})	Elements	Concentrations (mg L^{-1})
NH_4NO_3	114.3	N	40
KH_3PO_4	38.3	P	9
K_2SO_4	49.5	K	33
$CaCl_3 \cdot 2H_2O$	52.3	Ca	14
$MgSO_4 \cdot 7H_2O$	61.2	Mg	12
Fe-EDTA	26.4		

seedlings using a modified Saito's solution (Saito, 1977), the modification being to replace inorganic iron (Fe) with Fe-EDTA as shown in Table 25.1. Al was added as $AlCl_3$. Al concentrations used were 0, 13, and 26 ppm. NaOH or HCl was used to adjust the solution pH to 4.5, 4.0, and 3.5; therefore, a 3 × 3 factorial design was employed. Each pot was aerated by an air pump. There were 10 seedlings per pot. Seedlings were grown for four weeks from May 19 to June 16, 1995 in a naturally-lit greenhouse. Nutrient solutions were replaced every five days. The pH was also adjusted at the mid-point of each five-day period.

Ca/Al Molar Ratio Treatment

The solution culture method was the same as stated above. Ca concentrations varied with Al concentration and designed Ca/Al molar ratios (10, 1, 0.10). Al and Ca were added as $AlCl_3$ and $CaCl_3 \cdot 2H_2O$, respectively. The Al concentration used was 26 ppm. NaOH or HCl was used to adjust the solution pH to 3.50 for all the treatments. Ca concentrations in solution culture without Al were 0.3, 3, 30 mM. There were five seedlings per pot. The seedlings were grown for six weeks from August 17 to October 1 in a naturally-lit greenhouse.

Measurement

At the end of the experiment, needle lengths and needle chlorophyll contents were measured. At harvest, roots, stems, and leaves were rinsed in deionized water, dried at 80°C for a week, and weighed.

Statistical Analysis

A fully randomized design was used to assign Al concentration and solution pH treatments. All results are from a single representative experiment. Data were analyzed by analysis of variance (ANOVA) as a 3 × 3 factorial combination of pH treatments and Al concentrations. Variance for pH treatments (pH 4.5, 4.0, or 3.5) was partitioned into linear and quadratic components and variance for Al concentrations (0, 13, 26 ppm) was also partitioned into linear and quadratic components using orthogonal polynomial contrasts (Mize and Schultz, 1985; Chappelka and Chevone, 1989). For factors involved in a significant interaction, an examination of the average of a treatment over all factors can be misleading (Mize and Schultz, 1985). Therefore, the response to Al was examined for each pH treatment using orthogonal contrast. An orthogonal contrast using each fixed level of pH treatment across either linear or quadratic components of Al treatments was developed to determine if pH treatment influenced the effect of Al treatments, separately. A similar procedure of statistical analysis was conducted for responses to pH treatments.

RESULTS

Individual and Combined effects of Al and Low pH

Visible Foliar Injury

Leaves of red pine seedlings grown in 26 ppm Al solution with pH 4.0 and 3.5 exhibited purplish coloration, similar to that of phosphorus deficiency within two weeks after the initiation of the Al treatments. The seedlings grown in the solution of pH 3.5 containing Al at 13 ppm also exhibited slightly purplish leaves. However, there was no visible foliar injury at pH 4.5 over all the levels of Al.

Needle Elongation

The interactive effects of Al concentrations and low pH were significant and linearly synergistic (Table 25.2). The Al toxicity to current needle elongation was increased with reducing solution pH. In pH 4.5 treatment, Al had no toxic effects on the needle elongation, but in pH 4.0 or 3.5 treatment, needle elongation was linearly decreased with raising Al concentrations. On the other hand, Al toxicity at 26 ppm to the needle elongation was linearly enlarged with reducing pH from 4.5, via 4.0, to 3.5. Without Al, effects of solution pH values alone on needle elongation were not significant (Figure 25.1).

Dry Weights

Leaf, stem, and root dry weights of red pine seedlings were measured at the final harvest. Synergistic interactive effects of Al concentrations and pH in the whole plant dry weight were significant (Table 25.2). In pH 4.5 or 4.0 treatment, Al concentrations did not significantly affect whole plant dry weight. With reducing pH, Al toxicity was increased. Whole plant dry weight was linearly reduced with increasing Al concentration in pH 3.5 treatment (Figure 25.2).

Responses of leaf dry weight to Al and low pH were similar to those of needle elongation at stated before (Table 25.2 and Figure 25.3).

Without Al in solution, pH values in culture solution alone did not affect the responses of the parameters measured here (Figures 25.1 to 25.3).

Effects of Ca/Al Molar Ratio on Al Phytotoxicity

Root length and whole plant length were significantly linearly decreased at Ca/Al molar ratios of 0.1 with 26 ppm Al compared with Ca/Al molar ratios of 10 or 1 (Figure 25.4), and without Al. The Ca concentrations in solution culture without Al did not affect the root length and whole plant length of red pine seedlings (data not shown here).

DISCUSSION AND CONCLUSION

Purplish leaves were observed in red pine seedlings grown in the solutions containing Al with pH 4.0 or 3.5 in this study. Red spruce, white spruce, and black spruce treated with

Table 25.2. Mean Squares and Levels of Significance of Al and Low pH Effects in a Culture Solution.

Source of Variation	D.f.	Mean Squares[a]		
		Whole Plant DW	Leaf DW	Needle Length
Al treatment	2	0.370*	0.148*	22.975**
Linear (Lin)	1	0.325**	0.147**	20.700***
Quadratic (Q)	1	0.045	0.001	0.275
pH treatment	2	0.055	0.022	2.881
Lin	1	0.051	0.002	0.544
Q	1	0.004	0.020	2.337
Al x pH	4	0.275	0.061	11.333*
Lin x lin	1	0.069	0.039*	10.350**
Lin x Q	1	0.016	0.001	0.647
Q x lin	1	0.179*	0.010	0.320
Q x Q	1	0.026	0.011	0.016
Error	99	5.093	1.194	137.919

[a] Calculated F-values significant at 0.10, 0.05, 0.01, or 0.001 levels are denoted by *,**, or ***, respectively.

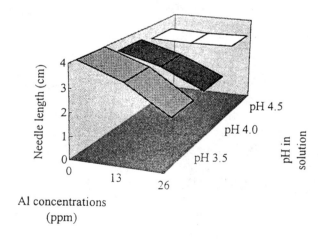

FIGURE 25.1. Combined effects of Al concentrations and low pH on current needle length.

phytotoxic levels of Al in solution culture exhibited brownish or blackened root systems, and yellowish or purplish needles characteristic (Hutchinson et al., 1986).

Al toxicity is more readily characterized by root morphology. Root length, rate of root elongation, and weight of plant tops are reliable measures of Al toxicity, but root weight is not (Adams, 1984; Adams and Lund, 1966; Mccormick and Steiner, 1978; Pavan et al., 1982; Steiner et al., 1984). In the present study, similar results were observed. Needle elongation, leaf dry weight, and especially root length and whole plant length are reliable and sensitive measures of Al toxicity. Therefore, we investigated the effects of Ca/Al ratio and Ca concentration without Al on root length and whole length in the second experiment.

No significant differences in the growing of red pine seedlings at pH 4.5, 4.0, and 3.5 without Al showed that red pine can tolerate acidic conditions and is relatively insensitive

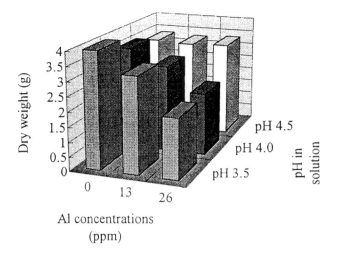

FIGURE 25.2. Combined effects of Al concentrations and low pH on whole plant dry weight.

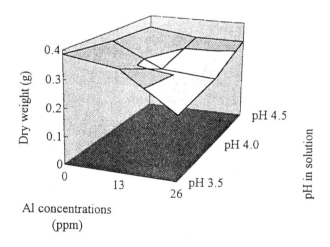

FIGURE 25.3. Combined effects of Al concentrations and low pH on leaf dry weight.

to low pH. However, synergistic interaction of low pH and the elevated Al concentrations was significant in current needle length, whole plant or leaf dry weight (Table 25.2, Figures 25.1 to 25.3). These results show that Al toxicity is increased with reduced pH. It suggests that soil pH reduction caused by acid deposition can result in increases of not only Al concentration but also Al toxicity. It was well documented that plant growth suppression in low pH soils is mainly a result of Al toxicity rather than H^+ toxicity except for extreme cases (Moore, 1974; Saigusa, 1991). These results show that high Al concentration was more closely related with the growth reduction induced by soil acidification than low soil pH, and the growth reduction of red pine was caused by the synergistic interaction of high Al concentration and low pH. Red pine is an intermediate species in sensitivity to Al. In nature, Al is common; therefore, low pH often becomes a problem together with Al. These data suggest that red pine may exhibit similar reductions in growth in acid soils with comparable Al and low pH.

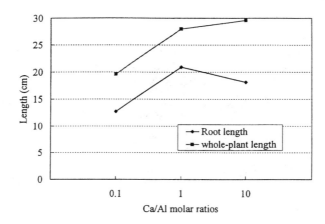

FIGURE 25.4. Effects of Ca/Al molar ratios on phytotoxicity of Al to red pine seedlings.

These solution culture experiments showed that Al phytotoxicity also varied with Ca/Al ratio and was reduced with elevated Ca/Al ratio. It has been reported that Al toxicity to forest trees with be manifested with Ca/Al molar ratio <1 (Ulrich, 1981 cited by Wold, 1990) or <2 (Cronan and Grigal, 1995). The Ca effects cannot be explained in terms of the alleviation of Ca deficiency, since in this study, Ca concentration elevated from 0.3 to 3 or 30 mM without Al did not increase root length and whole plant length of red pine seedlings. Farmers in many nations over the world have historically applied lime to acid soil to raise crop production. Such a favorable measure is due to not only raised soil pH, but also reduced dissolved Al concentration. Another more important factor is due to an elevated Ca/Al ratio. These results also suggest that liming is still an applicable measure to cope with soil acidification problems induced by natural or anthropogenic factors.

ACKNOWLEDGMENTS

The authors wish to express their appreciation to the Japan Society for the Promotion of Science (JSPS) for providing the JSPS Postdoctoral Fellowship to Dr. Yunfeng Shan and the Monbusho's Grant-in Aid for JSPS fellow. We would like to thank Mr. K. Horie of the Laboratory of Terrestrial Environment, Tokyo University of Agriculture and Technology, for his technical assistance.

REFERENCES

Adams, F. and Z.F. Lund. Effect of Chemical Activity of Soil Solution Aluminum on Cotton Root Penetration of Acid Subsoils. *Soil Sci.*, 101, pp. 193–198, 1966.

Adams, F. Ionic Concentrations and Activities in Soil Solution. *Soil Sci. Soc. Am. Proc.*, 35, pp. 420–426, 1984.

Chappelka, A.H. and B.I. Chevone. Two Methods to Determine Plant Responses to Pollutant Mixture. *Environ. Pollut.*, 61, pp. 31–45, 1989.

Cronan, C.S. and C.L. Schofield. Aluminum Leaching Response to Acidic Precipitation: Effects on High-Elevation Watersheds in the North East. *Sci.*, 204, pp. 304–306, 1979.

Cronan, C.S. and D.F. Grigal. Use of Calcium/Aluminum Ratios as Indicators of Stress in Forest Ecosystems. *J. Environ. Qual.*, 24, pp. 209–226, 1995.

Kohno, Y., H. Matsumura, and T. Kobayashi. Effects of Aluminum on the Growth and Nutrient Uptake in *Cryptomeria japonica* D. Don and *Chamaecyparis obtusa* Sieb. et Zucc. *J. Jpn. Soc. Atmos. Environ.*, 30(5), pp. 316–326, 1995.

Hutchinson, T.C., L. Bozic, and Munoz-vega. Responses of Five Species of Conifer Seedlings to Aluminum Stress. *Water, Air Soil Pollut.*, 31, 283–294, 1986.

Magistad, O.C. The Aluminum Content of the Soil Solution and Its Relation to Soil Reaction and Plant Growth. *Soil Sci.*, 20, pp. 181–225, 1925.

McCormick, L.H. and K.C. Steiner. Variation in Aluminum Tolerance Among Six Genera of Trees. *For. Sci.*, 24, pp. 565–568, 1978.

Miyaki, H., N. Kamei, T. Izuta, and T. Totsuka. Effects of Aluminum on the Growth of Hydroponically Grown Seedlings of *Cryptomeria japonica* D. Don. *Man. Environ.*, 17, pp. 10–16, 1991.

Mize, C.W. and R.C. Schultz. Comparing Treatment Means Correctly and Appropriately. *Can J. For. Res.*, 15, pp. 1142–1148, 1985.

Moore, D.P. Physiological Effects of pH on Roots, in *The Plant Root and Its Environment*, Carson, E.W., Ed., University Press of Virginia, Charlottesville, 1974, pp. 135–151.

Pavan, M.A., F.T. Bingham, and P.F. Pratt. Toxicity of Aluminum to Coffee in Ultisols and Oxisols Amended with $CaCO_3$, $MgCO_3$, and $CaSO_42H_2O$. *Soil Soc. Am. J.*, 46, pp. 1201–1207, 1982.

Saigusa, M. Plant Growth on Acid Soils with Special Reference to Phytotoxic Aluminum and Subsoil Acidity. *J. Jap. Soil Fertil. Soc.*, 62, pp. 451–459, 1991.

Saito, K. A Trial of Longterm Water Culture of *Cryptomeria japonica* Saplings. *Bull. Gov. For. Exp. Sta.*, 296, pp. 2–9, 1977.

Shan, Y., T. Izuta, Aoki, and T. Totsuka. Effects of Ozone and Soil Acidification, Alone and in Combination, on Growth, Gas Exchange Rate and Chlorophyll Content of Red Pine Seedlings. *Water, Air Soil Pollut.*, in press.

Steiner, K.C., J.R. Barbour, and L.H. McCormick. Response of *Polulus* Hybrids to Aluminum Toxicity. *Forest Sci.*, 30, pp. 404–410, 1984.

Ulrich, B., R. Mayer, and P.K. Khanna. Chemical Changes Due to Acid Precipitation in a Loess-Derived Soil in Central Europe. *Soil Sci.*, 130, pp. 193–199, 1980.

Wolt, J.D. Effects of Acidic Deposition on the Chemical Form and Bioavailability of Soil Aluminum and Manganese, in *Mechanisms of Forest Response to Acidic Deposition*, Lucier, A.A. and S.G. Haines, Eds., Springer-Verlag, 1990, pp. 62–107.

Chapter 26

The Effect of Heavy Metal Contamination on the Pigment Profiles of *Torreya* sp.

C. Penny, N.M. Dickinson, and N.W. Lepp

INTRODUCTION

Heavy metal contamination is derived from numerous sources, including the mining and smelting of nonferrous metals; metals emitted from these industries have been shown to accumulate in soil and vegetation in the vicinity of the industry (Buchauer, 1973; Helmisaari et al., 1995). This often leaves the surrounding areas derelict and devoid of vegetation. With the intensification of agriculture, there has been an increase in the addition of inorganic and organic manures in an attempt to maximize crop production. The addition of organic wastes and fertilizers based on sewage sludge, pig and poultry wastes may also add high levels of copper, zinc, and cadmium. Three such examples of heavy metal contamination are used to determine the effect of elevated soil metals on the pigment profiles of *Torreya californica* and *T. nucifera*.

When attempts are made to either revegetate derelict lands or to grow crops on agricultural lands contaminated by metals, plants often do not perform as well as others growing in uncontaminated soils. Stress response studies have been conducted on conifer seedlings to observe any adverse effects that elevated metals have on plant health and vigor. Various growth parameters have been used to assess the stress response. The genetic composition of conifers was shown to be disrupted in species growing near a zinc smelter (Prus-Glowacki and Nowak-Bzowy, 1992), with a lower observed heterozygosity in *Pinus sylvestris*. The phenolic compounds within the needles of *P. sylvestris* have been shown to be altered in the presence of Al, Cd, Mn, and Pb nitrates (Karolewski and Giertych, 1994). However, by far the most commonly used parameter is root growth. In *Pinus pinaster* and *P. pinea* a strong correlation was shown to exist between copper concentration in the roots and growth inhibition (Arduini et al., 1996). Exposure of *Pinus strobus*, *Acer rubrum*, and *Picea abies* to elevated levels of Cd and Zn resulted in reduced root initiation, poor development of lateral roots, chlorosis, dwarfism, wilting, and necrosis of the current season's growth (Mitchell and Fretz, 1977). It can be seen that, using root growth data, the effect of heavy metals on the growth of coniferous species can be determined; however, this method is difficult to employ in pot experiments and impossible to implement in the field.

It is known that factors which affect photosynthesis will invariably also affect pigments, owing to the close link between the condition and function of photosynthetic apparatus and its pigments (Young and Britton, 1990). Chlorophyll concentration has been used previously to assess the response of conifer species to elevated levels of metal ions (Schlegel et al., 1987). The concentration of chlorophyll in needles and cotyledons was shown to decrease when seedlings of *Picea abies*, growing in nutrient solution, were exposed to elevated concentrations of Cd, Zn, Hg, and Methyl-Hg. A later study discussed the possibility that the decreased rates of transpiration and lowered levels of chlorophyll in needles were the direct result of the presence of Hg, but concluded that these were responses to root damage that occurred, thus leading to a decrease in water supply and nutrient levels in the needles (Godbold and Hüttermann, 1988).

In this study, chlorophyll and carotenoid pigments of *T. californica* and *T. nucifera* were utilized to observe the stress responses exhibited by the two species when they were grown in soils contaminated by heavy metals.

MATERIALS AND METHODS

Species Used and Site Descriptions

Two species of *Torreya* were used, as they are both deemed to be threatened in their native habitats owing to their narrow endemic populations (P. Thomas, personal comm.), and also owing to their ease of propagation from cuttings. These were *T. californica* (California nutmeg tree) native to California, USA, and *T. nucifera* (Japanese Torreya), a native of Japan (Rushforth, 1987; Vidakovic, 1991).

Site Descriptions

Soils were collected from three different sites (Table 26.1) subjected to the following types of metal pollution:

1. aerial metal contamination, where the predominant ion was copper (Prescot)
2. metal contamination as a result of mining, where the predominant metal ion was zinc (Parys Mountain)
3. agricultural contamination, where there were only slightly elevated levels of both zinc and copper (Gateacre).

The heavy metal content of the soils was analyzed using an Atomic Absorption Spectrophotometer (AAS), after digestion in 5M nitric acid using a microwave under high pressure. The pH of the soil samples was also determined (Table 26.2).

Experimental

Torreya californica cuttings were grown in soils collected from the three different contaminated sites. Owing to the limited amount of rooted material, *T. nucifera* was only grown in Prescot soil. Each soil was "diluted" with ordinary potting compost, giving three different levels of contamination: 100% contaminated soil (high dosage), 50% contaminated soil mixed with potting compost (low dosage), and one 100% potting compost control for each species. All treatments were replicated five times and completely randomized in one block. Samples were collected for pigment analysis after 12 months' growth.

Pigment Analysis

Class 1 (newly formed) needles were removed from the same position on each plant, quickly immersed into liquid nitrogen, then stored at –20°C. Extraction of chlorophyll was

Table 26.1. Soil Collection Site Descriptions.

Site	Grid Reference	Contamination Source	Site Description
Parys Mountain	SH440905	Disused copper mine	Very sparse vegetation of *Calluna vulgaris*
Prescot	SJ465925	Aerial contamination from the former BICC copper refinery	*Acer pseudoplatanus* woodland
Gateacre	SD444872	Sewage sludge	Agricultural grassland

Table 26.2. Analysis of Soil Types (Mean ± SE).

Soil	Treatment	Metal Ion Concentration µg g^{-1}		Organic Matter (%)	pH
		Zinc	Copper		
Control	0%	199 ± 9	not traceable in the sample	17.5	6.33
Prescot	50%	253 ± 5	815 ± 31	14.2	5.44
	100%	338 ± 6	1482 ± 454	11.9	3.91
Gateacre	50%	303 ± 6	103 ± 49	5.9	6.48
	100%	332 ± 8	24 ± 15	4.1	5.90
Parys	50%	8187 ± 2607	594 ± 137	4.9	7.18
Mountain	100%	16540 ± 8680	791 ± 425	1.6	7.10

conducted in a darkened room. A known weight of needles were ground using a mortar and pestle in 100% redistilled acetone and then filtered under vacuum, ensuring that the remaining debris was white, indicating that all the chlorophyll had been removed. The supernatant was evaporated off using oxygen-free nitrogen gas and stored at –20°C. To analyze chlorophyll content, a known volume of 100% redistilled acetone was added to each sample and absorbancy measured using 1-cm cells and a UV/VIS spectrophotometer. Chlorophyll *a* was measured at a wavelength of 661.6 nm, chlorophyll *b* at 644.8 nm, and total carotenoids at 470 nm. The resultant absorbancy values were then converted to mg g^{-1} (fresh weight) using the equations for 100% redistilled acetone from Lichtenthaler (1987).

The composition of carotenoids was determined qualitatively using a Hewlett-Packard 1080 series High Pressure Liquid Chromatography (HPLC). Once the chlorophyll content had been analyzed quantitatively, the samples were then blown down again under oxygen-free nitrogen gas and resuspended in ethyl acetate, 20 µL was then injected onto a Spherisorb™* ODS2 column. The pigments were eluted using acetonitrile:water (9:1) and ethyl acetate solvent system.

RESULTS AND DISCUSSION

Chlorophyll Content

Prescot Soil Treatment

It is well documented that Cu is an essential micronutrient which plays an important in role in many plant processes, such as photosynthesis (Mehra and Farago, 1994). However,

---

* Registered trademark of Phase Separations Ltd., Deeside, Flintshire, United Kingdom.

in excess, it has been suggested that copper can induce chlorophyll destruction, which in turn will lead to carotenoid destruction (Lindon and Hendriques, 1991). The data for the concentration of chlorophyll pigments and total carotenoid concentration (Table 26.3) indicate that when growing in 100% Prescot soil which contains around 1480 mg kg^{-1} Cu (Table 26.2), *T. californica* exhibited a decrease in the total chlorophyll concentration from control levels. This was not observed for *T. nucifera* growing in the same treatment (Table 26.4); here, there was an increase in the total chlorophyll concentration compared to the control. In both cases the 50% treatment exhibited the greatest amount of total chlorophyll concentration of all three treatments (Tables 26.3 and 26.4). This may be due to the decrease in Cu in this treatment and the increase in organic matter and pH (Table 26.2), all of which contribute to the mobility and availability of Cu to the plant. A consideration which was not addressed here and should be considered is the nutrient status of the soil.

When the ratio of chlorophyll *a*: chlorophyll *b* in *T. californica* (Table 26.3) is observed, it can be seen that the value is similar for all three treatments, suggesting that although the total concentration of chlorophyll decreased in 100% Prescot soil treatment, the breakdown of chlorophyll *a* and chlorophyll *b* is at a similar rate, suggesting that although the levels of Cu may be a contributing factor in the breakdown of chlorophyll in this species, the metal is nonspecific to either chlorophyll pigment. The total concentration of carotenoids in the needles of *T. californica* (Table 26.3) indicates a breakdown in the carotenoid pigments when the species is grown in the 100% treatment; this suggests that the conditions offered by this soil are so adverse that they lead to a breakdown in both the chlorophyll and the carotenoid pigments within the plant, which in the long term would suggest the plant will experience poor growth and performance. *T. nucifera* did not exhibit any reduction in total carotenoid concentration in the 100% treatment when compared to the control (Table 26.4).

The effect of stressful conditions in relation to changes in pigmentation have focused on the xanthophyll cycle, with the majority of work centering on high light intensities (Adams et al., 1996; Thayer and Björkman, 1990). The xanthophyll cycle is comprised of three carotenoid pigments: zeaxanthin, violaxanthin, and antheraxanthin. Zeaxanthin is formed through the deep oxidation of violaxanthin to antheraxanthin catalyzed by the enzyme de-epoxidase. The reaction is reversible and the enzyme epoxidase reconverts zeaxanthin to antheraxanthin and violaxanthin; this reaction is light-dependent (Demmig-Adams, 1990). It is documented that zeaxanthin synthesized as protection in response to stressful conditions such as high light intensities (Demmig-Adams and Adams, 1992). It can be observed from Figure 26.1 that *T. californica* when grown in 100% Prescot soil synthesized zeaxanthin, with this pigment not being detected in any other treatment. Demmig-Adams (1990) suggests those plants that have the ability to produce zeaxanthin exhibit the ability to prevent overexcitation of photosystem II, and as a result reduce disruption caused by elements in excess such as copper. Figure 26.2, however, indicates that *T. nucifera* did not synthesize any zeaxanthin, suggesting that in conjunction with the chlorophyll data (Table 26.4), this species was not stressed when exposed to the elevated Cu and very low pH of the Prescot soil treatments.

The xanthophyll pool size refers to the sum of the xanthophyll cycle pigments, and an observed increase in the xanthophyll pool size has in the past been related to stressful conditions, such as high light intensity (Thayer and Björkman, 1990). The data in Figure 26.3 does not illustrate this, exhibiting a decrease in the xanthophyll pool size in relation to an increase in metal contamination for both species in the Prescot soil treatment.

Table 26.3. Chlorophyll Content (mg g^{-1} fwt) of *Torreya californica* Growing in Soils Contaminated with Heavy Metals (Mean ± Stdev).

Soil	Treatment	Chlorophyll a	Chlorophyll b	Total Chlorophyll	Total Carotenoid	Chl a: Chl b
Control	0%	10.4 ± 2.7	3.1 ± 1	13.5 ± 3.7	2.4 ± 0.8	3.5 ± 0.5
Prescot	50%	11.6 ± 1.9	3.6 ± 0.7	15.2 ± 2.5	2.8 ± 0.8	3.3 ± 0.3
	100%	7.2 ± 1.7	2.4 ± 0.9	9.6 ± 2.5	0.9 ± 1	3.2 ± 0.6
Gateacre	50%	13.6 ± 1.7	4.1 ± 0.6	17.7 ± 2.3	2.8 ± 0.9	3.3 ± 0.3
	100%	11.5 ± 1.7	3.3 ± 0.5	14.8 ± 2.2	2.2 ± 1.2	3.5 ± 0.2
Parys	50%	8.2 ± 2.9	2.9 ± 1.1	11.2 ± 3.7	1.7 ± 0.8	2.9 ± 0.6
Mountain	100%	4.7 ± 2.3	1.3 ± 0.8	6.0 ± 3	1.3 ± 0.7	4.6 ± 3.6

Table 26.4. Chlorophyll Content (mg g^{-1} fwt) of *Torreya nucifera* Needles Growing in Prescot Soil (Mean ± Stdev).

Soil	Treatment	Chlorophyll a	Chlorophyll b	Total Chlorophyll	Total Carotenoids	Chl a: Chl b
Control		8.0 ± 3	2.6 ± 1	10.6 ± 3.5	1.7 ± 1	3.4 ± 0.8
Prescot	50%	13.0 ± 1.5	3.8 ± 0.7	16.8 ± 2.1	2.7 ± 0.8	3.5 ± 0.3
	100%	9.5 ± 1.9	3.0 ± 0.5	12.4 ± 2.4	2.0 ± 0.8	3.2 ± 0.4

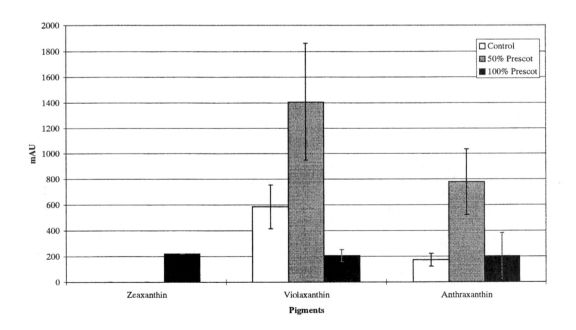

FIGURE 26.1. The xanthophyll cycle pigments measured in milliabsorbancy units (mAU) of *Torreya californica* growing in Prescot soil. (Mean ± Stdev)

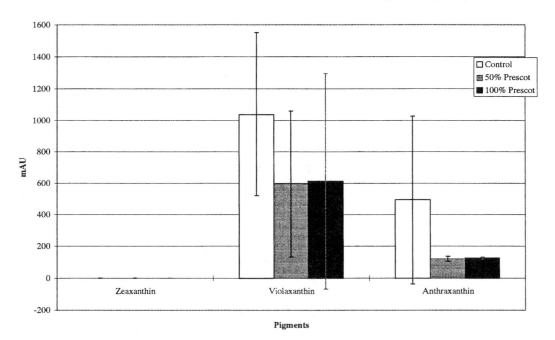

FIGURE 26.2. The xanthophyll cycle pigments (mAU) of *Torreya nucifera* growing in Prescot soil. (Mean ± Stdev)

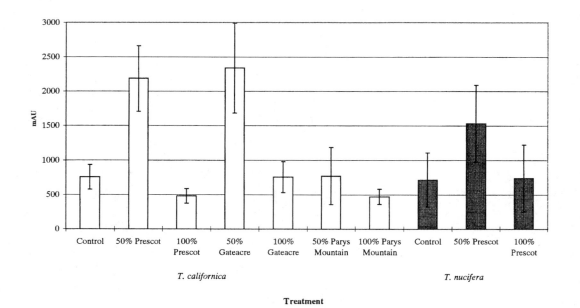

FIGURE 26.3. The xanthophyll pool size (mAU) of *Torreya sp.* growing in contaminated soils. (Mean ± Stdev)

Gateacre Soil Treatment

The Gateacre site was the least contaminated of the three sites (Table 26.2), with the concentrations of Cu and Zn only slightly exceeding background levels for agricultural soils of 20–30 mg kg^{-1} Cu (Baker and Senft, 1995) and 10–300 mg kg^{-1} Zn (Kiekens, 1995). As a result, there were no observed marked effects of the 100% contaminated soil treatment when compared to the control for the concentration of chlorophyll pigments and total concentration of carotenoids (Table 26.3) and no observed synthesis of zeaxanthin (Figure 26.4). However, as with the data for the *T. californica* and *T. nucifera*, the concentration of chlorophyll pigments and carotenoid concentration was greatest when the species were grown in 50% contaminated soils. Figure 26.3 also indicates a greater amount of xanthophyll cycle pigments for this treatment, indicating that the 50% treatment promotes photosynthesis and therefore the shoot growth of the plant.

Parys Mountain Soil Treatment

The Parys Mountain soil contained very high levels of Zn (Table 26.2), with concentrations of Cu just slightly below that of the 50% Prescot soil treatment, and in the 100% treatment the soil organic matter content is around 1% (dry weight), an order of magnitude less than the control (Table 26.2). Zinc, like copper, is termed an essential micronutrient playing a specific role within plant systems (Mehra and Farago, 1994), but is toxic to plants in elevated levels. The effect of Zn on plant pigments is not well documented; however, it has been suggested that excess Zn will interact with the thylakoids at site 2, but this reaction is light-dependent, with plants growing at low levels being less affected (Baker et al., 1982). In response to the elevated metals and soil conditions offered by the Parys Mountain soil it can be seen that the concentration of chlorophyll pigments and total carotenoid concentration is consistently less in the 50% and 100% treatments than the control (Table 26.3). The ratio of chlorophyll *a* : chlorophyll *b* is greater for plants growing in the 100% Parys Mountain treatment, suggesting that for this treatment, chlorophyll *b* is being broken down to a greater extent than chlorophyll *a*.

Figure 26.5 shows that zeaxanthin has been produced in response to the 100% contaminated soil treatment, but surprisingly, no zeaxanthin was produced in response to the 50% treatment. The xanthophyll pool size again indicates a decrease in xanthophyll pigments in response to an increase in contaminated soil dosage.

CONCLUSIONS

There is clear evidence that, in response to 100% Prescot and Parys Mountain soil treatments, the concentration of chlorophyll and carotenoid pigments decreases in *T. californica*. There was no observed effect of the Prescot soil treatment on *T. nucifera*, possibly indicating that this species is tolerant to Cu.

In the stressful conditions offered by the 100% treatments of Prescot and Parys Mountain soil, the synthesis of zeaxanthin was observed; however, in the less contaminated soils of the Gateacre treatments, no synthesis of zeaxanthin was observed. No zeaxanthin was synthesized in *T. nucifera* growing in the Prescot soil treatments, further suggesting that this species has the ability to tolerate elevated levels of Cu.

Therefore, it can be concluded that the analysis of the chlorophyll and carotenoid pigments can give an early indication of heavy metal stress. Further work is required to deter-

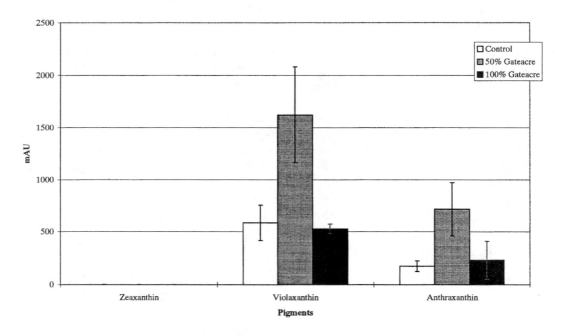

FIGURE 26.4. The xanthophyll cycle pigments (mAU) of *Torreya californica* growing in Gateacre soil. (Mean ± Stdev)

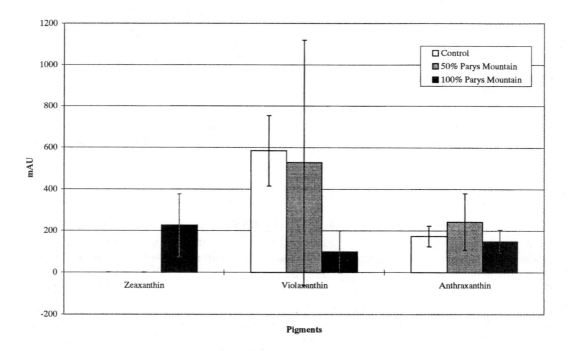

FIGURE 26.5. The xanthophyll cycle pigments (mAU) of *Torreya californica* growing in Parys Mountain soil. (Mean ± Stdev)

mine mechanisms of chlorophyll breakdown and changes in the carotenoid composition with respect to heavy metal contaminated soils. It is clear, however, that the method offers a potential nondestructive measure of the biological impact of metal contaminated soils on vascular plants.

ACKNOWLEDGMENTS

The authors would like to thank the HEFC for financial support for this work and the Commonwealth Science Council for a travel grant. The authors are indebted to M. Gardner and P. Thomas of the Royal Botanic Gardens, Edinburgh, for their cooperation and support. Many thanks are extended to Dr. K.O. Jones for his help and encouragement in the production of this chapter

REFERENCES

Adams, W.W., III, B. Demmig-Adams, D.H. Barker, and S. Kiley. Carotenoids and Photosystem II Characteristics of Upper and Lower Halves of Leaves Acclimated to High Light. *Australian J. Plant Physiol.* 23, pp. 669–677, 1996.

Arduini, I., D.L. Godbold, and A. Onnis. Cadmium and Copper Uptake and Distribution in Mediterranean Tree Seedlings. *Physiologia Plantarum*, 97, pp. 111–117, 1996.

Baker, N.R., P. Fernyhough, and I.T. Meek. Light-Dependent Inhibition of Photosynthetic Electron Transport by Zinc. *Physiologia Plantarum*, 56, pp. 217–222, 1982.

Baker, D.E. and J.P. Senft. Copper, in *Heavy Metals in Soils*, Alloway, B.J., Ed., Chapman and Hall, London, 1995, pp. 179–205.

Buchauer, M.J. Contamination of Soil and Vegetation Near a Zinc Smelter by Zinc, Cadmium, Copper and Lead. *Environ. Sci. Technol.*, 7(2), pp. 131–135, 1973.

Demmig-Adams, B. Carotenoids and Photoprotection in Plants: A Role for the *Xanthophyll zeaxanthin. Biochimica et Biophysica Acta*, 1020, pp. 1–24, 1990.

Demmig-Adams, B. and W.W. Adams, III. Carotenoid Composition in Sun and Shade Leaves of Plants with Different Life Forms. *Plant, Cell Environ.*, 15, pp. 411–419, 1992.

Godbold, D.L. and A. Hüttermann. Inhibition of Photosynthesis and Transpiration in Relation to Mercury-Induced Root Damage in Spruce Seedlings. *Physiologia Plantarum*, 74, pp. 270–275, 1988.

Helmisaari, H.-S., J. Derome, H. Fritze, T. Nieminen, K. Palmgren, M. Salemaa, and I. Vanha-Majamaa. Copper in Scots Pine Forests Around a Nearby-Metal Smelter in South-Western Finland. *Water, Air Soil Pollut.*, 85(3), pp. 1727–1732, 1995.

Karolewski, P. and M.J. Giertych. Influence of Toxic Metal Ions on Phenols in Needles and Roots, and on Root Respiration of Scots Pine Seedlings. *Acta Societatis Botanicorum Poloniae*, 63(1), pp. 29–35, 1994.

Kiekens, L. Zinc, in *Heavy Metals in Soils*, Alloway, B.J., Ed., Chapman and Hall, London, 1995, pp. 284–305.

Lichtenthaler, H.K. Chlorophylls and Carotenoids: Pigments of Photosynthetic Biomembranes. *Methods Enzymol.*, 148, pp. 350–382, 1987.

Lidon, F.C. and F.S. Henriques. Effects of Excess Copper on the Photosynthetic Pigments in Rice Plants. *Bot. Bull. Academia Sinica*, 33, pp. 141–149, 1992.

Mehra, A. and M.E. Farago. Metal Ions and Plant Nutrition, in *Plants and the Chemical Elements: Biochemistry, Uptake, Tolerance and Toxicity*, Farago, M.E., Ed., VCH, Weinheim and New York, pp. 67–86, 1994.

Mitchell, C.D. and T.A. Fretz. Cadmium and Zinc Toxicity in White Pine, Red Maple, and Norway Spruce. *J. Am. Soc. Horticultural Sci.*, 102(1), pp. 81–84, 1977.

Prus-Glowacki, W. and R. Nowak-Bzowy. Genetic Structure of a Naturally Regenerating Scots Pine Population Tolerant for High Pollution Near a Zinc Smelter. *Water, Air Soil Pollut.*, 62, pp. 249–259, 1992.

Rushforth, K. *Conifers.* Christopher Helm, London, 1987, p. 232.

Schlegel, H., D.L. Godbold, and A. Hüttermann. Whole Plant Aspects of Heavy Metal Induced Changes in CO_2 Uptake and Water Relations of Spruce (*Picea abies*) Seedlings. *Physiol. Plantarum*, 69, pp. 265–270, 1987.

Thayer, S.S. and O. Björkman. Leaf Xanthophyll Content and Composition in Sun and Shade Determined by HPLC. *Photosynthesis Res.*, 23, pp. 331–343, 1990.

Vidakovic, M. *Conifers, Morphology and Variation.* Graficki Zavod Hrvatske, p. 754, 1991.

Young, A.J. and B Britton. Carotenoids and Stress, in Stress Responses in Plants: Adaptation and Acclimation Mechanism, *Plant Biol.*, Vol. 12, Alscher, R.G. and J.R. Cumming, Eds., Wiley-Liss, Inc., New York, 1990, pp. 87–112.

Chapter 27

Colonization of Iron- and Zinc-Contaminated Dumped Filtercake Waste by Microbes, Plants, and Associated Mycorrhizae

T.M. Chaudhry, L. Hill, A.G. Khan, and C. Kuek

INTRODUCTION

Heavy metal contamination, caused by either natural processes or by human activities is one of the most serious environmental problems (Reedy and Prasad, 1990). Because plants function as the principal entry points of heavy metals into the food chain leading to animals and man (Rauser, 1990), it is very important to clean up soil and groundwater, which is a challenging proposition for a range of technical and economic reasons. Although several methods of cleanup have been developed, they are expensive and are only suitable for small areas of high commercial values (Negri and Hinchman, 1996). Furthermore, they restrict plant growth and render the soil biologically dead during the decontamination process.

Traditional methods for decontaminating soils have relied on high-impact technologies such as soil incineration or thermal desorption (to remove organic or volatile contaminants from soil), particle size separation (to remove the soil fraction that holds the greater portion of the contaminants), soil washing with selected chemicals (for the removal of organic or inorganic contaminants) or, as a last resort, the complete removal of the contaminated material, which is packed and sent to a landfill. However these technologies lead to other difficulties. Excavation of contaminated material and the need to dispose of it in suitable landfills causes environmental disruption. Systems that treat the soil usually generate secondary waste (e.g., wash solution) and produce soil that has lost its fertility along with its contaminants. In addition, these systems usually are expensive, energy-intensive, and vulnerable to equipment breakdown (Negri and Hinchman, 1996).

Scientists now are studying a new approach, called *phytoremediation*, which capitalizes on the ability of certain green plants to decontaminate the soil or water in which they are growing. Phytoremediation focuses on the power of nature to heal itself, somewhat like the proven process of bioremediation which uses bacteria to degrade soil and water contaminants. But phytoremediation is more proactive than bioremediation. It improves the appearance of sites more quickly and in some areas, can clean up sites that are not receptive to the use of bacteria (Bing, 1996).

Contaminated areas often support characteristic plant species, some of which are able to accumulate high concentrations of toxic metals in their tissues (Baker and Brooks, 1989). Many recent studies have demonstrated the feasibility of using such plants in decontamination of heavy metal contaminated soils (Baker et al., 1994; McGrath et al., 1996). But, as pointed out by Raskin et al. (1994), many heavy metal accumulating plants remain to be discovered. In addition, since metal uptake and tolerance depend on both plant and soil factors, we also require information on plant-soil interactions, especially soil microbes and their interactions with plants growing on such soils.

The role of symbiotic mycorrhizal fungi in such interactions is believed to be of great importance, given the various reports that these fungi are associated with roots of plants growing on contaminated soils (Griffioen et al., 1994; Khan and Chaudhry, 1996; Colpaert and van Assche, 1992). These findings have given a new dimension to phytoremediation because mycorrhizal roots can explore a greater soil volume than roots alone. Arbuscular Mycorrhizae are shown to enhance plant growth and increase biomass despite high levels of soil heavy metals (Galli et al., 1994; Shetty et al., 1995). Studies have indicated that certain AM fungal ecotypes from contaminated sites are more tolerant than those from noncontaminated sites (Gildon and Tinker, 1983; Griffioen et al., 1994). Gadd (1993) has reviewed interaction of fungi in general with toxic metals, but an overview of interactions of mycorrhizal fungi with toxic metals is lacking.

In view of the potential importance of micro- and macroflora to remediate heavy metal contaminated soil, this study was performed on a zinc waste contaminated filtercake dump area. The objectives of this study were to: (1) determine the microbial and fungal population of the research site; (2) identify the existing plant species; and (3) quantify the extent to which dominant plants are colonized by AM fungi.

MATERIALS AND METHODS

Location

The study area characterized in this study was located in the Flat Products Division of Broken Hill Propriety (BHP) at Port Kembla Steelworks, and it lies about 80 km south of Sydney. The contaminated site consisted of approximately 5 ha of iron-bearing filtercake stockpiled within an area known as Area 21. Filtercake consists of 51–68% ferrous oxide and 2–5% zinc (Thompson and Makin, 1991). The uncontaminated reference site consisted of nearby open forest.

Sampling

The soil and plants with roots were sampled to a depth of 15–20 cm from the contaminated filtercake and uncontaminated control areas along a transect which covered two principal age types. Nine samples of old (3–4 yr) filtercake were collected from sites A–I, and analyzed individually. Six samples from a small area containing freshly dumped filtercake (<2 yr) were taken and pooled for analysis as sample J. Associated vegetation samples, consisting of whole plants and their rhizospheres, were collected where possible. Soil, plants, and their rhizospheres were also collected from a nearby (<500 m) uncontaminated open forest woodland area for analysis as sample K.

Microbial and Fungal Population

The microbial populations of the contaminated (filtercake) and uncontaminated soil samples were determined by serial dilution and plating of soil suspensions on differential culture media. For total bacterial count, plate count agar (PCA) plates were used. Sabouraud agar was used to isolate and enumerate total fungal population from soil samples of the contaminated and uncontaminated sites (Harley and Prescott, 1993).

The API 120E system for Enterobacteriaceae and other nonfastidious gram-negative rods, was used for identification of bacteria of the contaminated and uncontaminated soil samples, grown on PCA plates.

Assessment of Mycorrhizal Infection

Roots of the dominant plant species growing on contaminated and reference uncontaminated sites were cleared and stained for AM fungal infection by using Philips and Hayman (1970) technique. The mycorrhizal infection of roots was recorded using a line-intersect method (Giovannetti and Mosse, 1980) and expressed as percentage infected root length. The presence of AM fungal structures such as arbuscules, vesicles, and intercellular hyphae, were also recorded. Rhizospheres were analyzed for the number of AM fungal propagules per 50 g soil by a wet sieving and decanting technique (Gerdemann and Nicolson, 1963).

Macroflora of the Study Sites

The plants growing on the contaminated and reference sites were studied by quadrat method (Krebs, 1985) and the frequency of occurrence of various plant species was determined as per the method of Colinvax (1973).

RESULTS AND DISCUSSION

The freshly dumped contaminated filtercake was devoid of any plant cover and contained fewer microbes compared to the other sites. Fresh filtercake soil contained the least number of colony-forming bacteria (0.27 cfu g^{-1}) as compared to 3–4 years old filtercake soil and the control soil, which contained 10.0 and 10.53 cfu g^{-1}, respectively (Figure 27.1). No growth of fungal colonies was observed on the plates inoculated with freshly dumped contaminated filtercake soil not supporting plant growth (Figure 27.2). These results may be due to the effects of the physical and chemical nature of the filtercake waste (Khan et al., 1997). The older contaminated and uncontaminated reference sites were able to sustain a reasonable environment for microbes due to physicochemical changes in rhizosphere by macroflora of that area.

The results of API-120E system employed for Enterobacteriaceae and other nonfastidious gram negative rods in the soil samples studied are summarized in Table 27.1. *Pseudomnas, Sphingomonas, Flavobacterium, Sphingobacterium, Bordetella,* and *Alcaligenes* species were present in both the contaminated and the uncontaminated soil samples, but *Xanthomonas* and *Chromobacterium* species were identified only in the contaminated soil samples (Table 27.1).

The presence of bacterial species on a contaminated site is possibly helpful in the biodegradation of hazardous chemical waste of the site. *Pseudomonas* strains are capable of degrading a wide array of chlorobenzoates and chlorophenols (Hartman et al., 1979; Reineke and Knackmuss, 1979). *Pseudomonas* species have the ability to biodegrade (2,4,5-T) trichlorophenoxy acetic acid (Chatterjee et al., 1982; Kilbane et al., 1982), 4-chlorophenyl (Pettigrew and Sayler, 1986), toluene (Jain et al., 1987), naphthalene (Blackburn et al.,

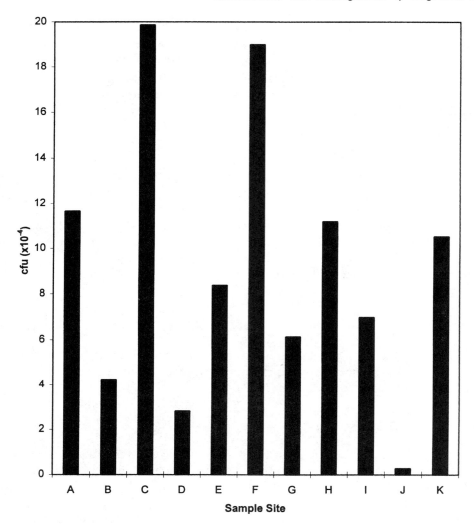

FIGURE 27.1. Mean bacterial colony-forming units g^{-1} soil from the sample sites. (Values are means of three replicates.)

1987), and are mercury-resistant bacteria (Barkay and Olson, 1986). *Alcaligenes* and *Bordetella* are ecologically interesting genera that grow well at temperatures over 70°C and have been isolated from hot springs and hot water tanks of homes and laundromats (Brock, 1981). The presence of *Alcaligenes* and *Bordetella* in two samples from the older filtercake probably indicates the biodegradation of phenolic substances present in the filtercake. The degradation of polychlorinated biphenyls (PCBs) and trichloroethylene is carried out aerobically by *Alcaligenes* (Bedard et al., 1987; Harker and Kim, 1990).

This study has also shown that the roots of the plants growing on the contaminated site were mycorrhizal, though to a lesser extent than those from the reference site (Table 27.2). These results are consistent with those of earlier workers (Diaz and Honrubia, 1990; Gildon and Tinker, 1981, 1983; Ietswaart et al., 1992; Sambandan et al., 1992; Weissenhorn

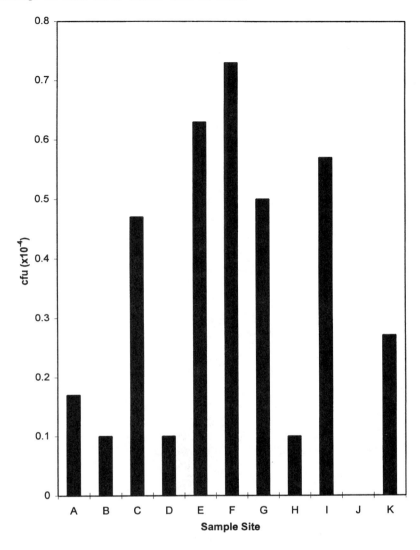

FIGURE 27.2. Mean fungal colony-forming units g^{-1} soil from the sample sites. (Values are means of three replicates.)

et al., 1991), who also reported arbuscular mycorrhizae in heavy metal contaminated sites. Studies by other workers have shown that AM fungal ecotypes from heavy metal contaminated sites seem to be more tolerant to heavy metals than reference strains from uncontaminated soils (Gildon and Tinker, 1983; Griffioen et al., 1994).

Rhizosphere of the plants from 3–4 yr old stockpiled filtercake contained AM fungal spores, mainly belonging to *Glomus* and *Gigaspora* spp. Soils from the reference site harbored various *Glomus* spp. but no *Gigaspora* spp. spores were recovered from the site. The freshly dumped contaminated filtercake site was devoid of AM fungal spores. Lack of plant establishment on this site may reflect its unfavorable physicochemical characteristics. Our results indicate that mycorrhizal infection, particularly with *Glomus* and *Gigaspora* strains, is necessary for soil reclamation.

Table 27.1. Identification of Bacteria of Zinc Waste Contaminated Filtercake and an Uncontaminated Control Site Using API 120E System.

Sample	Colony Type	Identification No.	Identified Bacteria
A3 (10^{-5})	yellow/raised	1002000	*X. maltophilia*
A3 (10^{-5})	white/raised	1002004	*F. meningosepticum*
B1 (10^{-5})	cream/raised/irregular	1003004	*Sphmon. paucimobilis* *Sphingo. multivorum* *F. meningosepticum*
B2 (10^{-5})	cream/flat/curled	2202004	*P. aeruginosa* *P. fluorescens/putida* *C. violaceum*
C3 (10^{-5})	bright yellow/raised	203004	*Sphmon. paucimobilis* *Sphingo. multivorum* *P. cepacia*
D2 (10^{-4})	pink/raised	0203004	*P. spp., F. odoratum* *P. fluorescens/putida* *Bordetella/A. spp.*
E1 (10^{-4})	brown, orange/raised	0202000	*X. maltophilia*
F3 (10^{-5})	pink/yellow/raised	1203004	*Sphmon. paucimobilis* *Sphingo. multivorum* *P. cepacia*
G2 (10^{-5})	cream/flat/irregular	1203004	as above
G2 (10^{-4})	orange/raised	2203004	*P. aeruginosa* *P. fluorescens/putida*
H1 (10^{-5})	black/convex	0203004	*P. spp., F. odoratum* *P. fluorescens/putida* *Bordetella/A. spp.*
H3 (10^{-5})	pink/raised	0203004	as above
I1 (10^{-5})	white, cream/convex	0202000	*X. maltophilia*
I2 (10^{-5})	yellow, white/raised	0202000	*X. maltophilia*
J2 (10^{-4})	yellow, cream/flat /irregular	5202004	*P. cepacia* *X. maltophilia*
K1 (10^{-6})	pink, orange/raised	0203004	*P. spp., F. odoratum* *P. fluorescens/putida* *Bordetella/A. spp.*
K2 (10^{-5})	white/raised	1203004	*Sphmon. paucimobilis* *Sphingo. multivorum* *P. cepacia*

X = Xanthomonas; P = Pseudomonas; F = Flavobacterium; Sphmon.= Sphingomonas; Sphingo. = Sphingobacterium; C = Chromobacterium; A = Alcaligenes

The population density and frequency of plant species are summarized in Table 27.3. The freshly dumped filtercake supported no plant growth. Among the plant species growing on the older filtercake, *Chrysanthemoides monilifera*, *Tagetes minuta*, *Foeniculum vulgare*, *Sonchus oleraceus*, and *Ricinus communis* were found to be abundant. All or some these plants may be suitable options for phytoremediation of heavy metal contaminated filtercake. The reference (control) site was also inhabited by various plant species, none of which were represented on the zinc contaminated filtercake stockpile. *Poa* spp. were very abundant on the reference site.

Table 27.2. AM-Colonization in Dominant Plant Species Growing on 3–4 Year Old Zinc Waste Contaminated Filtercake and a Reference Site.

Name of Plant Species	VAM-Colonization			
	Hyphae (%)	Vesicles (%)	Arbuscules (%)	Total Colonization (%)
Filtercake site				
Asteraceae				
Chrysanthemoides monilifera (L.) Norlindh	27.0	16.0	11.0	54.0
Tagetes minuta L.	12.0	24.0	8.0	44.0
Apiaceae				
Foeniculum vulgare Miller	26.0	22.0	5.0	53.0
Euphorbiaceae				
Ricinus communis L.	26.0	10.0	4.0	40.0
Reference site				
Poaceae				
Poa spp.	47.0	16.0	12.0	75.0

Table 27.3. Record of Data on Population Density and Frequency of Occurrence of Plant Species Growing on 3–4 Year Old Zinc Waste Contaminated Filtercake and a Reference Site.

Name of Plant Species	Population Density (no. m^{-2})	Frequency (%)	Frequency Class[a]
Filtercake site			
Asteraceae			
Chrysanthemoides monilifera (L.) Norlindh	5.0	67.0	A
Senecio quadridentatus Labill.	2.0	22.0	O
Sonchus oleraceus L.	4.0	44.0	F
Tagetes minuta L.	10.0	78.0	A
Apiaceae			
Foeniculum vulgare Miller	6.0	67.0	A
Euphorbiaceae			
Ricinus communis L.	12.0	89.0	A
Malvaceae			
Modiola caroliniana (L.) G. Don	2.0	44.0	F
Solanaceae			
Solanum nigrum L.	4.0	33.0	O
Reference site			
Asteraceae			
Ageratina adenophora (Spreng.) R.M. King and H. Robinson	8.0	50.0	F
Senecio madagascariensis Poiret	6.0	33.0	O
Verbenaceae			
Verbena bonariensis L.	7.0	50.0	F
Poaceae			
Poa spp.	68.0	100.0	VA

[a] O = Occasional (21–40%); F = Frequent (41–60%); A = Abundant (61–80%); VA = Very Abundant (81–100%); (Colinvax, 1973).

Further analytical as well as comparative studies of the morphological and physico-chemical variations between plants and AM fungal populations in their rhizosphere are in progress in order to identify heavy metal (especially zinc) tolerant and or hyperaccumulator ecotypes, and the mechanism of hyperaccumulation and tolerance involved. The potential phytoremediation of soil can be enhanced by inoculating plants perceived to be the most efficient hyperaccumulators with AM fungi appropriate for heavy metal contaminated sites.

REFERENCES

Baker, A.J.M. and R.R. Brooks. Terrestrial Higher Plants Which Hyperaccumulate Metallic Elements—A Review of Their Distribution, Ecology and Phytochemistry. *Biorecovery*, 1, pp. 81–126, 1989.

Baker, A.J.M., S.P. McGrath, C.M.D. Sidoli, and R.D. Reeves. The Possibility of In-Situ Heavy Metal Decontamination of Polluted Soils Using Crops of Metal-Accumulating Plants. *Resour. Conserv. Recycling*, 11, pp. 41–49, 1994.

Barkay, T. and B.H. Olson. Phenotypic and Genotypic Adaptation of Aerobic Heterotrophic Sediment Bacterial Communities to Mercury Stress. *Appl. Environ. Microbiol.*, 52, pp. 403–406, 1986.

Bedard, D.L., R.E. Wagner, and M.J. Brennan. Extensive Degradation of Arochlors and Environmentally Transformed Polychlorinated Biphenyls by *Alcaligenes eutrophus* H850. *Appl. Environ. Microbiol.* 53, pp. 1094–1102, 1987.

Bing, M. Back to Nature. *Mining Voice July/August 1996*. 1996, pp. 31–33.

Blackburn, J.W., R.L. Jain, and G.S. Sayler. Molecular Microbial Ecology of a Naphthalene-Degrading Genotype in Activated Sludge. *Environ. Sci. Technol.*, 21, pp. 884–890, 1987.

Brock, T.D. Extreme Thermophiles of the Genera *Thermus*, in *The Prokaryotes: A Handbook on Habitats, Isolation, and Identification of Bacteria*. Starr, M.P., H. Stolp, H.G. Truper, A. Ballows, and H.G. Schlegel, Eds., Springer-Verlag, Berlin, 1981, pp. 978–984.

Chatterjee, D.K., J.J. Kilbane, and A.M. Chakrabarty. Biodegradation of 2,4,5-Trichlorophenoxyacetic Acid in Soil by a Pure Culture of *Pseudomonas cepacia*. *Appl. Environ. Microbiol.*, 44, pp. 514–516, 1982.

Colinvax, P.A. The Commonness and Rarity of Species, in *Introduction to Ecology*. Herbert, D.P., Ed., John Wiley & Sons, New York, 1973.

Colpaert, J.V. and J.A. van Assche. Zinc Toxicity in Ectomycorrhizal *Pinus sylvestris*. *Plant Soil*. 143, pp. 201–211, 1992.

Diaz, G. and M. Honrubia. Infectivity of Mine Soils from South-East Spain. *Agric. Ecosystem Environ.*, 29, pp. 85–89, 1990.

Gadd, G.M. Tansley Review No. 47. Interaction of Fungi with Toxic Metals. *New Phytologist*, 124, pp. 25–60, 1993.

Galli, U., H. Schuepp, and C. Brunold. Heavy Metal Binding by Mycorrhizal Fungi. *Physiol. Plantarum.*, 92, pp. 364–368, 1994.

Gerdemann, J.W. and T.H. Nicolson. Spores of Mycorrhizal *Endogone* Species Extracted from Soil by Wet Sieving and Decanting. *Trans. British Mycological Soc.*, 46, pp. 235–244, 1963.

Gildon, A. and P.B. Tinker. A Heavy Metal Tolerant Strain of a Mycorrhizal Fungus. *Trans. British Mycological Soc.*, 77, pp. 648–649, 1981.

Gildon, A. and P.B. Tinker. Interactions of Vesicular-Arbuscular Mycorrhizal Infection and Heavy Metals in Plants. I. The Effect of Heavy Metals on the Development of Vesicular-Arbuscular Mycorrhizae. *New Phytol.*, 95, pp. 247–261, 1983.

Giovannetti, M. and B. Mosse. An Evaluation of Techniques for Measuring Vesicular-Arbuscular Mycorrhizal Infection in Roots. *New Phytol.*, 84, pp. 489–500, 1980.

Griffioen, W.A.L., I.H. Ietswaart, and W.H.O. Ernst. Mycorrhizal Infection of an *Agrostis capillaris* Population on a Copper Contaminated Soil. *Plant Soil*, 158, pp. 83–89, 1994.

Harker, A.R. and Y. Kim. Trichloroethylene Degradation by Two Independent Aromatic-Degrading Pathways in *Alcaligenes eutrophus* JMP134. *Appl. Environ. Microbiol.*, 56, pp. 1179–1181, 1990.

Harley, J.P. and L.M. Prescott. Basic Laboratory and Culture Techniques, in *Laboratory Exercises in Microbiology*, 2nd ed., Wm. C. Brown Publishers, Dubuque, 1993, pp. 14–46.

Hartmann, J., W. Reineke, and H.J. Knackmuss. Metabolism of 3-Chloro-, 4-Chloro-, and 3,5-Dichloro-benzoate by a Pseudomonad. *Appl. Environ. Microbiol.*, 37, pp. 421–428, 1979.

Ietswaart, J.H., W.A.J. Griffioen, and W.H.O. Ernst. Seasonality of VAM Infection in 3 Populations of *Agrostis capillaris* (Gramineae) on Soil With or Without Heavy Metal Enrichment. *Plant Soil*, 139, pp. 67–73, 1992.

Jain, R.K.G.S., J.T. Sayler, L.H. Wilson, and D. Pacia. Maintenance and Stability of Introduced Genotypes in Groundwater Aquifer Material. *Appl. Environ. Microbiol.*, 53, pp. 996–1002, 1987.

Khan, A.G. and T.M. Chaudhry. Effects of Metalliferous Mine Pollution on the Vegetation and their Mycorrhizal Association at Sunny Corner—A Silver Town of the 1880's, in *Abstracts of 1st International Conf. on Mycorrhizae.* p. 70. University of California, Berkeley, 1996.

Khan, A.G., T.M. Chaudhry., W.J. Hayes, C.S. Khoo., L. Hill., R. Fernandez, and P. Gallardo. Physical, Chemical and Biological Characterisation of a Steelworks Waste Site at Port Kembla, NSW, Australia. *Water, Air Soil Pollut.*, 104, pp. 389–402, 1998.

Kilbane, J.J., D.K. Chatterjee, J.S. Karns, S.T. Kellog, and A.M. Chakrabarty. Biodegradation of 2,4,5-Trichlorophenoxy Acetic Acid by a Pure Culture of *Pseudomonas cepacia*. *Appl. Environ. Microbiol.*, 44, pp. 72–78, 1982.

Krebs, C.J. Population Parameters, in *Ecology: The Experimental Analysis of Distribution and Abundance*, 3rd ed., Wilson, C.M. and H. Detgen, Eds., Harper & Row, Publishers, New York, 1985, pp. 157–172.

McGrath, S.P., S.J. Dunham, C.D.M. Sidoli, and F. Lodico. Four Years Growth of Hyperaccumulator Plants on Metal Polluted Soils. Extended Abstracts: *1st International Conf. on Contaminants and the Soil Environment in the Australasia-Pacific*. C.S.I.R.O., Division of Soils, Adelaid, Australia, 1996, p. 323.

Negri, M.C. and R.R. Hinchman. Plants That Remove Contaminants from the Environment. *Laboratory Medicine*, 27, pp. 36–40, 1996.

Pettigrew, C.A. and G.S. Sayler. The Use of DNA: DNA Colony Hybridization in the Rapid Isolation of 4-Chlorobiphenyl Degradative Bacterial Phenotypes. *J. Microbiol. Methods*, 5, pp. 205–213, 1986.

Phillips, J.M. and D.S. Hayman. 1970. Improved Procedure for Clearing Roots and Staining Parasitic and Vesicular-Arbuscular Mycorrhizal Fungi for Rapid Assessment of Infection. *Trans. British Mycological Soc.*, 55, pp. 158–161, 1986.

Raskin, I., P.B.A.N. Kumar, S. Dushenkov, and D.E. Salt. Bioconcentration of Heavy Metals by Plants. *Current Opinion in Biotechnology*, 5, pp. 285–290, 1994.

Rauser, W.E. Phytochelatins. *Ann. Rev. Biochem.*, 59, pp. 61–86, 1990.

Reedy, G.N. and M.N.V. Parsad. Heavy Metal-Binding Protein/Peptide: Occurrence, Structure, Synthesis and Function—A Review. *Environ. Expertl. Botany*, 30, pp. 251–264, 1990.

Reineke, W. and H.J. Knackmuss. Construction of Haloaromatic Utilizing Bacteria. *Nature* (London), 277, pp. 385–386, 1979.

Sambandan, K., K. Kannan, and N. Raman. Distribution of Vesicular-Arbuscular Mycorrhizal Fungi in Heavy Metal Polluted Soils of Tamil-Nadu, India. *J. Environ. Biol.* 13, pp. 159–167, 1992.

Shetty, K.G., B.A.D. Hetrick, and A.P. Schwab. Effects of Mycorrhizae and Fertilizer Amendments on Zinc Tolerance of Plants. *Environ. Pollut.*, 88, pp. 307–314, 1995.

Thompson, S.C. and S.L. Makin. *BHP Central Research Laboratories Report*, Shortland. 1991.

Weissenhorn, I., C. Leyval, and J. Berthelin. VA Mycorrhizal Colonisation of Maize in an Industrially Polluted Soil and Heavy Metal Transfer to the Plant, in Abstracts of Third European Symposium on Mycorrhizas. *Mycorrhizas in Ecosystem-Structure and Function*. Sheffield, UK. 1991, pp. 253.

Chapter 28

An Experimental Study on the Effectiveness of EDTA/HCl on the Removal of Pb, Cd, Cu, and Ni Soil

K.K. Chiu, S.C. Lee, and C.S. Poon

INTRODUCTION

Soil remediation or soil cleanup has become more common, especially in developed countries. The choice of extractant such as water, organic solvents, chelating agents, surfactants, acids, or bases depends on which types of contaminants are to be treated (Holden et al., 1989). Removal of metals by forming soluble complexes with the chelating agents is quite efficient, but the high cost of solvents has precluded their application in remediation of heavy metal-contaminated sites. Hydrochloric acid (HCl) is more effective in extracting metals than other acids, presumably because of the formation of soluble metal-chloride complexes (John et al., 1994). Ethylenediamine tetraacetic acid (EDTA) has been chosen as the representative of the group of chelating agents, since it can form very stable complexes in aqueous solution with Cd, Cr, Cu, Ni, Pb, and Zn ions (Tuin and Tels, 1990b). EDTA alone can remove over 90% of the lead and cadmium (Ellis et al., 1986). Two types of metal extraction agents, acids and chelating agents (Ehrenfeld and Bass, 1984; Rulkens and Assink, 1984), have been used for the removal of heavy metal from soil. Both of these extraction agents have shown a high efficiency in soil washing technology (Tuin and Tels, 1990a; Allen and Chen, 1993; Peters and Shem, 1992). However, after acid washing, the physical and chemical properties of the soil can drastically change from the natural condition. Farrah and Pickering (1978) found greater removal efficiencies for Cd and Pb with EDTA than with HCl in the case of artificially contaminated clay minerals. Norvell (1984) measured twice as large removal of Cu and Ni with 0.1 N HCl than with 0.005 M EDTA at pH=5.3; the amount of Cd extracted was about the same with HCl and with EDTA. Tuin and Tels (1990b) obtained a higher extraction percentage when 0.1 N HCl was applied before 0.005M EDTA, and the result could be further improved by using a higher EDTA concentration. Elliott et al. (1989) and Allen and Chen (1993) showed that with EDTA, pH does not influence the extraction efficiency. Peters and Shem (1992) found that over 64% of Pb could be removed by EDTA at the pH range of 4 to 12, with pH causing only a minor effect on the extraction efficiency.

In this study, we explored the effectiveness of : (1) in situ soil flushing, and (2) ex situ soil washing to remove inorganic contaminants in sandy soil and clay soil. The extractant solutions applied in this experiment were HCl and EDTA.

MATERIALS AND METHODS

Physical and Chemical Characterization of the Contaminated Soil

A sandy soil and a clay soil were obtained from a beach and a farming garden in the rural areas in Hong Kong, respectively. The physical characteristics of the soil are summarized in Table 28.1. Four metals (Pb, Ni, Cu, and Cd) were added to the sandy soil in order to make the concentration of the metals in the soil exceed the Dutch "C" value (Cariney, 1993). The soil was then air-dried, homogenized, and passed through a 2-mm sieve. The four metals (Pb, Ni, Cu, and Cd) nitrate were mixed with the soil for half an hour by using a portable concrete mixer. After mixing, the concentrations of metals in the soil were extracted and analyzed according to EPA Method 3051 (U.S. Environmental Protection Agency, 1986). The concentrations of metal in the soils are shown in Table 28.2.

Soil Flushing

Four PVC columns (0.5 m high and 0.2 m diameter) were constructed for the soil flushing test. A porous stone filter (50 mm thick 10-mm size aggregate) was placed at the bottom of each column to prevent the soil from being washed out. A detailed drawing is shown in Figure 28.1. The following soil packing procedure was used: each 100-mm soil layer was compacted by giving it 10 blows with a hammer to obtain a bulk density of about 1,035 kg m^{-3} and a porosity of about 0.7. The packed columns were kept in a vertical position and saturated with distilled water at a slow rate to remove the air bubbles trapped in the soil columns. The water-saturated soil column was allowed to drain under gravity as the inflow was the same as the outflow through the columns. The drained fluid was collected and analyzed for metals to calculate the residual metals remaining in the columns.

Two types of extraction agents (HCl and EDTA) were used. The extraction solvents were pumped into the columns by a Microprocessor Pump Drive® (Cole-Parmer Instrument Company through Masterflex Tygon Tubing size No. 18). The inlet flow rates for both columns were at 40 mL min^{-1}. The leachants were allowed to drain under gravity through the outlet to the glass collection bottles. Each day, half pore-volumes of the extraction solvents were pumped into the columns. The leachates were collected and the concentrations of the four metals determined by an Atomic Absorption Spectrometer (AAS). The removal efficiencies of heavy metals by soil flushing with different extraction solvents in synthetic contaminated sandy soil were determined. The distributions of metals concentrations in the soil profile with the depths after the soil flushing were also analyzed.

Soil Washing

Air-dried synthetic contaminated clay soil was placed in a glass beaker. At time zero, extractant solution was pipetted into the beaker and the suspension was mixed by means of a magnetic stirrer. 0.1M EDTA and 0.1N HCl were used as the extractant solutions for soil washing. The liquid/solid ratios were 5, 10, 100, and 200. After 1 h of extraction, the suspension was centrifuged and samples were taken from the supernatant liquid and analyzed for metal concentrations by AAS.

Two shaking methods (rotatory and orbital) were investigated under different extraction periods. 10 g of air-dried contaminated clay soil was placed in a 500-mL extraction bottle. 50 mL of 0.1N HCl was pipetted into the bottles and shaken in rotatory and orbital modes. The durations of the extractions were 1, 5, and 24 h. The revolution rate of both

Table 28.1. Physical Properties of the Soil.

	Sandy Soil	Clay Soil
Particle size Distribution[a]	96% sand, 4% gravel	100% clay
Median diameter[a]	0.42 mm	—
Organic matter[b]	0.82%	8.87%
pH value[c]	8.76	6.55

[a] By the ASTM, 1987, Method D 422-63.
[b] By *Standard Methods for the Examination of Water and Wastewater*, 18th edition, 1992, Method 2540E.
[c] By American Standard Method of Soil Testing 2e, part 2.

Table 28.2. Metals Concentrations (mg kg^{-1}) in Natural Soil and Synthetic Contaminated Soil.

Element	Soil Flushing Natural Sandy Soil	Soil Washing Contaminated Sandy Soil	Natural Clay Soil	Contaminated Clay Soil	Dutch "A" Value	Dutch "C" Value
Cadmium	below mdl.[a]	24	below mdl.[a]	28	1	20
Copper	53	940	48	1031	50	500
Lead	178	1060	46	826	50	600
Nickel	31.3	978	23	929	50	500

[a] mdl = minimum detection limit.

the orbital and rotatory shakers was 30 cycles/min. After the extraction, the suspension was centrifuged at 1600 rpm and samples were taken from the supernatant liquid and analyzed for the metals concerned by AAS.

RESULTS AND DISCUSSION

Soil Flushing

Removal Efficiency of Metals

- For lead, as shown in Figure 28.2a, the efficiencies of removal were high for both extractants. Both cases could achieve a removal rate above 94% after 10 pore-volume (PV) of flushing. Their removal efficiencies were similar. But in the initial stage (1–5 PV), the removal rate of 0.1 M EDTA was much higher than that of 0.1 N HCl.
- For nickel, the removal efficiencies were over 89% in both columns, and the higher efficiency (94%) was with 0.1 N HCl (Figure 28.2b). In the first 2.5 PV flushing, the rates of nickel leached out from both columns increased significantly, and after then, the extents of increase became insignificant.
- For copper, the results of removal in each column were plotted in Figure 28.2c. In the first 3.5 PV, the removal rate of copper with 0.1 M EDTA was higher than that of the others. After 3.5 PV, additional flushing did not cause any significant increase in leaching, and it was similar to the case of lead.
- For cadmium, the removal efficiencies for each column were about 98% (Figure 28.2d). Similar to that of Ni, the rates of Cd leached out from both columns were rapid in the first 2.5 PV.

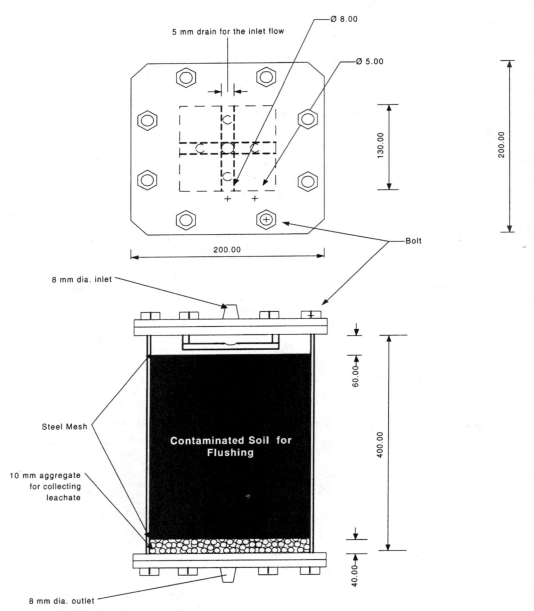

FIGURE **28.1.** The sketch of the contaminated soil reactor.

Extractability of Metals

The extractability of the four metals from the synthetic contaminated sandy soil decreased in the order: Cd > Pb > Cu > Ni. Harter (1983) and Tuin and Tels (1990b) also reported a lower extraction percentage for Ni compared to Cu, Pb, and Cd in soil-washing technology. For the artificially contaminated clay soil, sequential extraction measurements proved greater Ni contents in the residual fractions (Tuin and Tels, 1990a). Pb and Cd

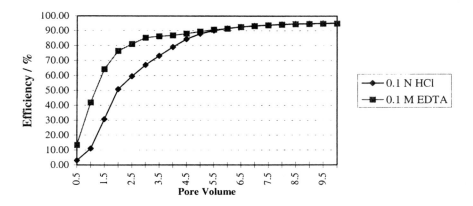

FIGURE 28.2a. Removal rate of Ph with different solvents soil flushing.

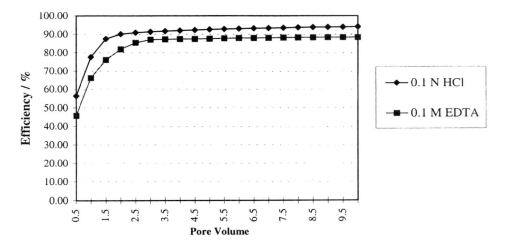

FIGURE 28.2b. Removal rate of Ni with different solvents soil flushing.

were not strongly bound to the sandy soil. Cu and Ni could be bound to interlayer sites which are resistant to removal by the extractant. The ionic radii of the four metal ions are shown in Table 28.3 (Weast et al., 1982). For Pb and Cd ions, they are too big to bind to the interlayer sites. In the case of extraction by HCl, Pb and Cd can form relatively soluble chloride complexes (Kragten, 1978), which may increase their removal efficiencies. With EDTA, the high Pb and Cd removal efficiencies could be explained by their physical ionic radii. These ions are generally bound to the more exterior sites in the soil as compared with Cu and Ni. These exterior sites can easily be reached by the EDTA, and the metal ions may form soluble complexes which are more amenable to extraction.

Comparison Between HCl and EDTA

In the first 5 PV flushing, the extraction efficiencies of Pb and Cu by 0.1 M EDTA were greater than that of 0.1 N HCl. Also, in the first PV, Pb and Cu removed by 0.1 M EDTA were twice as much as that of 0.1 N HCl.

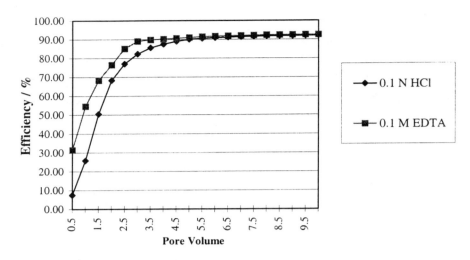

FIGURE 28.2c. Removal rate of Cu with different solvents soil flushing.

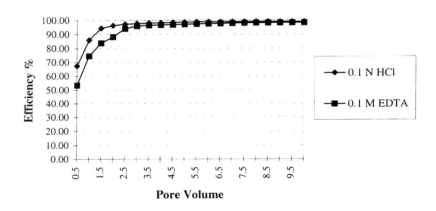

FIGURE 28.2d. Removal rate of Cd with different solvents soil flushing.

Table 28.3. Ionic Diameters of the Four Metal Ions.

Metal Ion	Pb^{2+}	Cd^{2+}	Cu^{2+}	Ni^{2+}
Crystal ionic radius (Å)	1.20	0.97	0.72	0.69

Soil Washing

Removal Rate by EDTA and HCl

Among the four different metals washed by HCl/EDTA solvents, higher removal rates of Ni and Cd were obtained with 0.1N HCl (Ni: 26% and Cd: 100%) than that with 0.1M EDTA (Ni: 21% and Cd: 95%). For Pb and Cu, the removal rates of metals were higher with 0.1M EDTA (Pb: 80% and Cu: 72%) than with 0.1N HCl (Pb: 78% and Cu:

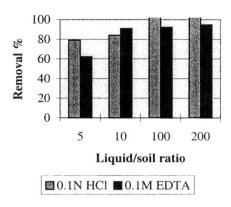

FIGURE 28.3a. Removal rate of Cd with EDTA and HCl.

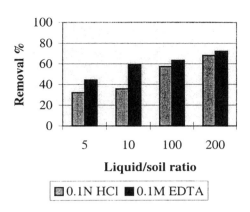

FIGURE 28.3b. Removal rate of Cu with EDTA and HCl.

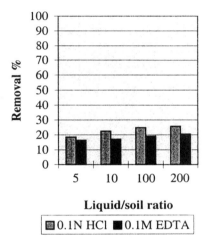

FIGURE 28.3c. Removal rate of Ni with EDTA and HCl.

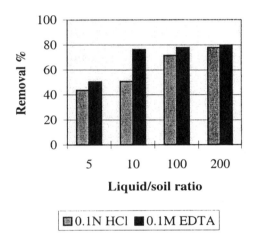

FIGURE 28.3d. Removal rate of Pb with EDTA and HCl.

68%). The results were similar to those obtained in the soil flushing experiment. The removal rates of Pb and Cu with EDTA were about twice that of HCl in the initial phase of the flushing.

The Effect of Liquid Solid Ratio

The removal rates of the metals with EDTA and HCl in different liquid/solid ratios were plotted and shown in Figure 28.3a to Figure 28.3d. For both EDTA and HCl, the removal rates of all four metals increased as the liquid/solid ratio increased. But for Ni, the effect of the liquid/solid ratio was not significant. When higher liquid/solid ratios were used, little improvement in Ni removal was found. The atomic radius of Ni was the smallest compared with the other three metals. So it was probably bound more strongly

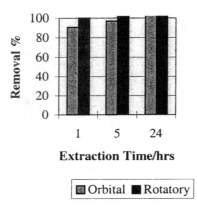

FIGURE 28.4a. Orbital vs. rotatory shaking (Cd).

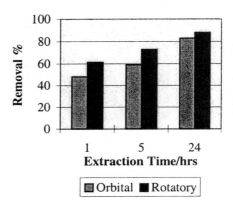

FIGURE 28.4b. Orbital vs. rotatory shaking (Cu).

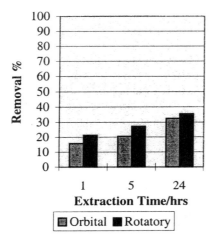

FIGURE 28.4c. Orbital vs. rotatory shaking (Ni).

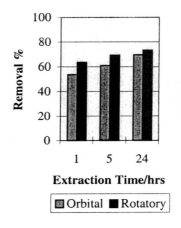

FIGURE 28.4d. Orbital vs. rotatory shaking (Pb).

to the clay components, and the extraction efficiency did not increase with the liquid/solid ratio.

The Effect of Shaking (Orbital and Rotatory) Mechanism

The removal rates of metals with orbital and rotatory shaking at different shaking times are shown in Figure 28.4a to Figure 28.4d. In this experiment, 0.1N HCl was used as the extractant and the rate of revolution (30 cycles min^{-1}) was the same. The rotatory shaking achieved a higher removal rate with the four metals than the orbital shaking. When the time of extraction was increased from 1 hour to 24 h, the difference between the removal efficiencies was reduced. The results showed that rotatory shaking provided better mixing and more contact area between the extractant and the soil, which reduced the time required to reach equilibrium. With orbital shaking, the desorption rates of the metals in the soil were much slower and the equilibrium can only be reached when time of extraction was increased.

Extractability of the Four Metals

The extractability of the four metals in the synthetic contaminated clay soil decreased in the order: Cd > Pb > Cu > Ni, which is similar to that of the soil flushing results.

CONCLUSIONS

In this study, the removal efficiencies of four metals (Pb, Ni, Cu, and Cd) with different extraction solvents were investigated using soil flushing and washing.

Soil Flushing

- The removal efficiencies of heavy metals (Pb, Ni, Cu, and Cd) by EDTA and HCl were above 90% after 10 PV (pore-volume) flushing. No significant difference was observed for the two solvents after 10 PV, but in the first 1–3 PV, the removal efficiencies of EDTA with Cu and Pb were nearly two times higher than that of HCl.
- The extractability of the four metals from the synthetic-contaminated sandy soil decreased in the order: Cd > Pb > Cu > Ni.

Soil Washing

- The removal efficiencies of Pb and Cu with EDTA were 80% and 72%, respectively, and were better than that of HCl (Pb: 78%, Cu: 68%). But for Cd and Ni, the removal efficiencies by HCl (Cd: 100%, Ni: 26%) were better than by EDTA (Cd: 95%, Ni: 21%).
- When the liquid/solid ratio increased, the removal rate of all four metals increased, although a smaller increase was observed with Ni. It may due to the fact that Ni binds more strongly to soil components.
- At the same revolution rate, rotatory shaking provided a better mixing and contact area between the extractant and the soil than orbital shaking, but with increasing extraction time, the differences between rotatory and orbital shaking reduced.

REFERENCES

Allen, H.E. and P.H. Chen. Remediation of Metal Contaminated Soil by EDTA Incorporating Electrochemical Recovery of Metal and EDTA, *Environ. Progress*, 12(4), p. 86, 1993.

Cairney, T. *Contaminated Land Problems and Solutions*, Blackie Academic & Professional, United Kingdom, 1993.

Ehrenfeld, J. and J. Bass. 1993. *Evaluation of Remedial Action Unit Operations at Hazardous Waste Disposal Sites*, Noyes Publications, Park Ridge, NJ, 1993.

Elliott, H.A., J.H Linn, and G.A. Shields. Role of Fe in Extractive Decontamination of Pb-Polluted Soils, *Haz. Waste Haz. Matter*, 6, pp. 223–229, 1989.

Ellis, W.D., T.R. Fogg, and A.N. Tafuri. Treatment of Soils Contaminated with Metals, *Proceedings of the 12th Annual Research Symposium: Land Disposal, Remedial Action, Incineration and Treatment of Hazardous Waste*, pp. 201–207, 1986.

Farrah, H. and W.F. Pickering. Extraction of Heavy Metal Ions Sorbed on Clays. *Water Air Soil Pollut.*, 9, pp. 491–498, 1978.

Harter, R.D. Effect of Soil pH on Adsorption of Lead, Copper, Zinc and Nickel. *Soil Sci. Soc. Am. J.*, 47, pp. 47–51, 1983.

Holden, T. et al. How to Select Hazardous Waste Treatment Technologies for Soils and Sludges. *Pollut. Technol. Rev.*, Noyes Data Corporation, Park Ridge NJ, 1989, p. 163.

Kragten, J. *Atlas of Metal-Ligand Equilibria in Aqueous Solution*, Ellis Horwood, Halsted Press, 1978.

Norvell, W.A. Comparison of Chelating Agents as Extractants for Metal in Diverse Soil Materials. *Soil Sci. Soc. Am. J.*, 48(6), pp. 1285–1292, 1984.

Peters, R.W. and L. Shem. *Use of Chelating Agents for Remediation of Heavy Metal Contaminated Soil, in Environmental Remediation.* G.F. Vandegrift, D.T. Reed, and I.R. Tasker, Eds., ACS Symposium Series 509, American Chemical Society, Washington, DC, 1992, pp. 70–84.

Rulkens, W.H. and J.W. Assink. Extraction as a Method for Cleaning Contaminated Soil: Possibilities, Problems and Research, *Proceedings of the 5th National Conference on Management of Uncontrolled Hazardous Waste Sites*, Washington, DC, 1984, pp. 576–583.

Tuin, B.J.W. and M. Tels. Removing Heavy Metals from Contaminated Clay Soils by Extraction with Hydrochloric Acid, EDTA or Hypochlorite Solutions. *Environ. Technol.*, 11, pp. 1039–1052, 1990b.

Tuin, B.J.W. and M. Tels. Distribution of Six Heavy Metals in Contaminated Clay Soils Before and After Extractive Cleaning. *Environ. Technol.*, 11, p. 935, 1990a.

U.S. Environmental Protection Agency. *Test Methods for Evaluating Solid Waste: Physical/Chemical Methods*, 3rd ed., U.S. Environmental Protection Agency, Office of Solid Waste and Emergency Response, Washington, D.C., Office of Solid Waste and Emergency Response, 1986.

Van Benschoten, J.E., B.E. Reed, M.R. Matsumoto, and P.J. McGarvey. Metal Removal by Soil Washing for an Iron Oxide Coated Sandy Soil. *Water Environ. Res.*, 66(2), p. 168, 1994.

Weast, R.C. et al. *Handbook of Chemistry and Physics*, 62nd ed. CRC Press, Boca Raton, FL, 1982.

Chapter 29

Vetiver Grass for Erosion Control, Hong Kong and Guizhou, China

R.D. Hill and M.R. Peart

INTRODUCTION

Both Hong Kong and Guizhou lie in China's southern erosion region (Wen, 1993). This region is characterized by steep slopes, mostly at moderate elevations, deeply-weathered soil parent materials, annual rainfall totals mainly in the range of 1100–2500 mm, and high rainfall intensities, especially in the coastal zone open to typhoons, during which 500–600 mm of rain in two or three days are not uncommon. The original forest has long ago been removed from the uplands, for timber and fuel, as well as by forms of shifting cultivation. To some degree, especially in Hong Kong, native forest has been replaced by planted forests of *Acacia*, *Pinus*, and *Cunninghamia*, with scrub, grass, and *Dicranopteris* fern forming widespread secondary formations, their status being maintained by more or less regular cutting and burning. Contrary to general opinion, it seems possible that surface erosion under such secondary vegetation communities is quite limited, provided that the cover is closed. Cultivation is of much greater erosional significance outside Hong Kong. In Guizhou Province, for instance, more than half of the cultivated land area is sloping, much of that fraction growing rice or maize in summer and wheat or broad beans in winter, sometimes on slopes as steep as 35°, a practice technically illegal but one imposed by poverty and a lack of feasible alternatives. Some 40% of the land area in the province is regarded as 'eroded,' a proportion that is steadily increasing. This contrasts with a mere 4%, 4400 ha, of eroded badland in Hong Kong, where some 67.5% of the land area is reasonably well protected by forest plantings, scrub, grassland, and *Dicranopteris* fern (Hong Kong, 1995). However, the work of Peart (1995, 1997) shows that small areas of bare land such as those resulting from roadworks and powerline corridors can greatly influence sediment concentrations in streams.

Authorities in both regions have attempted, not always successfully or quickly, to control sediment generation on slopes. These attempts have been both biological and nonbiological, the latter being exemplified by the use of *chunam* and 'shortcrete' on landslip failures. Biological control can be classified into barrier and cover approaches. This chapter, in so far as Vetiver grass is concerned, addresses the former approach: the use of barriers to check runoff and cause sediment deposition. In contrast, the cover approach using trees for erosion control has a number of disadvantages. First, on rapidly-eroding substrates, especially those which are nutrient-poor, trees are quite ineffective. Failure to thrive is common, not merely because of soil-nutrient problems but also because the relatively slow growth rate of trees permits erosion to cause root exposure. This further exacerbates the problem. As Lam's data for Hong Kong badlands indicate, about 1.5 cm of soil loss annually is not uncommon (Lam, 1978). In such circumstances, where trees survive, as at Tai Lam Chung, it takes several years for a closed canopy to develop. Second, even where a closed cover develops, the species commonly used, coupled with even-aged stands of fairly uniform tree-heights, often lead to rather sparse litter and ground cover. Significant erosion, notably by drip, may continue even under the closed canopy. As Young (1989) notes: "... a canopy of trees more than a few meters high is not expected to substantially reduce erosion, other than by the litter which falls from it." In a word, trees do not necessarily control erosion and are slow to act in a controlling role. Finally, trees can also be difficult to establish on sites such as ridge crests which can be prone to water shortage, especially if there are pronounced dry seasons.

These considerations have, in the wider region of Southeast Asia, led to the use of quick-growing perennial herbs and grasses. In Malaysia, for example, leguminous creepers such as *Pueraria*, *Calopogonium*, and *Centrosema* are widely used in vulnerable rubber and other tree-crop plantations (Burkill, 1966). In Hong Kong, especially on slopes cut in the course of civil engineering works, hydroseeding of grass seed, fertilizer, and an organic 'carrier' (usually macerated waste paper), onto textile-covered slopes is well-established as a method of choice. Both approaches have limitations. Quick-growing creepers cannot be established in dry weather and there is always the risk of high-intensity rainfall, in southern China usually originating from troughs or upper-air disturbances in late spring, before a good cover is established. Creepers have further disadvantages in that they give no direct economic return and, though effective in controlling runoff and erosion, they may quickly 'swamp' trees that may be planted or at the very least, require regular cutting back or herbicide application to prevent smothering. Hydroseeding is expensive and basically limited to areas accessible to truck-mounted pumps. While covering slopes with geotextile does offer some protection to slopes, there is nevertheless a 'window of vulnerability' before the grass becomes established and even then, grass is ineffective in trapping gravel and larger materials because those used are fairly flaccid. In addition, after the grasses flower, abundant seed may be produced and this is inappropriate in an agricultural context because it adds to potential weed problems.

With these matters in mind, Vetiver grass, *Vetiveria zinzanioides* L., was introduced to Hong Kong (by Hill) in 1990 with the objective of devising an appropriate planting strategy for badland areas, most of which are beyond the physical reach of hydroseeding. Subsequently, with Chen Xuhui, Hill was able to set up a small erosion control project at Luodian, southern Guizhou province, where it had already been shown that Vetiver grass would grow satisfactorily. At both locations the major task has been to develop planting strategies and at Luodian to test the grass in an agricultural context, in the interest of lowering costs, preferably without fertilizer application.

Vetiver grass is by origin a grass of seasonal swamps in the Indian subcontinent. Its leaves and culms are notably close-growing, stiff, erect, and nonspreading, the last an important characteristic in agricultural contexts. Having been taken into cultivation (for its essential oil) long ago and being commonly reproduced asexually, most cultivars do not set fertile seed, again making it appropriate for agricultural areas. Its use for erosion control probably goes back some hundreds of years but in more recent times it has been used to reduce runoff and sediment loss, especially on cultivated clayey soils such as those derived from volcanics in Fiji where sloping lands are used for sugar cultivation. Vetiver grass hedges used in such situations have built up terraces several meters high. It is hardy, though growth naturally slows under cold or under stress of dryness, easily established and cheap to maintain. It recovers quickly following burning but is not shade-tolerant, though it will survive under moderate shade.

While proven effective in reducing runoff and erosion, it has further benefits. Vetiver grass performs better if periodically cut back to a height of 30–50 cm from the roughly 2 m to which it will otherwise grow. Cutting promotes tillering and reduces the shading effect upon crops nearby. The cut material is a medium-quality feedstuff for animals and fish such as grass carp. It can be used for thatching material and in handicrafts such as making sun-hats, bags, and the like. Work (by Chen Rongjun) at the Kadoorie Agricultural Research Centre in Hong Kong has demonstrated that as fuel, dried Vetiver grass has a similar calorific value to the better native grasses of the region.

MATERIALS AND METHODS

In 1990, a stock of Vetiver grass slips was obtained from the South China Botanical Institute. The slips were planted out on a level terrace consisting of a sandy loam soil at the Kadoorie Agricultural Research Centre (K.A.R.C.) in the New Territories of Hong Kong. From this stock sufficient material was generated to undertake some field trials to determine whether Vetiver could be used for control erosion in the territory. The various trials in Hong Kong are described in Table 29.1 and their location given in Figure 29.1.

It was also desired to assess the performance of Vetiver grass in an agricultural context. Consequently, in conjunction with Professor Chen Xuhui of the Guizhou Academy of Agricultural Science, field trials were undertaken at Luodian, Guizhou, P.R.C. (for location see Figure 29.1). Multiple double-row hedges were planted in April 1995 on slopes of around 20°. Soils were angular, pebbly, fine sandy clay loam to clay loam on weathered shales. Herbaceous fodders were planted between the Vetiver hedges. Eight fertilizer treatments were compared with one control, five replicates of each, (a) on farmers' land and (b) on abandoned, bare, eroded land. Further treatment details can be found in Chen Xuhui and Hill (1996) and are summarized in Table 29.1.

RESULTS

In Hong Kong the success of the Vetiver grass trials was judged according to the strike rate of the planting, their long-term survival, and the ability of the hedge to trap sediment and form a terrace. The various trials are summarized in Table 29.2, and it can be concluded that some success was obtained in the use of Vetiver.

For example, in Hong Kong all the Vetiver grass hedges developed terracettes consequent to sediment deposition, some of which were over 20 cm high. At 2 out of the 7 sites the terracettes became unstable due to the use of single-row planting which permitted breaks in the hedge to occur too easily. Measured strike rates of planted material in general exceeded 83%. The trial to stabilize gullies at Jordan Valley was not successful, with much of the Vetiver grass being washed out.

At Luodian, Guizhou Province, the trials show that on farmers' land Vetiver grass hedges, as measured by dry-matter production, do best without fertilizer, a surprising but welcome finding for those erodible shale soils. Even on abandoned degraded land, Vetiver grass without fertilizer produces 270 g m^{-2}, about 30% less than with the best fertilizer regime, three split applications of NPK (340 g m^{-2}), or farmyard manure (293 g m^{-2}). However, performance was measured by dry-matter production and the less productive

Table 29.1. Summary of Vetiver Trials Methodology.

A. **Kadoorie Agricultural Research Centre**
   Site 1: Four double-row hedges planted 7/91. Cut slope 40°. Bouldery sandy clay loam on
   old volcanics. No fertilizer applied.
   Site 2: Four double-row hedges planted 8/91. Cut slope 40°. Bouldery sandy clay loam. No
   fertilizer applied.

B. **Jordan Valley**
   Site 1: Two double-row hedges planted 10/92. NPK + Alginure at planting, no maintenance
   fertilizer. 2 m below ridge, 12–15°. Sandy clay loam on weathered granite (subsoil),
   hard-pan at 20 cm. With *Acacia confusa* 3 m apart.
   Site 2: Three single-row hedges planted 6/93. NPK + Alginure on planting, no maintenance
   fertilizer. Rilled slope 18°–20°. Sandy clay loam on weathered granite (subsoil). With
   Acacia confusa, A. mangium, Casuarina, 3 m apart.
   Site 3: Four single-row hedges planted 6/93. NPK + Alginure on planting, no maintenance
   fertilizer. Cut slope 2°–5°. Sandy to coarse sandy clay loam on weathered granite.
   With *Acacia confusa, A. mangium,* 3 m apart.
   Gullies: Four gully-mouths double-row planted, 10/92 and 6/93. NPK + Alginure on planting,
   no maintenance fertilizer. Loose alluvial sand to coarse sand from weathered granite.
   No trees planted.

C. **Wan Chuk Yuen** - trial to test post-fire survival. 300 slips in 3 hedges on burnt-over slope of
   10° on a ridge, planted 1-93.

D. **Shau Kei Wan** Three single-row hedges planted 8/94, doubled up 9/95. NPK + Alginure on
   planting and maintenance NPK (13, 13, 21%) 10 g m^{-1} 4/96. Slope 15–25°. Sandy clay to
   coarse sandy clay on weathered granite. With *Acacia confusa* and *A. mangium* 3 m apart.

E. **Mount Parker** Four double-row hedges planted 5/96. NPK + Alginure on planting and
   maintenance NPK (13, 13, 21%) to be applied at 10 g m^{-1}. Slope 15°–20°. Pebbly, bouldery,
   sandy clay loam on crusted weathered granite (subsoil). No trees.

F. **Luodian Guizhou, PRC.** Fertilizer trial (see Chen and Hill, 1996). Multiple double-row hedges
   planted 4/95. Eight fertilizer treatments with one control, five replicates of each, on (a)
   farmers' land (b) abandoned, bare, eroded land. No Alginure. Slope 20°. Angular pebbly fine
   sandy clay loam to clay loam on weathered shale. Herbaceous fodders planted between
   hedges.

hedges were nevertheless highly effective in trapping sediment. The findings are summarized in Table 29.2.

DISCUSSION

A number of general considerations in the use of Vetiver grass have also emerged from
the trials. The use of Vetiver grass necessitates production of young, but not too young,
vigorous slips for planting out. Experience has shown that timing and methods of production in the nursery are crucial in obtaining good strike-rates after planting. However, since
costs vary according to production method, transportation, planting and labor costs, no
single approach will invariably be optimal.

Yoon (1995) warns against bare-root planting, noting that new roots are only formed
from new tillers or from the nodes of old culms. He says, "Any planting using slips with
cut-roots would be very slow to establish and grow." This has not been our experience in
Hong Kong, where all plantings including the eroding site trials have been of this nature.
The first planting on a cut slope at K.A.R.C. had negligible failures. A planting at Jordan
Valley, H.K., into residual soil moisture in October 1993 had a 99% strike rate 6 months

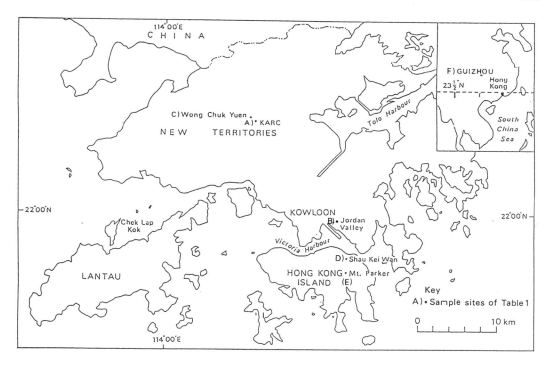

FIGURE 29.1. Location of study sites.

after planting. A planting at Mt. Parker early in the 1996 season had only 83% strike rate but that experienced a period of no rain for two weeks following planting. That so many survived in a very difficult location is an indication of Vetiver's hardiness. The secret is to use young tillers. Clumps raised in the nursery are simply dug up and split into smaller clumps containing 3–4 tillers.

In engineering applications, Yoon (1995) advises the use of polybags, filled with good soil, of a minimum size of 10×15 cm, with three tillers per slip. Alternatively, foam (under mist) may be used and this reduces subsequent transportation costs by 75%. Other potting materials and other types of containers were found by him to be generally disappointing. In Hong Kong good soil is difficult to obtain and expensive and the polybag approach, being labor-intensive, is thus too expensive for general use.

Age of slips seems to have a significant impact on the success of Vetiver grass establishment. Yoon (1995) found that subsequent root development was best in slips 16–23 weeks old, older material proving slow to tiller. Our work in Hong Kong confirms this, one planting at Jordan Valley probably suffering from using over-aged slips.

In terms of transport to the site, once Vetiver grass bare-root slips, each containing three tillers, have been lifted from the nursery, surplus soil can be knocked off to save space and weight, a major consideration on sites difficult to access. Standard sources such as Vetiver Information Network (1993) suggest trimming both shoots and roots before transportation, but neither is necessary or desirable. The slips are quite hardy and if kept shaded and moist, will last several days if circumstances prevent same-day planting. Polybag slips, of course, will survive longer. Where access is very difficult, clean bare-root slips can be air-dropped for subsequent planting without damage.

Table 29.2. Results of Vetiver Trials.

A. Kadoorie Agricultural Research Centre
Site 1: 100% strike-rate. Mostly destroyed by rodents 1-4/94 but stable terracettes formed.
Successful

Site 2: 100% strike-rate. Survives, stable terracettes present.
Successful

B. Jordan Valley
Site 1: 99% strike-rate. One hedge burnt, 1994 dry season - rapid regeneration. 1996 -
hedges quiescent under closed cover of *Acacia* 2.5–3.0 m high. Litter 3–5 cm deep.
Stable terracettes developed.
Successful

Site 2: Strike rate not measured. 1996, Vetiver, poor shoot growth, good root growth; trees
0.5–2.0 m high. No litter, terracettes unstable, slopes between edges eroded, rills
deeper and more extensive.
Unsuccessful

Site 3: Strike rate not measured. 1996: Vetiver—very poor growth, stunted, many gaps; trees
0.5–1.5 m, stunted, roots exposed. No litter, terracettes unstable, surface wash
continuing.
Unsuccessful

Gullies: Strike rate not measured. 1996: Many gaps though some survivals, slips washed out
bodily. Gullies still active.
Unsuccessful

C. Wan Chuk Yuen: Trial to test post-fire survival. Survival at 29/3/93: 18% blown out, 39%
dead in ground, 44% growing, moderate growth. Burnt 9-10/12/95. Regenerates as well as
native *Arundinella* and better than native *Dicranopteris* (fern) and *Imperata* sp.
Successful

D. Shau Kei Wan: Strike rate >99%. 1996: Vetiver well-grown giving dense, continuous hedges,
trees 0.5–1.5 m, stunted, roots exposed. Terracettes stable, 20–30 cm high, some colonizing
adventives.
Successful

E. Mount Parker: Strike rate 83%. 10/96: fair growth, some minor gaps (filled by pinning
culms), some adventives colonizing stable terracettes 10–20 cm high.
Fairly successful

F. Luodian Guizhou, PRC: 100% strike rate. Stable terracettes 15–25 cm high. Fertilizer unnecessary on farmers' land, acceptable performance on bare land without fertilizer. Vetiver
hedges without fertilizer only about 12% less effective than growing it with the most effective
fertilizer treatment, namely NPK in three split applications during the growing season.
Successful

The trials reveal that the key to successful erosion control is correct planting of good-quality material. Particularly on rapidly-eroding substrates it is false economy to plant in single rows, though current international practice is to use this pattern in order to make planting material go as far as possible. All of the early plantings in Hong Kong were in double rows, on the contour, at 15–20 cm spacing. The double-row plantings have gener-ally been successful whereas single-row planting have not (e.g., Jordan Valley sites 2 and 3). In single-row plantings, the failure of a single slip may result in concentration of the flow of water and entrained sediment, thereby breaking the barrier.

Hedges can be laid out on the contour by eye, although the use of an A-shaped frame or a level made of transparent plastic pipe filled with water will assist accuracy. Slips should be planted in rows 15–20 cm apart, with individual slips being planted every 15–20 cm within each row. It is recommended by both the Vetiver Information Network (1993) and the U.S. National Research Council (1993) that hedges be set 1 m apart, but the work at K.A.R.C. and other locations suggests this may be conservative. However, it is better to be conservative and prevent concentrated overland flow from developing because it has the ability to erode and destroy the hedge. Retroactive restoration will be expensive.

On nutrient-poor substrates, as in Hong Kong, NPK (13-13-22) slow-release, granulated fertilizer is applied at 5 g per bare-root slip, together with a small handful of 'Alginure,' a proprietary water-retaining gel. The latter is of some importance since dry spells of 1–2 weeks frequently occur even during the rainy season. Our practice has been to plant untrimmed slips but, once planted, to trim the shoot back to a height of about 30 cm. This reduces transpiration losses and may stimulate root growth. The trimmings are not discarded but are placed immediately upslope of the newly-planted hedge, where they assist in the trapping of moving sediment and after decaying, which takes 12–15 months, add organic matter to the soil.

The timing of planting out is not particularly crucial, other than to avoid extended drought. The main consideration in tropical and subtropical areas is soil moisture. Short dry spells of 3–5 days seem to have little effect upon strike-rates but longer dry spells clearly do. In southern China and Southeast Asia, data on soil moisture, by lysimetry, is very scanty and for hill soils is virtually nonexistent. In areas affected by the dry monsoon it seems likely that the period from November to April is too dry for planting unless hand-watering is possible. In Hong Kong, October plantings, on residual soil moisture, have been perfectly satisfactory. Temperature is less of a problem. Although growth is slow or nonexistent below about 15°C, Vetiver grass will survive below this threshold and will also survive short frost periods or near-frost conditions. Provided that water supply is adequate, heat stress is not a problem. Since the growing shoot is semisubterranean, regrowth following severe stress, even fire, is usually rapid and vigorous, as our experience in Hong Kong has shown.

Several trials were also made at Jordan Valley to try to control runoff at the mouths of gullies, using short double-row hedges. These were not successful, though isolated plants did survive because the slips were washed out by the force of water. Deep planting, the shoot being placed 10 cm below the surface at planting, was also unsuccessful because growth was poor.

Vetiver grass hedges need little maintenance. Most crucial is the early filling of any gaps. This is particularly important on rapidly-eroding substrates since a gap will concentrate flow, whereas the whole objective of the hedge is to spread it. Gaps can be filled either by replacement planting with new material or by bending culms over and pinning them to the ground to promote rooting at the internodes. The latter can also be used readily to extend a hedge up or down a slope. Since trimming is a maintenance operation, it is obvious that if culm-pinning is to be used, trimming should be postponed. Trimming of Vetiver grass hedgerows has the effect of promoting tillering. Once a year is a minimum. Trimmings should be gathered and placed upslope of the hedge, as with the original planting.

On nutrient-poor substrates, once or twice yearly maintenance application of a slow-release NPK fertilizer is highly desirable, although as organic matter builds up along the hedge this probably becomes less crucial with time. On other soils, fertilizer application may not be necessary. The Guizhou trial showed that on farmers' land, growth without any fertilizer at all was marginally better than performance under the most-effective fertilizer application regimes, NPK in split applications, and farmyard (organic) manure. On bare, eroding land, the performance of Vetiver grass without fertilizer was somewhat inferior to most fertilizer treatments, which on average gave 30% higher fresh-grass yields. Nevertheless, the performance of unfertilized Vetiver, so far as erosion control was concerned, was wholly satisfactory in this trial (Chen and Hill, 1996).

The trials at Jordon Valley, Sites 2 and 3, were without fertilizer beyond the initial application, and involved the simultaneous planting of trees, mainly Acacias and Casuarinas. Shoot growth was very poor though root growth was good. It is possible that the soil nutrient status was so marginal that trees compete with the Vetiver grass for what little is available, to the detriment of the latter. However, the combination of good root growth and very poor shoot growth remains a puzzle (currently under investigation). Soil depth is not a problem, in contrast to Site 1 at Jordan Valley where the soil, also on decomposed granite, is very hard at a depth of 20 cm (average penetrometer reading 4.5 g cm^2).

Vetiver grass has generally been found to be remarkably free from pests and other problems, though a stem borer has been reported in China. The Guizhou trial, now in its second year, has not shown any problems, using double-row planting as described earlier. One of the two original plantings on a steep cut slope at the Kadoorie Agricultural Research Centre was severely damaged (by rodents), but only after a closed cover of adventive plants had developed. The slope had been, and remains, stabilized. (That damage was probably due to house rats being deprived of their usual food waste from a nearby kitchen, a nearby planting entirely escaping damage.)

While Vetiver grass has had some success it may not be suitable for all applications. For example, it could be argued that Vetiver hedges, while preventing erosion, would be rather unsightly in the Country Parks of Hong Kong. It may also be less desirable than trees in terms of habitat.

An important consideration in erosion prevention is cost, and in general this is very much dependent upon the availability and cost of labor and planting material. K.A.R.C. charges HK $4 per clump, from each of which 3–4 slips for transplanting can be obtained. A local contractor charged $4.75 per slip for the Mt. Parker site and this included collection and division of clumps, transport to site (with some hand-carrying up a rough, steep slope), cutting slip tops to 30 cm, purchasing, mixing, and applying slow-release fertilizer and alginate, planting double rows at 20-cm centers, using a pick-ax to prepare the "soil," and spreading cut tops upslope of hedge. The total cost (excluding maintenance) is thus about HK $87.50 per lineal meter; i.e., U.S. $11.00. At a 2-m vertical interval between the hedges, 1 lineal meter of hedge would protect 2 m^2 on a 45° slope, some 4 m^2 on a 23° slope, and about 8 m^2 on an 11° slope. Labor cost is the major component and in the Chinese context is about one-fiftieth that of Hong Kong (roughly 8 Yuan per worker per day in Guizhou compared with HK $350 per worker per day—at a minimum—in Hong Kong).

In the Hong Kong context, it is clear that using Vetiver grass hedges has no advantage over hydroseeding so far as cost is concerned. However, Vetiver hedges can be planted in places inaccessible to hydroseeding. Moreover, Vetiver hedges begin to trap sediment the moment they are planted. This gives them an advantage over hydroseeding and, especially, over trees, which, of course, are equally readily planted in inaccessible areas. Experience at Jordan Valley (sites 2 and 3) and Shau Kei Wan shows that on rapidly-eroding, virtually nutrient-free slopes, trees do not perform well. Root exposure and stunting are common. Three years after planting, no litter layer has developed and the vegetative cover remains far from complete. The surface is still substantially bare and subject to surface wash and mobilization of sediment by raindrop impact. Using *Acacias* and *Casuarina* it is doubtful that sediment mobilization is necessarily reduced to low levels, for under them the ground cover may be quite scanty, the litter thin, and the soil surface is still vulnerable to drip from branches and to concentration of water flow across the surface. Vetiver grass hedges spread out the surface flow, trap sediments and probably the seeds of many adventive plants as well. In agricultural contexts Vetiver hedges function in the same way. Since tillage is by far the most powerful mobilizer of sediment, a means of trapping much of it is essential to the improvement of soil moisture regimes on cultivated slopes and to the prevention of soil degradation and loss. Unlike solid structures which can be over-topped, Vetiver grass grows up as the materials trapped behind it form a terracette, at the same time providing direct benefits as fodder, fuel, thatch, and bedding material for livestock.

CONCLUSIONS

The data (Table 29.2) indicate that on difficult, virtually nutrient-absent substrates, the method of using Vetiver grass now practiced has a good chance of success, subject to the vagaries of the weather at planting. The basic elements of this method are as follows:

- young, green material not over about 23 weeks old
- 3 tillers per slip
- place cut tops directly upslope from newly-planted hedge
- slow-release NPK + water-retaining alginate
- staggered double-row planting, on the contour, at maximum 20-m centers
- multiple-row planting where flow is concentrated (or use sand-bags + Vetiver)
- tree-planting is unnecessary as adventives colonize
- 2–3 m vertical distance between successive hedges on slope (or closer if needed to prevent concentrated overlandflow)
- encourage tillering by cutting, preferably before culm formation and flowering
- apply maintenance dressing(s) of NPK.

On reasonably good soils; i.e., those with adequate nutrients and moisture, neither fertilizer nor water-retaining alginate are required, thus lowering costs. Single-row planting, though economizing on planting material, is not recommended for application in the region where rainfall intensities and runoff potential is extremely high and runoff very rapid. Though bare-root planting is advised against by authorities such as Yoon (1995), who prefers polybag or foam for slips, our experience is that bare-root planting is satisfactory, with alginate and fertilizer, though possibly more risky should a dry spell follow planting. Overall, the method is very promising and thoroughly deserving of more trials on a greater variety of substrates. Maintenance application of NPK also needs further investigation. In China, a clear need is to extend the use to agricultural croplands where the potential for soil erosion and degradation is great.

ACKNOWLEDGMENTS

The assistance of Dr. Richard Webb, lately of the Territories Development Department, Hong Kong Government, is gratefully acknowledged. So, too, is the fine work of Professor Chen Xuhui, Guizhou Academy of Agricultural Sciences, who was responsible for the Luodian fertilizer study briefly reported here. Research funding for the Hong Kong studies came from the Territories Development Department (H.K. Government), the Shell Company (HK) Ltd., with assistance from the Kadoorie Agricultural Research Centre. The Guizhou study is funded by the Hui Oi Chow Fund, and CUPEM, both at the University of Hong Kong and by the Potash Institute, Hong Kong (Director, Dr S. Portch).

REFERENCES

Burkill, I.H. *A Dictionary of the Economic Products of the Malay Peninsula*. Ministry of Agriculture and Co-operatives, Kuala Lumpur, 1966.

Chen, X. and R.D. Hill. Effect of Applying Fertilizer to Vetiver Grass on Sloping Lands. Paper for International Conference on Vetiver, Chiangmai, January 1996.

Hong Kong. *Annual Digest of Statistics*. Government of Hong Kong, 1995, p. 281.

Lam, K.C. Soil Erosion, Suspended Sediment and Solute Production in Three Hong Kong Catchments. *J. Tropical Geography*, 47, pp. 51–62, 1978.

Peart, M.R. Fingerprinting Sediment Sources: An Example from Hong Kong, in *Sediment and Water Quality in River Catchments*. Foster, I.D.L. et al., Eds., John Wiley, London, 1995, pp. 179–186.

Peart, M.R. The Human Impact Upon Sediments in Rivers: Some Examples from Hong Kong, in *Human Impact on Erosion and Sedimentation*, Walling, D.E. and J.L. Probst, Eds., Proceedings of the Rabat Symposium. I.A.H.S. Publication No. 245, I.A.H.S. Press, Wallingford, 1997, pp. 111–118.

United States, National Research Council, *Vetiver Grass, a Thin Green Line Against Erosion*. National Academy Press, Washington, DC, 1993.

Vetiver Information Network. *Vetiver Grass: Technical Information Package*, 2 vols., Washington, DC, 1993.

Wen, D. Soil Erosion and Conservation in China, in *World Soil Erosion and Conservation*, D. Pimental, Ed., Cambridge University Press, Cambridge, 1993, pp. 63–85.

Yoon, P.K. 1995. Important Biological Considerations in Use of Vetiver Grass Hedgerows (VGHR) for Slope Protection and Stabilization, *Vetiver Newslett.*, 13, pp. 29–32, 1993.

Young, A. *Agroforestry for Soil Conservation*, C.B.B. International, Wallingford, 1989.

Chapter 30

Landfill Leachate Used as Irrigation Water on Landfill Sites During Dry Seasons

J. Liang, J. Zhang, and M.H. Wong

INTRODUCTION

Sanitary landfilling is the most common method employed in disposing of municipal solid waste (MSW) in many countries of the world (Schrab et al., 1993; Shrive et al., 1994). There has been a trend in recent years toward the use of completed sanitary landfills as parks and golf courses, all of which require establishment of vegetation (Saint, 1992). However, difficulties in vegetating landfill sites have been experienced over the last decade in many countries (Ettala, 1987; Wong, 1988; Wong and Leung, 1989). Drought stress is one of the major factors limiting plant growth and development on landfill sites which results, at least in part, from the difficulty for water availability arising through the refuse to reach the cover soil during the dry seasons due to the low rate of capillary rise of water (unpublished data). Moreover, the cover soil is usually highly compacted during construction, which limits infiltration and may lead to excessive rainwater runoff. In addition, the shallow root systems that usually developed in landfill sites due to the high concentration of carbon dioxide around roots and the high compaction of soil make the plants even more susceptible to water stress (Chan et al., 1991).

Landfill leachate is a by-product of solid waste decomposition during and after the landfilling period, and is a mixture of rainwater and the soluble portions of waste and its degradative products containing high concentrations of ammonium, nitrate, organic chemicals, and heavy metals (Devare and Bahadir, 1994). A high concentration of leachate is usually toxic to seed germination and vegetation establishment, mainly due to its high concentrations of toxic substances (e.g., high concentrations of heavy metals, high ammonia-N content, etc.) (Devare and Bahadir, 1994; Gordon et al., 1989; Wong and Leung, 1988, 1989). However, trees growing on the lower slope of landfills, receiving more leachate, seemed to behave much better than those growing on the upper slope (Ettala, 1987; Wong and Leung, 1989). Shrive et al. (1994) showed that the photosynthetic rate and growth of two broad-leaf tree species, *Acer rubrum*, and *Populus* spp., were stimulated when irrigated with municipal landfill leachate and that differences in responses to leachate treatments existed between different species. Our previous results showed that, at an early stage of irrigation with a low concentration of leachate, tree growth was stimulated and this stimulation was related to the increase in the photosynthetic rate. However, a high concentration of leachate exerted an adverse effect (unpublished results).

The main objectives of the present study were to investigate the effects of landfill leachate on the growth and other physiological aspects of trees growing on landfill, and to investigate the possibility of landfill leachate being used as an irrigation source, especially during dry seasons when trees were subjected to water stress. A further objective was to assess whether the experimental outcomes could provide useful information for the development of guidelines for the recirculation of landfill leachate as a means of irrigation.

MATERIALS AND METHODS

Leachate Collection and Analysis

Landfill leachate was collected from Shuen Wan and Pillar Point landfill sites of Hong Kong and stored in a cold room for further analysis and use as irrigation water. pH values and electrical conductivity of leachate were measured with a pH meter (Model Orion 420A, Orion Research Inc., Boston, USA) and a digital electrical conductivity meter (PW 9526 digital conductivity meter, Philip Co., The Netherlands), respectively. Osmolarity was measured using a vapor pressure osmometer (Model 5500, Wescor Inc., Utah, USA). The total nitrogen, NH_4-N and NO_3-N and total phosphorus were determined with the Automated Ion Analyzer (Quikchem AE, Lachat Instruments, USA). The concentrations of K, Cu, Ni, Pb, Fe, and Al were measured with a Varian AA-20 atomic absorption spectrophotometer, operated with an air-acetylene flame, after acid digetstion with concentrated sulfuric acid. The machine was calibrated with a series of standard solutions prepared from 1000 mg L^{-1} metal standard solutions from Merck and recalibrated every 50 to 100 determinations of samples.

Ecological Surveys

Ecological surveys were carried out from October to December, 1994, on the Pillar Point landfill site, Hong Kong, which started operating in 1983 and still continues as an active site. On the upper part of the slope, 50 three-month-old, uniform-sized seedlings (for each species) of *A. confusa*, *S. jambos*, *E. torelliana*, and *E. robusta* were transplanted in July 1994. The following treatments were imposed on 20 October, 1994: (1) watering treatment, i.e., 1000 mL of tap water per plant were applied weekly to trees growing on the upper and lower slopes; (2) fertilizer applying and watering treatment, i.e., 5 g NPK complete fertilizer dissolved in 1000 mL tap water were applied to trees growing on the upper of slopes and then supplemented with 1000 mL water weekly as treatment (1); and (3) leachate-irrigating treatment: 1000 mL of 50% leachate collected from Pillar Point landfill were applied weekly to trees growing on the lower slopes and (4) control (unwatered). One month later, the survival rate and plant height increase of trees growing on the upper slope were recorded.

Physiological Investigations

On the lower part of the slope, vegetation is well established and includes *Acacia confusa*, *A. magium*, *Syzygium jambos*, *Eucalyptus torelliana*, *E. robusta*, *Leucaena leucocephala*, *Sesbania cannabina* (Retz) Pers. and some grass species. Five well-grown and uniform-sized plants for each species of *A. confusa*, *E. torelliana*, and *L. leucocephala* were chosen and similar treatments but 10 g NPK fertilizer applied as described above imposed on them.

Soil Sampling and Measurement of Soil Water Content

A depth of 10 cm of surface soil was discarded using a spade, and soil samples at 10 to 20 cm depth were collected with a soil sampler (stainless steel tube: 1 m in length and 10 cm in diameter) and placed into plastic boxes and sealed immediately. After fresh weight determination, the samples were oven-dried at 100°C for dry weight measurement. Five samples from each treatment were collected.

Measurements of Photosynthetic Rate and Chlorophyll Fluorescence Emission Efficiency

Photosynthetic rate (A) of the youngest, fully expanded leaves of *A. confusa, E. torelliana,* and *L. leucocephala* was measured with a gas-exchange system (CIRAS-1, PP system, Hitchin, UK) at a CO_2 concentration of 350 mm m^{-3} and a natural light intensity (about 1200 µmol m^{-2} s^{-1}). Leaf chlorophyll fluorescence emission was measured on dark-adapted (at least for 30 min) leaves with a plant efficiency analyzer (PEA system, Hansatech, Norfolk, UK) (Liang et al., 1997). The parameter measured and computed by the equipment was F_v/F_m, a ratio of maximal variable fluorescence out of a fully light-saturated, peak fluorescence. Variable fluorescence is subtracted from peak fluorescence with a constant fluorescence of dark-adapted leaves. Dark-adaptation was achieved with the specially designed, light-proof clips attached on leaves for at least 30 min. At least 10 and 30 measurements were conducted for photosynthesis and chlorophyll fluorescence emission, respectively.

Measurements of Leaf Conductance and Leaf Water Potential

Leaf conductance (g_s) was determined using a steady state porometer (Model Li-1600, Li-Cor Inc., Lincoln, Nebraska, USA) as described previously (Liang et al., 1996). For each treatment, at least 10 youngest, fully expanded leaves from four plants of each of three species were tested. Leaf water potential was measured following the measurements of photosynthesis and leaf conductance. Leaf discs (6 mm in diameter) were sampled and quickly sealed in thermocouple chambers which were connected to a dew point microvoltmeter (HR-33T, Wescor Inc., Logan, Utah, USA). After 2 h incubation at 25°C for thermal and humidity equilibrium, readings of microvolts at dew-point mode were taken and calibrated into water potential units (Liang et al., 1996).

Leaf Sampling for Abscisic Acid (ABA) Analysis

Leaves of each treatment were also sampled for ABA analysis after measurements of photosynthesis and water potential. The youngest, fully expanded leaves were harvested and plunged into liquid nitrogen. Freeze-dried leaf samples were stored in a desiccator pending assay.

ABA analysis was carried out using the radio immunoassay (RIA) method (Menser et al., 1983). Highly specific monoclonal antibody (MAC225) was provided by Dr. S.A. Quarrie (John Innes Centre, Norwich, UK). Briefly, about 50 mg of ground leaf sample was weighed to a plastic vial, to which 4 mL of distilled and deionized water was added. The suspension was shaken at 4°C for 24 h and 0.05 mL of the supernatant obtained by centrifuging the extract at 3,000 g for 15 min without further purification was mixed with 0.2 mL phosphate-buffered saline (pH 6.0), 0.1 mL ^{3}H-ABA (about 20,000 dpm) and 0.1 mL diluted antibody solution. The mixture was incubated at 4°C for 45 min and ^{3}H-ABA bound with the antibody was precipitated by adding 0.5 mL saturated $(NH_4)_2SO_4$ solution to the mixture and washed once with 50% saturated $(NH_4)_2SO_4$ solution. The radioactivity was measured with a liquid scintillation counter (Beckman L-1900, USA). The immunoreactive contamination in crude leaf extracts was tested using a spike-dilution test (Jones, 1987). Fractionally diluted extracts were added with different, known amounts of synthetic ABA and assayed for ABA the same way as normal samples. The lines, obtained by plotting the measured values against the added values of ABA for each dilution, were statistically parallel to each other, and therefore showed no significant nonspecific interference. The assay sensitivity (as low as 0.4 pmol ABA per assay vial) and reliability were the same as reported earlier (Zhang and Davies, 1990).

RESULTS

Table 30.1 shows the major physical and chemical properties of landfill leachate collected from Shuen Wan and Pillar Point, where the landfilling is still in progress. No significant differences in physical and chemical properties of the leachates were observed between the two sites. The leachates contained high concentrations of salts (high electrical conductivity and osmolarity) and NH_4-N, but low phosphorus content. They were also very alkaline (pH value >8.0) and the concentration of heavy metals was comparable to an earlier report (Chan et al., unpublished data). Because of their similar properties, only the leachate collected from the Pillar Point landfill site was used in the present study.

Figure 30.1 shows the major meteorological parameters between October and December, 1994, the period during which the experiment was carried out. The rainfall was nearly zero from the beginning of October to early December, but the total evaporation was high (around 4 mm). The relative humidity of this period also dropped to the lowest point of the year. The daily average air temperature was about 22°C and the solar radiation about 12 MJ m^{-2}.

The gravimetric soil water content of unwatered controls decreased steadily near to the permanent wilting point (near to 0.07 g H_2O g^{-1} dried soil) by the end of the experiment. Watering treatments maintained a soil water content at around 0.12 g H_2O g^{-1} dried soil throughout the experimental period (Figure 30.2).

The survival rate of trees in response to soil drying varied greatly among species. *A. confusa* was the most tolerant to soil drying, with the control maintaining a survival rate of 100%. However, soil drying resulted in a high mortality of *S. jambos* seedlings, with a control survival rate as low as 5%. The survival rate of *E. robusta* was also substantially affected by soil drying (Table 30.2). Irrigation with either water, fertilizer solution, or 50% landfill leachate all significantly improved the plant survival rate, especially for *S. jambos*. It increased from 5% before irrigation to over 75% after one month of irrigation, and there was no significant difference among the treatments (Table 30.2). The survival rate of *E. torreliana* in the unwatered control was lower than that of *A. confusa*, although irrigation with either water, fertilizer solution, or 50% landfill leachate did not lead to any significant improvement in terms of survival rate, implying that other factors rather than the water stress itself cause the high mortality (Table 30.2). Tree growth was dramatically reduced under the condition of soil drying, but this reduction varied greatly among species. *A. confusa* maintained a relatively high growth rate even under the soil drying condition, although the growth rates of *S. jambos*, *E. torelliana*, and *E. robusta* were very low

Table 30.1. Physical and Chemical Properties of Leachate Collected from Shuen Wan and Pillar Point Landfill Sites.

	Pillar Point	Shuen Wan
pH	8.69	8.06
Electrical conductivity (ms cm^{-1})	9.62	12.95
Osmolarity (mmol kg^{-1})	243.33	273.0
COD (mg L^{-1})	766.1	850.7
BOD (mg L^{-1})	210.3	189.5
Total nitrogen (mg L^{-1})	1195.2	1051.1
NH_4-N (mg L^{-1})	623.3	918
NO_3-N (mg L^{-1})	4.8	1.42
NO_2-N (mg L^{-1})	7.67	4.47
Total phosphorus (mg L^{-1})	1.19	1.03
Cu (mg L^{-1})	0.375	0.22
Fe (mg L^{-1})	7.21	39.19
Ni (mg L^{-1})	0.296	0.305
Pb (mg L^{-1})	0.247	0.198
Al (mg L^{-1})	0.038	0.021
K (mg L^{-1})	968.87	1022

under the same conditions. The watered-treatment significantly stimulated tree growth, although the stimulatory effect was much less significant than that of fertilizer-application and leachate irrigation (Table 30.3).

As the soil water deficit progressed, leaf water potential steadily decreased, and reached the lowest point at about –2.5, –2.0, and –1.8 MPa for *L. leucocephala*, *E. torelliana*, and *A. confusa*, respectively (Figure 30.3). The leaf water potential of all three species tested increased soon after all three treatments were imposed. One week after the treatments commenced, the leaf water potential increased from about –1.7 MPa to above –1.55 MPa for *A. confusa* (Figure 30.3A); from –2.25 to above –2.0 MPa for *L. leucocephala* (Figure 30.3B) and from –1.7 to above –1.45 MPa for *E. torelliana* (Figure 30.3C). No obvious differences in leaf water potential were observed among treatments.

The stomatal conductance of *A. confusa*, *Leucaena leucocephala*, and *E. torelliana* steadily decreased as soil drying increased, but significant increases in stomatal conductance were observed after either watering, fertilizer application, or irrigation with 50% of landfill leachate (Figure 30.4). The stimulations in stomatal conductance following fertilizer application and leachate irrigation were more significant than that of the watering treatment alone for *A. confusa* (Figure 30.4A) and *L. leucocephala* (Figure 30.4B), especially at the early stage of treatments. Furthermore, some differences in the responses of stomatal opening to treatments were observed among three of the species, *L. leucocephala* and *E. torelliana* being relatively more responsive than *A. confusa* (Figure 30.4).

Contrary to the behavior of stomatal movement, leaf abscisic acid (ABA) content increased continually as the water deficit increased. Upon treatment, leaf ABA content decreased significantly initially, after which the rate of decrease slowed down (Figure 30.5). For example, for *A. confusa*, more than a threefold decrease in leaf ABA content was observed one week after the start of the treatments (Figure 30.5A). No obvious differences were observed in the effects on leaf ABA contents among treatments (Figure 30.5).

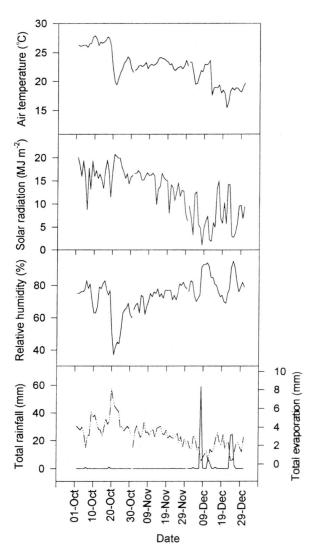

FIGURE 30.1. Changes of meteorological parameters during the experimental period. Data were obtained from the Royal Observatory Station, Hong Kong.

Further investigation on the effects of watering, fertilizer application, and leachate irrigation were conducted on photosynthetic properties (Figures 30.6 and 30.7). Figure 30.6 shows the changes of photosynthetic rate for three tree species; i.e., *A. confusa*, *L. leucocephala*, and *E. torelliana*, under different treatments. The results indicated that all treatments tended to have a positive effect on photosynthetic rate for the three tree species investigated, with the increases in *L. leucocephala* being significant (Figure 30.6). There were no obvious differences in photochemical activity of chloroplasts (F_v/F_m) measured between control and treatments, indicating that the photochemical activities of chloroplasts were relatively insensitive to soil drying (Figure 30.7).

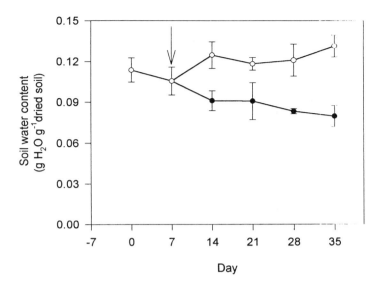

FIGURE 30.2. Changes in soil water content during the experimental period. Filled circle (●): control (unwatered) and open circle (○): watering. The arrow indicates the starting date of treatment. The value of each point is the average of five samples ± SD.

Table 30.2. The Survival Rate of Tree Seedlings Growing on Pillar Point Landfill Site (%). (Tree seedlings were transplanted in July 1994. After one month of treatment, the survival rate of seedlings was surveyed.)

Species	Unwatered Treatment	Watered Treatment	Fertilizer Application	50% Leachate Irrigation
Acacia confusa	100	100	100	100
Syzygium jambos	5	75	76	78
Eucalyptus torelliana	88	92	85	84
Eucalyptus robusta	60	78	73	71

Table 30.3. The Height Increase of Tree Seedlings Growing on Pillar Point Landfill Site. (Seedlings were transplanted in July 1994. The height was measured with a ruler before and after one month of treatment (cm) (n=5). Asterisks indicate values significantly higher than the respective control (P≤0.01).)

Species	Unwatered Treatment	Watered Treatment	Fertilizer Application	50% Leachate Irrigation
Acacia confusa	2.8 ± 0.37	3.7 ± 1.14**	6.7 ± 0.55**	6.9 ± 1.90**
Syzygium jambos	0.8 ± 0.15	1.8 ± 0.38**	3.1 ± 0.19**	3.1 ± 0.71**
Eucalyptus torelliana	0.6 ± 0.13	2.5 ± 0.50**	4.7 ± 0.85**	2.6 ± 0.29**
Eucalyptus robusta	0.8 ± 0.20	1.5 ± 0.25**	2.7 ± 1.05**	2.0 ± 0.31**

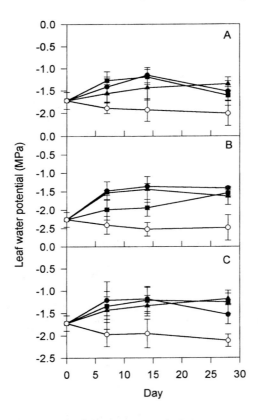

FIGURE 30.3. Effects of treatments on leaf water potential of three tree species. A: *Acacia confusa*, B: *Leucaena leucocephala*, and C: *Eucalyptus torelliana*. Water potential was measured using a dew point microvoltmeter (HR-33T, Wescor Inc., USA). Open circle (O): control (unwatered); filled circle (●): irrigated with water; filled square (■): fertilizer application and filled triangle (▲): irrigated with 50% leachate. The value of each point is the average of 4 measurements from 4 plants ± SD.

DISCUSSION

Water stress is one of the major limiting factors to the successful vegetation of sanitary landfill sites. The possible causes are the poor development of root systems under landfill environments, which lead to difficulty in accessing as well as absorbing water from soil. The landfill soil environment is unfavorable to root growth and development because of the high content of CO_2 and high mechanical resistance. The high CO_2 concentrations in landfill soil significantly inhibits root growth and root activity (Chan et al., 1991; Qi et al., 1994). During the landfilling period, the municipal solid wastes and the cover soil were compacted, layer by layer, by steel wheeled compactors to maximize landfill capacity and to minimize land settlement, so that the degree of compaction of landfill "soil" is much greater than that of agricultural soil. In Hong Kong, the bulk density of landfill soil is between 1.75 and 1.87 g cm^{-3} (unpublished data), which is much higher than that of typical agricultural soils (1.0 to 1.4 g cm^{-3}). High soil compaction severely limits the penetration of roots into deeper layers and most of the root mass occurs in the surface 30 cm (unpublished data). The shallow root systems thus tend exacerbate the problems resulting

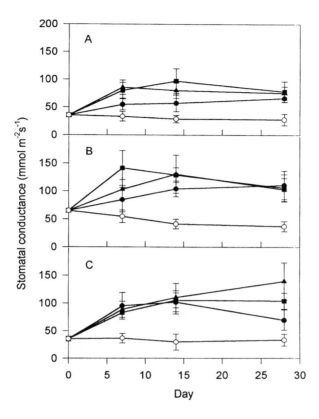

FIGURE 30.4. Changes of stomatal conductance of three tree species after treatment application. A: *Acacia confusa*, B: *Leucaena leucocephala*, and C: *Eucalyptus torelliana*. Stomatal conductance was measured using a steady state porometer (LI-1600, Li-cor Inc., Nebraska). Open circle (○): control (unwatered); filled circle (●): irrigated with water; filled square (■): fertilizer application and filled triangle (▲): irrigated with 50% leachate. The value of each point is the average of ten measurements from four plants ± SD.

from water deficit, problems compounded by other factors such as the high concentrations of CO_2 and CH_4 and the lack of O_2.

In Hong Kong, over 80% of rainfall occurs in the period between June and September. Outside this period, rainfall is low and erratic, but the evaporation is very high because of the high solar radiation and low relative humidity (Figure 30.1). Therefore, plants are usually exposed to a long period of dry conditions which can lead to a serious inhibition of plant growth and development. It is of great importance to alleviate the impacts of soil drying on plant growth and development in order to vegetate the landfill sites successfully.

The properties and composition of landfill leachate vary greatly and are determined not only by the nature of the waste disposed of in the landfill site and its design, but also by the climate and age of the site. Landfill leachate poses a serious environmental threat to groundwater reserves and plant growth because of the high contents of many hazardous materials (Gordon et al., 1989; Cheung et al., 1993; Schrab et al., 1993; Ernst et al., 1994). Mensor et al. (1983) showed that landfill leachate had a profound toxic effect on tree growth and lead to a significant mortality. Wong and Yu (1989) reported about 25% inhibition of the growth of *A. confusa* after 50 days of irrigation with landfill leachate. However, the toxic

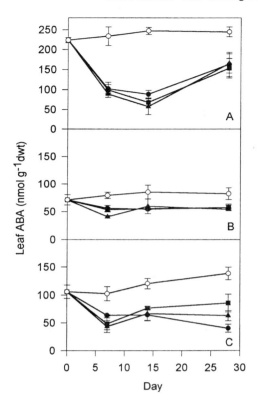

FIGURE 30.5. Changes of leaf ABA contents of three tree species following treatments. Symbols are the same as those in Figure 30.3. The value of each point is the average of four measurements from two plants ± SD.

effects of landfill leachate on plant growth and development may depend on its concentration and the way of application (Gordon et al., 1989). If the levels of hazardous substances were reduced to a reasonable level by either dilution or decreasing the application rate, the leachate may be used as an irrigation water source to alleviate the plant water deficits. The preliminary results reported in this study showed that, at a low concentration, landfill leachate could stimulate the growth *of A. confusa* at the early stage of application and have an adverse effect on plant growth at a high concentration of leachate (data not shown). Recently, Cureton et al. (1991) found that irrigation of poplar and willow for two seasons with landfill leachate stimulated height growth by 42% and 141%, respectively. Shrive et al. (1994) compared the effects of different irrigation type, mode of application, and rate of application on photosynthesis and growth of red maple and hybrid poplar and found that the mean seasonal photosynthetic rates increased for irrigated plants of both species relative to rain-fed controls. In the present experiment, comparisons were made of the effects of irrigation with water, fertilizer solution, and 50% leachate on plant survival, growth and physiological aspects during the dry season. In spite of the differences in tolerance to soil water deficits among tree species, irrigation with 50% leachate once a week, similar to irrigation with water or fertilizer solution, could improve the survival rate and stimulate the growth rate, especially for those species less tolerant to water deficits, such as *E. torelliana* (Tables 30.2 and 30.3).

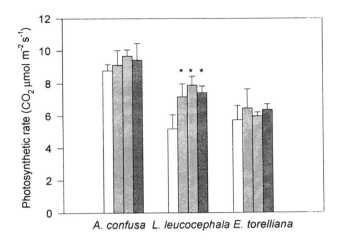

FIGURE **30.6.** Changes of photosynthesis rate of *A. confusa*, *L. leucocephala*, and *E. torelliana* subjected to different treatments for one month. Blank bars: control (unwatered); hatched bars: watered treatment; inverted hatched bars: fertilizer application and crossed bars: irrigated with 50% leachate. Treatments were imposed on October 20, 1994, and photosynthetic rate was measured using a gas exchange system (CIRAS-1 PP system, UK) at $350 \, l \, L^{-1} \, CO_2$ on November 20. The value of each point was the average of four measurements from two plants $\pm$ SD. Asterisks indicate values significantly different from the respective control ($P \leq 0.01$).

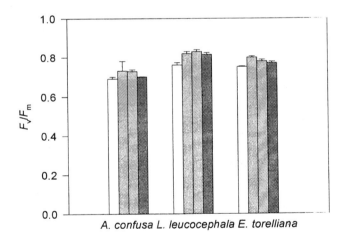

FIGURE **30.7.** Changes of chlorophyll fluorescence of *A. confusa*, *L. leucocephala*, and *E. torelliana* subjected to different treatments for one month. Chlorophyll fluorescence was determined on dark-adapted leaves using a Plant Efficiency Analyzer (PEA) (see Materials and Methods section for details). Symbols were the same as in Figure 30.6. The value of each point was the average of 10 measurements from four plants $\pm$ SD.

Similar to the effect of watering and fertilizer-application treatments, leachate irrigation lessened the inhibitory effects of soil drying on stomatal conductance and photosynthesis (Figures 30.4 and 30.6). After irrigation with leachate, leaf water potential of all tree species tested increased (Figure 30.3) and the leaf ABA content decreased significantly

(Figure 30.5). These results implied that the stimulation of tree growth by leachate irrigation might be related to the improvement of plant water relations and the stimulation of photosynthesis or alleviation of inhibitions imposed by water deficit. Therefore, the landfill leachate could be used as an irrigating water source and alleviate the limitations of water deficits on plants, especially during the dry seasons.

ACKNOWLEDGMENTS

The authors thank the Research Grants Council of the University Grants Committee for financial support.

REFERENCES

Chan, G.Y.S., M.H. Wong, and B.A. Whitton. Effects of Landfill Gas on Subtropical Woody Plants. *Environ. Manag.*, 15, pp. 411–431, 1991.

Cheung, K.C., L.M. Chu, and M.H. Wong. Toxic Effect of Landfill Leachate on Microalgae. *Water, Air, Soil Pollut.*, 69, pp. 337–349, 1993.

Cureton, P.M., P.H. Groenuelt, and R.A. McBride. Landfill Leachate Recirculation: Effects on Vegetation Vigor and Clay Surface Cover Infiltration. *J. Environ. Qual.*, 20, pp. 17–24, 1991.

Devare, M. and M. Bahadir. Biological Monitoring of Landfill Leachate Using Plants and Luminescent Bacteria. *Chemosphere*, 28, pp. 261–271, 1994.

Ernst, W.R., P. Hennigar, K. Doe, S. Wade, and G. Julien. Characterization of the Chemical Constituents and Toxicity to Aquatic Organisms of a Municipal Landfill Leachate. *Water Pollut. Res. J. Canada*, 29, pp. 89–101, 1994.

Ettala, F.B. Influence of Irrigation with Leachate on Biomass Production and Evapotranspiration on a Sanitary Landfill. *Aqua. Fennica*, 17, pp. 69–86, 1987.

Ettala, M.O. Short Rotation Tree Plantations at Sanitary Landfills. *Waste Manag. Res.*, 6, pp. 291–302, 1988.

Ettala, M.O., K.M. Yrjoen, and E.J. Rossi. Vegetation Coverage at Sanitary Landfills in Finland. *Waste Manag. Res.*, 6, pp. 281–289, 1988.

Gordon, A.M., R.A. McBride, and A.J. Frisken. Effect of Landfill Leachate Irrigation on Red Maple (*Acer rubrum* L.) and Sugar Maple (*Acer saccharum* Marsh) Seedling Growth and on Foliar Nutrient Concentrations. *Environ. Pollut.*, 56, pp. 327–336, 1989.

Jones, H.G. Correction for Non-Specific Interference in Competitive Immunoassays. *Physiol. Plant*, 70, pp. 146–154, 1987.

Liang, J., J. Zhang, and M.H. Wong. Stomatal Conductance in Relation to Xylem Sap ABA Concentration in Two Tropical Trees, *Acacia confusa* and *Litsea glutinosa*. *Plant Cell Environ.*, 19, pp. 29–36, 1996.

Liang, J., J. Zhang, and M.H. Wong. Can Stomatal Closure Caused by Xylem ABA Explain the Inhibition of Leaf Photosynthesis under Soil Drying? *Photosynth. Res.*, 51, pp. 149–159, 1997.

Menser, H.A., W.H. Winant, and O.L. Bennet. Spray Irrigation with Landfill Leachate. *Biocycle*, 24, pp. 22–25, 1983.

Qi, J., J.D. Marshall, and K.G. Mattson. High Soil Carbon Dioxide Concentrations Inhibit Root Respiration of Douglas Fir. *New Phytol.*, 128, pp. 435–442, 1994.

Quarrie, S.A., P.N. Whitford, N.E.J. Appleford, T.L. Wang, S.K. Cook, L.E. Henson, and B.R. Loveys. A Monoclonal Antibody to (s)-Abscisic Acid: Its Characterization and Use in a Radioimmunoassay for Measuring Abscisic Acid in Crude Extracts of Cereal and Lupin Leaves. *Planta*, 173, pp. 330–339, 1988.

Saint Fort, R. Fate of Municipal Refuse Deposited in Sanitary Landfills and Leachate Treatability. *J. Environ. Sci. Health, Part A: Environ. Sci. Eng.*, 27, pp. 369–401, 1992.

Schrab, G.E., K.W. Brown, and K.C. Donnelly. Acute and Genetic Toxicity of Municipal Landfill Leachate. *Water, Air Soil Pollut.*, 69, pp. 99–112, 1993.

Shrive, S.C., R.A. McBride, and A.M. Gordon. Photosynthetic and Growth Responses of Two Broad-Leaf Tree Species to Irrigation with Municipal Landfill Leachate. *J. Environ. Qual.*, 23, pp. 534–542, 1994.

Suter, G.W., III, R.J. Luxmoore, and E.D. Smith. Compacted Soil Barrier at Abandoned Landfill Sites Are Likely to Fail in the Long Term. *J. Environ. Qual.*, 22, pp. 217–226, 1993.

Wong, M.H. Soil and Plant Characteristics of Landfill Sites Near Merseyside, England. *Environ. Manage.*, 12, pp. 491–499, 1988.

Wong, M.H. and C.K. Leung. Phytotoxicity of Landfill Leachate (Gin Drinkers' Bay Landfill), in *Water Pollution Control in Asia*, Panswad, T., C. Polprasert, and K. Yamamoto, Eds. Pergamon Press, Oxford, 1988, pp. 707–716.

Wong, M.H. and C.K. Leung. Landfill Leachate as Irrigation Water for Tree and Vegetable Crops. *Waste Manage. Res.*, 7, pp. 341–357, 1989.

Wong, M.H. and C.K. Yu. Monitoring of Gin Drinkers' Bay Landfill, Hong Kong II. Gas Contents, Soil Properties, and Vegetation Performance on the Side Slope. *Environ. Manage.*, 13, pp. 753–762, 1989.

Zhang, J. and W.J. Davies. Changes in the Concentration of ABA in the Xylem Sap as a Function of Changing Soil Water Status Can Account for Changes in Leaf Conductance and Growth. *Plant Cell Environ.*, 13, pp. 271–285, 1990.

Chapter 31

Growth and Mineral Nutrition of *Casuarina Equisetifolia* on a Pulverized Fuel Ash-Rich Substrate

S.K.S. Lam and Y.B. Ho

INTRODUCTION

Coal-fired electricity generation was introduced to Hong Kong in 1982 to stabilize electricity prices (Thornely, 1988). By 1993 about 3,000 tonnes of pulverized fuel ash (PFA) was produced daily (HKEPD, 1995). Only a part of the PFA is utilized, mainly in cement and concrete production, in land reclamation and quarry restoration, and in various construction projects (HKEPD, 1993). The surplus PFA is stored in ash lagoons located in Tsang Tsui, in the New Territories.

Proper management of the ash lagoons, including establishment of a vegetation cover, is required before the site can be put to its final, recreational use. However, PFA is pozzolanic, has high alkalinity, high salinity, and high boron content, and is deficient in nitrogen. This makes soil development and colonization of natural vegetation on PFA sites very slow. Hodgson and Townsend (1973) showed that only a limited number of plant species could establish successfully on a PFA site in the first five or six years. Similarly, Shaw (1992) reported that the colonization of unamended PFA site by self-grown woodland took about 25 years. Therefore, it is necessary to improve the condition of the PFA substrate in order to shorten the time for vegetation establishment.

The revegetation of many industrial wastes is based on the application of a top layer of soil which can blanket many undesirable properties of the wastes that suppress plant growth and establishment. For example, Hodgson and Buckley (1975) showed that applying a 15-cm layer of soil to PFA aided the successful establishment and growth of shrubs and trees. Other studies demonstrated that organic amendments, including composts, livestock manure, and sewage sludge, gave a greater ameliorating effect than soil on PFA (Rippon and Wood, 1975). The use of livestock manure compost not only improved the textural properties of the PFA, it also significantly increased the nutrient supply to plants and the chance of successful vegetation establishment on ash disposal sites (Lee and Wigmore, 1988; Mulhern et al., 1989; Menon et al., 1992).

In this study, livestock manure compost is used as an organic amendment for planting on a PFA-rich substrate. *Casuarina equisetifolia* L. ex Forst. is the plant selected since it is tolerant to high salinity and alkalinity (Clemens et al., 1983) and it grows symbiotically with nodule bacteria to fix nitrogen and hence helps to improve soil fertility. The aims of this study are therefore, first, to study the growth of and mineral accumulation by *C.*

equisetifolia on a PFA/livestock manure compost-amended soil, and, secondly, to evaluate the potential use of this tropical tree to revegetate ash lagoons.

MATERIALS AND METHODS

Sample Collection and Analysis

Weathered PFA was taken from an ash lagoon of China Light & Power Co. Ltd. at Tsang Tsui, and livestock manure compost was collected from the Sha Ling Composting Plant at Yuen Long in 1992. Decomposed granitic sandy soil was bought from a commercial company at Yuen Long. Samples of these three substrates were air-dried, ground, sieved through a 2-mm screen, and stored in separate plastic containers. They were analyzed for their pH and electrical conductivity (EC). Total N and P were extracted by sulfuric acid-peroxide digestion. The amounts of NH_4-N and PO_4-P in the digest were analyzed by an autoanalyzer using the indophenol-blue and ascorbic acid method, respectively (Allen, 1989). Extractable N and P were similarly analyzed after extraction (150 rpm for 30 min) of the samples by 1M KCl and 1M $NaHCO_3$ at pH 8.5, respectively. Organic carbon in livestock manure compost was determined by loss-on-ignition (550°C, 2 h), while that of PFA and decomposed granitic sand was analyzed by a rapid titration method (Walkley and Black, 1934). Total metals in the samples were determined by atomic absorption spectrometry after wet digestion (16M HNO_3:18M H_2SO_4=5:1 at 150°C). Extractable metals were similarly analyzed after extraction (150 rpm for 30 min) with 1M pH 7 ammonium acetate (K and Na), or 0.005M pH 7.3 DTPA-TEA (Cu, Fe, Mn and Zn). Hot-water boron was extracted by boiling the sample with deionized water in a Teflon* flask under reflux for five minutes. The extractant was decolorized with activated charcoal and the filtrate measured by a modified azomethine-H method (John et al., 1975).

Plant Growth Experiment

The field trial was conducted from August 1992 to May 1993 at the Kadoorie Agricultural Research Centre of The University of Hong Kong. Three substrate treatments; namely, mixed soil (MS) with PFA:manure compost:sandy soil at a 3:4:3 ratio, sandy soil (S), and sandy soil overlaying 15 cm of PFA (SB) were included. Each of the three treatments, in triplicate, were laid out in plots each measuring 1m x 2m, and a depth of 0.45 m.

Ten 6-month-old *Casuarina equisetifolia* saplings, bought from the government nursery at Tai Tong and measuring about 50 cm tall, were transplanted to each plot. One week later, an inorganic fertilizer was applied to the S and SB treatments at the rate of 50 kg N ha^{-1}, 30 kg P ha^{-1}, and 50 kg K ha^{-1}. The trees were grown for a period of nine months, during which air temperature ranged from 9.4 to 32.3°C. The plants were watered whenever necessary.

Survival rate, height, and basal stem diameter (5 cm above soil surface) of the trees were measured one week after transplanting, and then after 3, 7, and 9 months. Initial sapling size did not differ significantly among the treatments. At each measurement session, young green tissue from each sapling was collected for chemical analysis. After 9 months, three trees from each plot were chosen randomly for dry weight determination. The aerial part of the trees was cut at soil surface and the root was subdivided into fine (diameter < 2 mm) and coarse (> 2mm) fractions. The roots were washed thoroughly with tap water to remove

---

*Registered trademark of E. I. du Pont de Nemours and Company, Inc., Wilmington, Delaware.

soil particles. The harvested plant materials were rinsed with deionized water and then oven-dried at 80°C for 72 h. The dry weights were then recorded.

Soil and Plant Analysis

Prior to planting, samples of the prepared substrates; i.e., MS, S, and SB, were randomly taken. At the end of the experiment, two subsamples of the surface soil (0–15 cm) from each plot were also collected. The pH, EC, organic C, B, total and extractable macronutrient (N, P and K), and metal (Cu, Fe, Mn, Na and Zn) contents in the substrates were determined by the methods as detailed above.

Dried plant materials collected from each plot were pooled before two subsamples, each weighing about 20 g, were finely ground for tissue composition analysis. Macronutrients were analyzed as for the soil samples. For metals and boron, 1 g of each dried ground subsample was ignited at 550°C for 2 h, the ash was then dissolved in 6M HCl before the elements were analyzed by the same methods as for soil analysis.

Statistical Treatment of Data

All data were subjected to one-way ANOVA tests to compare the means of different treatments. If an F test of the variance ratio was significant at $P < 0.05$, individual means were tested by Duncan's multiple range test.

RESULTS AND DISCUSSION

Chemical Properties of the Substrates

Table 31.1 gives the chemical properties of the substrates. PFA was low in organic matter, N, P, and K levels. However, its pH, B, and extractable Na levels were significantly higher ($P < 0.05$) than the other two substrates. In contrast, manure compost was a fertilizer, as it was rich in macro- and micronutrients, whereas the reverse applied to the sandy soil. The acidity and low salt level of the sandy soil made it a useful bulking material for lowering the pH and salt levels of the PFA and manure compost.

Revegetation success is dependent on nutrient input (Schoneholtz et al., 1992). Thus the addition of nutrient-rich manure compost to PFA would give the task of revegetation a greater chance of success. Besides, pH influences the availability of micronutrients in the soil (Miller et al., 1990), the reduced pH of the MS substrate increased their bioavailability. Further, a marked decrease in B and extractable Na levels occurred after PFA and compost were mixed with sandy soil (Table 31.1).

Table 31.2 shows the pH, EC, and nutrient contents of the substrates at the start and finish of the growth experiment. After nine months the total N and P contents in the MS substrate were still higher ($P < 0.05$) than S and SB, thus indicating that manure compost could provide a long-term nutrient supply. However, the levels of extractable N, P, K, and B, the total and extractable Cu and Na in MS were significantly lowered over the nine months, and this was also reflected in the EC value. The reduction might be due to absorption by the plant and to leaching. The amount of organic matter of MS also decreased significantly after nine months as a result of decomposition by soil organisms.

There was a significant increase in the final total N and Cu, and extractable N and P levels over the initial in the S substrate (Table 31.2). The overall increase in the soluble salt

Table 31.1. General Properties of the Growth Substrates.[a]

Parameter	PFA	Compost	Sandy Soil	MS
pH	8.89[a]	6.87[c]	5.02[d]	7.64[b]
E.C. (μS cm^{-1})	4580[a]	4505[b]	21.7[d]	3120[c]
C (%)	0.45[b]	42.9[a]	0.45[d]	6.86[c]
Total N (%)	0.015[c]	2.28[a]	0.002[c]	0.52[b]
Extractable N (μg g^{-1})	7.86[c]	3480[a]	1.92[d]	771[b]
Total P (%)	0.46[b]	2.15[a]	0.006[c]	0.52[b]
Extractable P (μg g^{-1})	24[c]	2210[a]	nd	97[b]
Total K (%)	0.17[c]	3.65[a]	0.22[c]	0.94[b]
Extractable K (μg g^{-1})	151[c]	16500[a]	nd	3670[b]
Total element content				
Cu (μg g^{-1})	50[c]	299[a]	nd	86[b]
Fe (%)	1.94[a]	0.71[c]	2.03[a]	1.46[b]
Mn (μg g^{-1})	404[b]	779[a]	450[c]	452[b]
Na (%)	0.39[c]	0.81[a]	0.01[d]	0.35[b]
Zn (μg g^{-1})	64[c]	555[a]	55[c]	298[b]
Extractable element content				
B (μg g^{-1})	63.2[a]	18.3[c]	1.18[d]	33.5[b]
Cu (μg g^{-1})	0.14[b]	1.18[a]	nd	0.06[b]
Fe (μg g^{-1})	2.16[b]	10.4[a]	0.62[d]	0.81[b]
Mn (μg g^{-1})	2.07[d]	10.6[b]	13.0[a]	6.06[c]
Na (μg g^{-1})	2740[a]	2420[b]	13[d]	1130[c]
Zn (μg g^{-1})	nd	6.60[a]	0.04[c]	0.93[b]

[a] Means within each row followed by the same letter are not significantly different at 5% level according to Duncan's multiple range test.
nd: not detectable.

content of the substrate was also apparent from its high EC value. Such an increase might partly be due to the addition of inorganic fertilizer to the soil after transplantation of the tree saplings and to soil improvement after growing the N-fixing *C. equisetifolia*. A similar small increase in nutrients was also apparent for the SB substrate.

Plant Growth Performance

Initially, the saplings growing on the MS treatment showed 10–20% mortality. Further, both shoot elongation and stem diameter increase were significantly lower ($P < 0.05$) than S and SB (Figure 31.1). The high B content of over 20 μg g^{-1} (Townsend and Gillham, 1975) in PFA and salinity of over 4 mS cm^{-1} (Townsend and Hodgson, 1973) for both PFA and manure compost were potentially toxic and might have accounted for the initial mortality and stunted growth. Clemens et al. (1983) reported similar results for *Casuarina* growing in saline soil. After nine months, the EC and B content of the MS substrate were very much reduced (Table 31.2) and growth inhibition was no longer apparent, as evidenced by the similar shoot and root biomass of the saplings between the MS and S treatments (Figure 31.2).

Figure 31.3 presents the distribution of root biomass down the soil profile. Generally, the bulk of the root biomass was in the top 15 cm layer of soil, and coarse roots constituted about 90% of the total root biomass. The MS tended to have less coarse root but higher

Table 31.2. Initial and Final Chemical Properties of the Growth Substrates.

Parameter	MS Initial	MS Final	S Initial	S Final	SB Initial	SB Final
pH	7.64	7.67	5.02	5.24	5.02	4.96
E.C. (μS cm^{-1})	3120	121a	21.7	40.5[a]	21.7	30.8
C (%)	6.86	5.19[a]	0.15	0.14	0.15	0.11
Total N (%)	0.515	0.580	0.002	0.011[a]	0.002	0.010
Total P (%)	0.522	0.566[a]	0.006	0.007	0.006	0.004
Total K (%)	0.937	0.220[a]	0.216	0.146	0.216	0.206
Extractable N (μg g^{-1})	771	49.8	1.92	4.71[a]	1.92	4.08[a]
Extractable P (μg g^{-1})	97.4	50.5[a]	nd	0.81[a]	nd	0.41[a]
Extractable K (μg g^{-1})	3670	340[a]	49.3	44.2	49.3	63.5
Total elemental content						
Cu (μg g^{-1})	85.8	101[a]	nd	12.2[a]	nd	2.10[a]
Fe (μg g^{-1})	1.46	1.33	2.03	1.25[a]	2.03	2.09
Mn (μg g^{-1})	452	485	450	471	450	648
Na (μg g^{-1})	0.35	0.05[a]	0.01	0.01	0.01	0.01
Zn (μg g^{-1})	298	360[a]	55.3	70.9	55.3	46.8
Extractable						
B (μg g^{-1})	33.5	4.45[a]	1.18	1.47	1.18	1.58
Cu (μg g^{-1})	0.06	0.33[a]	nd	0.05[a]	0.62	0.34
Fe (μg g^{-1})	0.81	nd[a]	0.62	nd[a]	0.62	0.34
Mn (μg g^{-1})	6.06	3.31	13.0	12.0	13.0	11.1
Na (μg g^{-1})	1130	26.3[a]	13.0	11.3	13.0	8.77
Zn (μg g^{-1})	0.93	1.22	0.04	0.23[a]	0.04	0.17

[a] The final mean is significantly different from the initial reading at $P<0.05$ *t*-test.

fine root biomass than the other two treatments. This indicated that the MS substrate had a good supply of nutrients. More fine roots, making up to about 50% of the total root biomass at greater soil depth, were developed by the tree to absorb nutrients. The fine roots (diameter < 2 mm) are the absorbing roots of a tree. Conversely, the S and SB substrates did not have a sustained supply of nutrients and the tree produced more coarse roots to search for nutrients. As the biomass and distribution of both the coarse and fine roots between S and SB were similar, this indicated that the underlying PFA layer did not affect root growth and penetration.

Elemental Composition of the Plant Tissue

Table 31.3 presents the N, P, K, B, Mn, Cu, Zn, Fe, and Na contents in the root of the three treatments after nine months of growth. Figure 31.4 shows the temporal changes of the mineral elements in the green tissue. *C. equisetifolia* grown on MS had higher P and K contents in both the root and green tissue than the plants on the other two substrates, and this reflected the good supply of these two elements in MS. The low P content of less than 0.15% in the green tissue showed that the S and SB substrates were deficient in the element, although through its mycorrhizal association the plant had the ability to absorb rather low levels of P from the soil (Walker et al., 1993) and hence was able to establish successfully on nutrient-poor substrates.

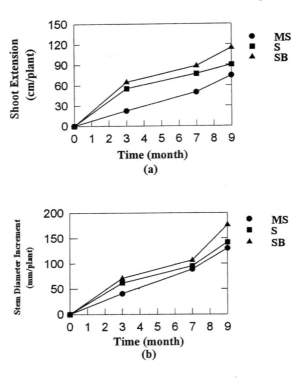

FIGURE 31.1. Temporal changes in shoot extension (a) and stem diameter increment (b) of the saplings.

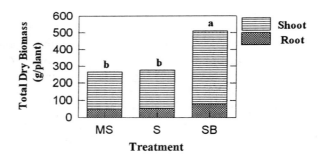

FIGURE 31.2. Shoot, root, and total dry biomass of the saplings after a growth period of 9 months.

For the first seven months, plants grown on MS accumulated higher level of B in the green tissue than those of the other two treatments. But the difference became insignificant by the end of the experiment since the B content in the MS substrate decreased as a result of leaching (Table 31.2). However, the root B content remained high in the MS treatment (Table 31.3). Elevated B content in plant tissue and/or growth suppression due to exposure to PFA had also been reported in other studies (Townsend and Hodgson, 1973; Townsend and Gillham, 1975; Page et al., 1979; El-Mogazi et al., 1988; Carlson and Adriano, 1991; Wong, 1995).

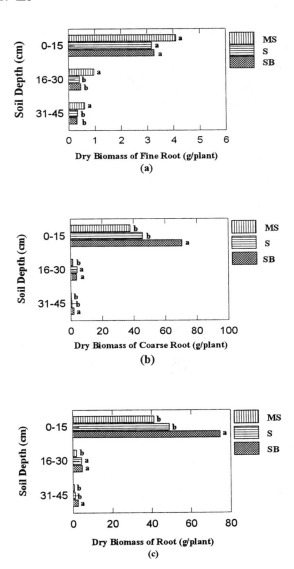

FIGURE 31.3. Dry biomass of fine root (a), coarse root (b), fine and coarse root (c) of the saplings at different soil depths.

The Mn content of both the root and leaf of MS was significantly lower than those of the other two treatments (Table 31.3 and Figure 31.4). This was due to the relatively high pH of the MS substrate which reduced the Mn supply and was reflected by the lower extractable Mn content of the MS substrate (Table 31.2). Manganese bioavailability is pH-dependent and deficiency commonly occurs in alkaline soil (Carlson and Adriano, 1991; Kabata-Pendias and Pendias, 1992). The root of the SB treatment also had less Mn content than S and this might be due to its penetration into the PFA layer, resulting in reduced Mn availability toward the latter part of the growth period.

Table 31.3 Elemental Composition in the Fine Root Tissue of
Casuarina equisetifolia at Month 9.

Parameters	MS	S	SB
Nitrogen (%)	1.15[a]	0.89[b]	1.00[ab]
Phosphorus (%)	0.16[a]	0.04[b]	0.04[b]
Potassium (%)	0.51[a]	0.30[b]	0.36[b]
Boron (μg g^{-1})	48.2[a]	8.21[b]	10.5[b]
Cu (μg g^{-1})	15.1[a]	7.31[b]	5.02[b]
Fe (μg g^{-1})	426[ab]	543[a]	340[b]
Mn (μg g^{-1})	58.9[c]	180[a]	110[b]
Na (%)	0.05[a]	0.05[a]	0.06[a]
Zn (μg g^{-1})	137[a]	102[ab]	77.5[b]

Note: Means within each row followed by the same letter are
not significantly different at 5% level according to Duncan's
multiple range test.

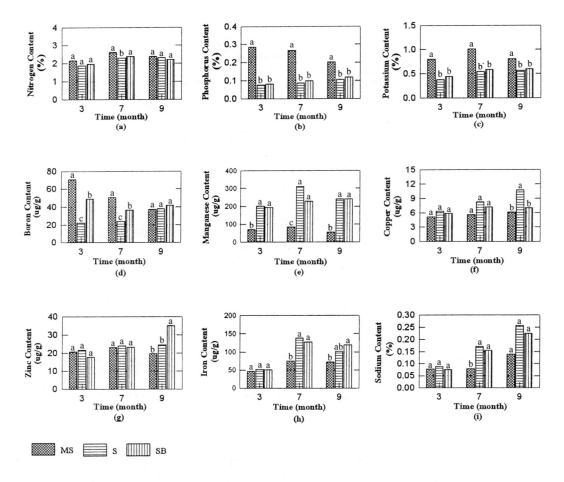

FIGURE 31.4. Contents of nitrogen (a), phosphorus (b), potassium (c), boron (d), manganese (e), copper (f), zinc (g), iron (h), and sodium (i) in the green tissue of the saplings during the experimental period.

Manure compost contained high levels of Cu and Zn (Table 31.1) and resulted in a high level of these two metals in the MS substrate. Thus, roots growing in this substrate accumulated higher Cu and Zn concentrations than those of the S and SB treatments (Table 31.3).

CONCLUSION

This study showed that a layer of 45-cm thick mixed soil (PFA/manure compost/sandy soil) or sandy soil with inorganic fertilizer could support a satisfactory growth of *Casuarina equisetifolia*. In nine months, the height of the saplings increased by 75 to 100 cm, and after three and a half years the trees measured some 6 to 9 m. Due to leaching, the initial high Na and B contents of the mixed soil decreased considerably within the experimental period and thus their effects on the plant was only temporary. Manure compost provided a rich source of nutrients, especially N, P, and K to the soil and was hence beneficial to the growth of the plant.

We believe that *C. equisetifolia* is suitable for the revegetation of a PFA-rich environment, such as ash lagoons. The plant has the ability to grow in saline and alkaline soil. Through its symbiotic relationships with nodule bacteria and mycorrhizal fungi, the plant can establish and grow in a substrate deficient in nitrogen and phosphorus. Since the ash lagoons at Tsang Tsui are planned to become an amenity area to the public, we suggest that additional suitable plants, including grass, shrubs, and trees, should be screened for a revegetation program. Further, apart from manure compost, other organic nutrient sources such as sewage sludge should be tested for their suitability as organic fertilizers and soil amendments for supporting revegetation.

ACKNOWLEDGMENTS

We would like to thank China Light and Power Co. Ltd. for the provision of PFA; and Professor D.K.O. Chan, Director of the Kadoorie Agricultural Research Centre of The University of Hong Kong, for permission to use the facilities at the Centre; and sponsorship of the Environmental and Conservation Fund for attending the Conference. This work formed part of the thesis of SKSL submitted to The University of Hong Kong for the award of a M. Phil. degree.

REFERENCES

Allen, S.E. *Chemical Analysis of Ecological Materials*, 2nd ed., Blackwell Scientific Publications, Oxford, 1989, p. 368.

Carlson, C.L. and D.C. Adriano. Environmental Impacts of Coal Combustion Residues. *J. Environ. Qual.*, 22, pp. 227–247, 1993.

Clemens, J., L.C. Campbell, and S. Nurisjah. Germination, Growth and Mineral Ion Concentrations of *Casuarina* Species Under Saline Conditions. *Australian J. Botany*, 31, pp. 1–9, 1983.

El-Mogazi, D., D.J. Lisk, and L.H. Weinstein. A Review of Physical, Chemical, and Biological Properties of Fly Ash and Effects on Agricultural Ecosystems. *Sci. Total Environ.*, 74, pp. 1–37, 1988.

Hodgson, D.R. and G.P. Buckley. A Practical Approach Towards the Establishment of Trees and Shrubs on Pulverized Fuel Ash, in *The Ecology of Resource Degradation and Renewal*, Chadwick, M.J. and G.T. Goodman, Eds., Blackwell Scientific Publications, Oxford, 1975, pp. 305–329.

Hodgson, D.R. and W.N. Townsend. The Amelioration and Revegetation of Pulverized Fuel Ash., in *Ecology and Reclamation of Devastated Land*, Vol. 2, Hutnik, R.J. and G. Davis, Eds., Gordon and Breach, London, 1973, pp. 247–270.

Hong Kong Environmental Protection Department. *Environment Hong Kong 1993, A Review of 1992*. Hong Kong Government, Hong Kong, 1993, p. 183.

Hong Kong Environmental Protection Department. *Monitoring of Municipal Solid Waste 1993 and 1994*. Hong Kong Government, Hong Kong, 1995, p. 44.

John, M.K., J.J. Chauah, and J.H. Neufield. Application of Improved Azomethine-H Method to the Determination of Boron in Soils and Plants. *Analytical Lett.*, 8, pp. 559–568, 1975.

Kabata-Pendias, A. and H. Pendias. *Trace Elements in Soils and Plants*, 2nd ed., CRC Press, Boca Raton, FL, 1992, p. 365.

Lee, M.H.P. and J.M. Wigmore. Planting Trials Using PFA as a Bedding Material, in *Proceedings of Polmet 1988: Pollution in the Metropolitan and Urban Environment, Hong Kong*, Hong Kong Institution of Engineers, Hong Kong, 1988, pp. 113–118.

Menon, M.P., G.S. Ghuman, J. James, and K. Chandra. Effects of Coal Fly Ash-Amended Composts on the Yield and Elemental Uptake by Plants. *J. Environ. Sci. Health, Part A-Environmental Science*, 27, pp. 1127–1139, 1992.

Miller, R.W., R.L. Donahue, and J.U. Miller. *Soils—An Introduction to Soils and Plant Growth*, 6th ed., Prentice-Hall International, Inc., London, 1990, p. 768.

Mulhern, D.W., R.J. Robel, J.C. Furness, and D.L. Hensley. Vegetation of Waste Disposal Areas at a Coal-Fired Power Plant in Kansas. *J. Environ. Quality*, 18, pp. 285–292, 1989.

Page, A.L., A.A. Elseewi, and I.R. Straughan. Physical and Chemical Properties of Fly Ash from Coal-Fired Power Plants with Reference to Environmental Impacts. *Residue Rev.*, 71, pp. 83–120, 1979.

Rippon, J.E. and J. Wood. Microbiological Aspects of Pulverized Fuel Ash, in *The Ecology of Resource Degradation and Renewal*, Chadwick, M.J. and G.T. Goodman, Eds., Blackwell Scientific Publications, Oxford, 1975, pp. 331–349.

Schoenholtz, S.H., J.A. Burger, and R.E. Krech. Fertilizer and Organic Amendment Effects on Mine Soil Properties and Revegetation Success. *Soil Sci. Am. J.*, 56, pp. 1177–1184, 1992.

Shaw, P.J.A. A Preliminary Study of Successional Changes in Vegetation and Soil Development on Unamended Fly Ash (PFA) in Southern England. *J. Appl. Ecol.*, 29, pp. 728–736, 1992.

Thornely, J.H.A. PFA-The Versatile Material. *Hong Kong Constructor*. May–June, pp. 15–22, 1988.

Townsend, W.N. and E.W.F. Gillham. Pulverized Fuel Ash as a Medium for Plant Growth, in *The Ecology of Resource Degradation and Renewal*. Chadwick, M.J. and G.T. Goodman, Eds., Blackwell Scientific Publications, Oxford, 1975, pp. 287–304.

Townsend, W.N. and D.R. Hodgson. Edaphological Problems Associated with Deposits of Pulverized Fuel Ash, in *The Ecology of Resource Degradation and Renewal*, Hutnik, R.J. and G. Davis, Eds., Blackwell Scientific Publications, Oxford, 1973, pp. 45–56.

Walker, R.B., P. Chowdappa, and S.P. Gessel. Major Element Deficiencies in *Casuarina equisetifolia*. *Fertilizer Res.*, 34, pp. 127–133, 1993.

Walkley, A. and C.A. Black. An Examination of the Degtjareff Method for Determining Soil Organic Matter and a Proposed Modification of the Chromic Acid Titration Method. *Soil Sci.*, 37, pp. 29–38, 1934.

Wong, J.W.C. The Production of Artificial Soil Mix from Coal Fly Ash and Sewage Sludge. *Environ. Technol.*, 16, pp. 741–751, 1995.

Chapter 32

A Constructed Wetland for the Treatment of Urban Runoff

A.S. Mungur, R.B.E. Shutes, D.M. Revitt, M.A. House, and C. Fallon

INTRODUCTION

Constructed wetland systems have been shown to successfully treat municipal, industrial, and agricultural effluents (Cooper et al., 1996). They are now increasingly being used to treat urban runoff, and several studies have shown that pollutant removal rates are sufficiently high to allow discharge of runoff into receiving waters. There are currently no established design and performance criteria for constructed wetlands for the treatment of urban and highway runoff. This study assesses the removal of Cd, Cu, Pb, and Zn by a full-scale constructed wetland and its potential for treatment as the system becomes more established and metal loadings discharging into the system increase as the residential development is completed.

LOCATIONS AND METHODS

The wetland is located adjacent to a residential development (188.18 ha) in Great Notley Garden Village, Essex, which is currently being constructed by Countryside Properties PLC. The development will provide 2,000 new homes and low density business space, together with shopping, recreational, and social facilities within a garden village environment by the year 2000. A bypass road that discharges runoff into the wetland was opened in March 1996, although only one lane is currently operating (Figure 32.1).

The constructed wetland (7900 m²) has been designed to provide treatment and act as a balancing pond to store surface water runoff from the catchment and discharge it into an adjacent recreational pond (16,000 m²) via six interconnecting pipes at a controlled rate. The estimated gross area of the catchment including the pond is 38.58 ha, of which 23.11 ha is impermeable surface (Mungur et al., 1994). The Environment Agency (EA) is responsible for approving the discharge rate, as well as setting standards for the quality of the discharged water.

Water and sediment samples from up to 14 sites in the two ponds were collected and analyzed for Cd, Cu, Pb, and Zn. Specimens of *Typha latifolia*, *Iris pseudacorus*, and *Scirpus lacustris* were collected at Sites 8, 9, and 10, respectively (Figure 32.1) for tissue heavy metal analysis. Macrophyte sampling took place initially in autumn 1993, prior to runoff discharging into the wetland, to establish background tissue concentrations. Further macrophyte samples were collected in the growing seasons of 1994 and 1995. *Phragmites aus-*

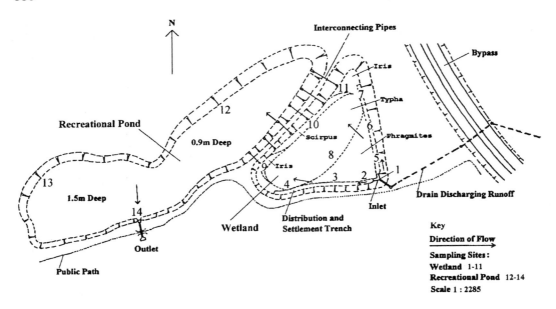

FIGURE 32.1. The constructed wetland and recreational pond showing the 14 sampling sites.

tralis was not planted in the wetland until the spring of 1995 (collected near Site 8), and thus only one sample set has been analyzed.

The plant samples were washed with tap water to prevent tissue damage and to remove all traces of attached sediment, and separated into two parts comprising the aerial component (stem and leaf) and subsurface component (roots and rhizome). These plant components (after grinding) were oven-dried at 100°C for 24 h and digested with concentrated nitric acid. The sediment samples were also oven-dried at 100°C for 24 h, sieved to the fraction less than 250 μm, and digested with a concentrated nitric acid and perchloric acid mixture (9:1 by volume). Duplicate samples were analyzed for Cd, Cu, Pb, and Zn by atomic absorption spectrophotometry using flame atomization. Water samples were digested with concentrated nitric acid, and duplicate samples were analyzed for Cd, Cu, Pb, and Zn using inductively coupled plasma atomic emission spectroscopy

RESULTS AND DISCUSSION

Water Metal Concentrations from the Inlet and Outlet of the Wetland System

The ranges of Cd, Cu, Pb, and Zn water concentrations at the inlet pipe (Site 1) (Table 32.1) show levels that are comparable with the concentrations recorded in highway drainage from major highways (Mungur et al., 1995). The area has a low traffic density and the high levels are probably due to flushes of pollutants that had accumulated upon the road surface during the construction of the bypass. The bypass surface was cleaned for the opening and the metal concentrations were lower following its opening.

The Pb and Zn concentrations found in the waters in the inlet were higher than the Cd and Cu concentrations (Table 32.1). Construction vehicles were heavily used in the vicinity of the wetland during the construction of the bypass. Zinc in the inflowing runoff was

Table 32.1. Comparison of Quality of UK/European Highway Drainage with Runoff from the Great Notley Site.

Metal	Concentration Ranges (μg L^{-1})				
	Motorways[1]	Suburban Roads[a]	A406 Road[b]	Inlet (before the opening of the bypass)	Inlet (after the opening of the bypass)
Total Cd	—	—	2–12	3–16	1–3
Total Cu	50–690	10–120	10–40	12–72	26.3–31
Total Pb	340–2410	10–150	40–160	37–152	64.7–73
Total Zn	170–3550	20–1900	10–100	11–86	4.4–50

[a] Data from Hedley and Lockey (1975).
[b] Data from Mungur et al. (1995).

probably derived from the oil and grease of the construction vehicles (Cd and Cu are not significant constituents). The Pb concentrations can be similarly explained.

Inlet concentrations were generally higher than those in the outlet and the highest inlet concentrations of Cd (51 μg L^{-1}), Cu (72 μg L^{-1}), Pb (152 μg L^{-1}), and Zn (305 μg L^{-1}) can be partially explained by shock loads due to regular rainfall events (i.e., high monthly totals) prior to the sampling dates (Figure 32.2). The higher Cd, Cu, Pb, and Zn concentrations in water in the inlet generally coincide with the autumn/winter period of 1994 and the spring of 1995 when the bypass was being constructed, reflecting the increased vehicle activity near the wetland site.

Outlet metal concentrations in water are generally lower than the inlet concentrations in the case of all the metals. They are occasionally higher than those in the inlet in the winter months (Figure 32.2). This may reflect shock loads in the runoff from the catchment area around the recreational pond (and the outlet area) which, at those times, had not been planted or developed and thus could not attenuate surface runoff.

The inlet water concentrations of the Great Notley wetland are comparable to, but generally higher than the levels seen in other wetland systems which receive urban runoff (Table 32.2). This reflects the higher metal loads discharged into the wetland due to the construction activities occurring in the vicinity throughout the majority of the monitoring period. The lower level of treatment (based on comparisons of the inlet and outlet concentrations) may be attributed to surface runoff from the then undeveloped catchment area around the recreational pond which could not attenuate surface runoff efficiently.

Metal Concentrations in Sediment from the Inlet and Outlet of the Wetland System

The sediment metal concentrations show similar trends to the water concentrations. With the exception of Cd, they were generally lower at the outlet than at the inlet. The sediment concentrations recorded at the inlet and outlet (2.2 and 2.4 μg kg^{-1}, 14.2 and 13.9 μg kg^{-1}, 41.5 and 28.4 μg kg^{-1} and 49 and 39.3 μg kg^{-1} for Cd, Cu, Pb, and Zn, respectively) were generally lower than the levels reported for other wetland systems and are in fact comparable to the concentrations seen in an unpolluted wetland (2, 20, 40, and 35 mg g^{-1} for Cd, Cu, Pb, and Zn, respectively) (Zhang et al., 1990). These results represent baseline levels and show that the Great Notley wetland is still in its early stages and has the

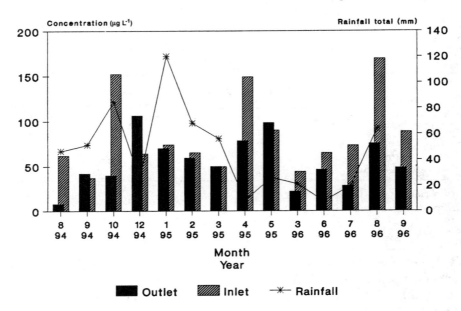

FIGURE 32.2. Temporal variation in inlet and outlet concentrations of Pb in water.

Table 32.2. Comparison of Average Inlet/Outlet Metal Water Concentrations ($\mu g \ L^{-1}$) in the Great Notley Wetland System and Other Wetland Systems Receiving Runoff.

	Cd		Cu		Pb		Zn	
Description	Inlet	Outlet	Inlet	Outlet	Inlet	Outlet	Inlet	Outlet
Great Notley wetland, UK	7.5	7.7	27.1	19.7	78.7	57.3	62.8	56.0
Sandford swamp, FL[a]	3.9	—	19.9	—	24.7	—	3.9	—
Freshwater marsh, Orlando, FL[a]	—	—	8.0	1.0	18.0	3.0	75.0	25.0

[a] Kadlec and Knight (1996).

potential to become an efficient sink for heavy metals when the housing development is fully established.

Metal Concentrations in Water from the Wetland and Recreational Pond

Water and sediment samples from a further 12 sites in the wetland and recreational pond were collected over the monitoring period. The sites were grouped into the five categories for a clearer representation of variation in metal concentrations within the system over the monitoring period (Table 32.3).

Table 32.3 shows that there were no clear trends in the water metal concentrations as the water flows from the inlet pipe to the outlet flume via the settlement trenches, the interior of the wetland and its interconnecting pipes, and the recreational pond (Figure 32.1). Cadmium, Cu, and Zn water concentrations were fairly consistent through the sys-

Table 32.3. Average Water Metal Concentrations and Standard Deviations ($\mu g\ L^{-1}$) in the Grouped Areas of the Wetland System.

Metal	Group I Inlet	Group II Sediment Traps	Group III Wetland Interior	Group IV Recreational Pond	Group V Outlet
Cd	7.5±5.5 (2.2)	6.8±4.0 (5.3)	5.4±3.8 (4.9)	7.4±5.4	7.7±8.6 (5.3)
Cu	27.1±18.0 (24.1)	28.9±26.7 (15.3)	25.0±6.3 (48.8)	24.0±12.5	19.7±9.2 (11.2)
Pb	78.7±42.8 (60.6)	96.2±51.2 (60)	65.4±21.3 (50.2)	66.7±32.1	57.3±33.1 (32.0)
Zn	62.8±94.5 (20.8)	64.8±103.2 (23.1)	23.3±8.3 (28.6)	19.9±12.7	56.0±95.9 (15.7)

( ): Average concentrations after the opening of the bypass.

tem over the monitoring period. Lead water concentrations generally show more variation through the system and concentrations in the sediment trenches, the wetland interior, and recreational pond generally vary with increases and decreases in the inlet and outlet Pb water concentrations.

Metal Concentrations in the Sediment from the Wetland and Recreational Pond

Table 32.4 shows that there was little variation within the metal concentrations through the system over the monitoring period. The opening of the bypass appears to have had no clear effect on the concentrations of all the heavy metals through the wetland, although this may also be due to low rainfall contributing to less runoff into the wetland. The highest sediment concentrations for each metal (4.8, 49.2, 76.7, and 72.8 $\mu g\ kg^{-1}$ for Cd, Cu, Pb, and Zn, respectively) were generally found in the sediment trenches following periods of consistent rainfall.

Performance of the Wetland System in Metal Removal

It is clear from Figure 32.3 that the metal loading rates into the wetland decreased markedly between October 1994 and January 1995. This is interpreted as a direct result of the reduction in construction activity through the winter of 1994/1995. Figure 32.3 also shows that the highest loading rates tend to follow regular rainfall events. However, the higher loads seen in August and October 1994 were probably associated with drainage derived from the construction activities during the building of the housing development. The very low loads seen in January 1995 coincide with a decrease in heavy construction activity. The higher loadings seen in September 1996 occurred after a relatively dry summer and are probably due to a shock load during the storm event which was in progress when the measurements were taken.

Loading rates out of the wetland at the interconnecting pipes between the wetland and the recreational pond and the outlet (Sites 9 to 11 and Site 14, respectively; see Figure 32.1) were measured in March and September 1996. Table 32.5 lists the removal efficiencies that were calculated at the various dates. The heavy metal removal efficiencies of the wetland for a nonstorm event and a storm event between the inlet (Site 1) and outlet (Site 14) range between 9.9 and 99.3%. These results are comparable to the removal efficiencies reported in full-scale wetlands by various workers (Table 32.6). The loadings at the interconnecting pipes were combined to represent the total load of metals transferring to the recreational pond. The heavy metal removal efficiencies between the inlet (Site 1) and the recreational pond range between –271.6 and 84.7%.

Cd removal in wetlands is thought to occur because of the formation of its sulfide and subsequent sedimentation of the metal (Hendry et al., 1979; CH2M HILL, 1992). The wide range of Cd removal in the Great Notley wetland (Table 32.6) may be attributed to the low concentrations of Cd usually found in runoff. Thus even small fluctuations in concentration will affect the removal efficiency greatly. Cu, Pb and Zn removals appear to be correlated with inflow concentrations with removal efficiency increasing with increasing inflow concentrations (Kadlec and Knight, 1996). The high affinity of Cu for peat and humic substances (Kadlec and Keoleian, 1986) and, to a lesser degree, macrophyte tissues (Sinicrope et al., 1992; Zhang et al., 1990) appears to play an important role in its removal

Table 32.4. Average Sediment Metal Concentrations and Standard Deviations ($\mu g\ kg^{-1}$) in the Grouped Areas of the Wetland System.

Metal	Group I Inlet	Group II Sediment Traps	Group III Wetland Interior	Group IV Recreational Pond	Group V Outlet
Cd	2.2±1.3	2.5±1.3 (2.7)	2.0±1.9 (5.0)	1.7±1.0	2.4±2.5 (3.4)
Cu	14.2±8.8	25.9±11.8 (10.5)	23.8±19.4 (25.8)	16.8±5.0	13.9±6.7 (7.5)
Pb	41.5±17.9	44.0±14.1 (24.7)	31.0±10.3 (27.2)	32.4±7.0	28.4±12.6 (15.0)
Zn	49.0±19.9	62.1±7.8 (47.3)	65.0±49.7 (57.6)	41.0±9.2	39.3±14.6 (21.3)

( ): Average concentrations after the opening of the bypass.

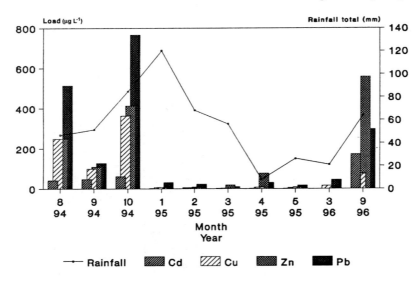

FIGURE 32.3. Temporal variation of heavy metal loading rates at the inlet (Site 1).

Table 32.5. Removal Efficiencies for the Inlet (Site 1)/Recreational Pond and Inlet (Site 1)/ Outlet (Site 14), Based on Metal Loads.

Date	Removal Efficiency (%) Sites	Cd	Cu	Pb	Zn
03/96	Site 1/Pond	−170.0	−46.7	24.6	−271.6
Nonstorm event	Site 1/Site 14	10.0	94.1	89.0	9.9
09/96[a]	Site 1/Pond	80.0	−281.6	26.8	84.7
Storm event	Site 1/Site 14	99.3	97.4	97.1	99.2

[a] Storm event—water samples collected and flow measurements made at the outlet (Site 14) represent the first flush of water out of the system (3 hours after initial sampling at the inlet (Site 1)—i.e., the retention time).

Table 32.6. Comparison of the Heavy Metal Removal Performance of the Constructed Wetland System at Braintree with Other Full-Scale Wetlands.

Metal	% Removal range between the inlet and outlet in the Great Notley wetland system	% Removal range between inlet and outlet in other full-scale wetland systems (Kadlec and Knight, 1996)
Cd	10.0–99.3	0–98.7
Cu	94.1–97.4	38–96
Pb	89.0–97.1	−181–83.3
Zn	9.9–99.2	33–89.5

in wetlands and may explain the high Cu removal seen in the Braintree wetland. Lead removal appears to be essentially due to the formation of insoluble compounds followed by subsequent sedimentation. The wide range for Zn removal in the Braintree wetland (Table 32.6) compared to Pb may be attributed to Zn being present in water as a predominantly

Table 32.7. Maximum Heavy Metal Concentrations (μg kg^{-1}) in Wetland Macrophytes.

Metal	Typha latifolia		Iris pseudacorus		Scirpus lacustris		Phragmites australis	
	ss[a]	as[b]	ss	as	ss	as	ss	as
Cd	1.7	0.8	7.6	1.4	5.2	1.4	3.0	0.6
Cu	16.2	5.4	24.1	9.2	138.2	7.0	19.4	7.0
Pb	8.0	30.1	18.1	9.5	35.5	16.0	7.0	5.0
Zn	52.6	38.0	68.6	50.4	98.2	18.0	212.8	31.2

[a] ss: subsurface tissue.
[b] as: above surface tissue.

soluble bioavailable metal ion or weak complex (Revitt and Morrison, 1987). Sediment association/uptake for Zn is generally lower than Pb since the latter exhibits little remobilization once deposited (Meiorin, 1989) and this may affect the overall removal efficiency of Zn. However, this was not seen in the Braintree wetland where the average sediment Zn concentrations in the system are higher than those of Pb (Table 32.6). Another possibility is that Zn removal varies more than Pb removal because of more variable inflow concentrations.

Metal Concentrations in Macrophytes in the Wetland

Metal analyses of the macrophytes indicate bioaccumulation of Cd, Cu, Pb, and Zn (Table 32.7; Figures 32.4 to 32.6). Tissue concentrations are generally higher in the roots and rhizomes compared to the leaves and stems. This is consistent with the results of several other studies (Mungur et al., 1995). There is an increasing trend with time in heavy metal concentrations of Zn in the subsurface tissues of *Iris* (Figure 32.4). Pb in the leaves and stems of *Scirpus* (Figure 32.5) and Cu and Zn in all tissues of *Typha* show similar increases (Figure 32.6). Zn appears to be preferentially accumulated over Pb, Cu, and Cd. This is consistent with other studies on metal uptake by macrophytes in wetlands receiving urban runoff (Meiorin, 1989; Simpson et al., 1983; Zhang et al., 1990) and may be explained by Zn being present in water as a predominantly soluble bioavailable metal ion (Revitt and Morrison, 1987) which may allow it to be taken up by the tissues more rapidly than the other metals.

The higher Zn concentrations may also be explained by the formation of an iron plaque on the roots of the macrophytes (Crowder and St.-Cyr, 1991). Iron plaque consists mainly of iron hydroxy-oxides (Mendelssohn and Postek, 1982) and there is a possibility that plants forming an iron plaque could be at an advantage with regard to the uptake of metals due to the adsorption and immobilization of heavy metals by the iron plaque (Taylor and Crowder, 1983), although the tolerance mechanism is as yet unclear (Ye et al., 1994). The plaque seems to slow Zn transport to the above-surface tissue but not to reduce Zn uptake into the roots, thus concentrating Zn in this area.

The maximum metal concentrations in the macrophyte tissues (Table 32.7) are considerably less than the concentrations seen in the macrophytes of the natural wetland near the Brent reservoir, which receives runoff from a major highway. This is to be expected, since the macrophytes in the Great Notley wetland were sampled when the plants were relatively young and had not been completely established.

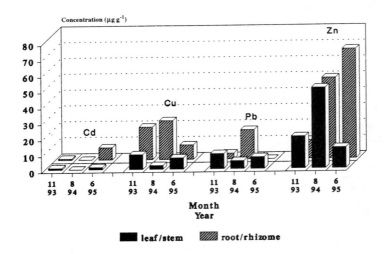

FIGURE 32.4. Temporal variation of heavy metals in the tissues of *Iris pseudacorus*.

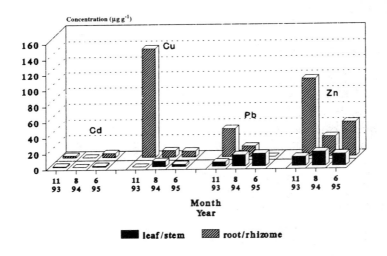

FIGURE 32.5. Temporal variation of heavy metals in the tissues of *Scirpus lacustris*.

This is clearly shown by a comparison of the maximum metal concentrations in *Typha latifolia* in the Great Notley wetland with those in a natural wetland and in a receiving basin in North London which both receive urban stormwater runoff (Figure 32.7) (Zhang et al., 1990). The comparison shows that the concentrations are lowest in the *Typha latifolia* present in the Great Notley wetland. Since macrophytes can accumulate high levels of heavy metals, as shown by the *Typha* in the natural wetland, it is envisaged that macrophyte tissue metal concentrations in the Great Notley wetland will increase over successive growing seasons as the system becomes more established and the wetland receives runoff more consistently.

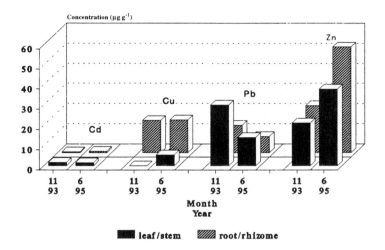

FIGURE 32.6. Temporal variation of heavy metals in the tissues of *Typha latifolia.*

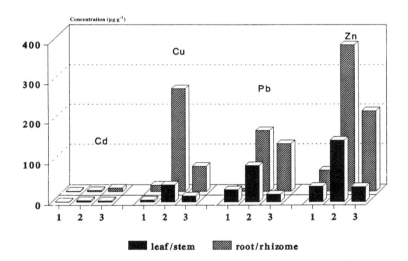

FIGURE 32.7. Comparison of maximum heavy metal concentrations in the tissues of *Typha latifolia* in (1) the Great Notley wetland, (2) a natural wetland, and (3) a receiving basin.

CONCLUSIONS

The results provide a baseline study for the assessment of the heavy metal removal performance of the constructed wetland system at Great Notley Garden Village. Delays in construction of the bypass and the housing development have meant that only limited results could be obtained after the opening of the bypass, which is the main source of runoff discharged into the wetland. These initial results show a variable metal removal performance by the constructed wetland. An improvement in performance is expected with the growth to maturity of the plants and the accumulation of a litter layer; the reduction and eventual completion of construction activity; and the regular removal of sediment from the settlement trenches.

The wetland is located in a newly created country park which will attract wildlife and enhance the original environment of arable fields, and is planted with four species of macrophyte which improve its aesthetic value. It is also associated with an irregular-shaped ornamental lake which complements its environmental and aesthetic value. Furthermore, the baseline monitoring results indicate that the runoff treatment performance should improve as the system becomes more established. However, the wetland has a regular triangular shape which decreases its aesthetic value and the flow is possibly shortcircuiting in the wetland and thus reducing treatment potential.

Heavy metal removal efficiencies between the inlet and outlet ranged from 9.9 to 99.3%, and more consistent results are expected with time. A continuation of the study would provide a valuable case study of the performance of the wetland system before and after the completion of the bypass and the housing development. The results will also influence the adoption of constructed wetlands in future residential developments in the UK.

ACKNOWLEDGMENTS

The authors wish to thank Countryside Properties PLC for their agreement to the monitoring program.

REFERENCES

CH2M HILL Carolina Bay Natural Land Treatment Program. Final report. Volume I—Data Summary. Volume II—Appendices. Prepared for Grand Strand Water & Sewer Authority, 1992.

Cooper, P.F., G.D. Job, M.B. Green, and R.B.E. Shutes. *Reed Beds and Constructed Wetlands for Wastewater Treatment*. Water Research Centre Publications, Medmenham, UK, 1996.

Crowder, A.A. and L. St.-Cyr. Iron Oxide Plaque on Wetland Roots. *Trends Soil Sci.*, 1, pp. 315–329, 1991.

Hedley, G. and J.C. Lockley. Quality of Water Discharged from an Urban Motorway. *J. Water Pollut. Control*, 74, pp. 659–674, 1975.

Hendry, G.R., J. Clinton, K. Blumer, and K. Lewin. Lowland Recharge Project Operations, Physical, Chemical and Biological Changes 1975–1978. Final Report to the Town of Brookhaven, Brookhaven National Laboratory, Brookhaven, New York, 1979.

Kadlec, R.H. and G.A. Keoleian. Metal Ion Exchange on Peat. *Peat and Water*, Fuchsman, C.H, Ed., Elsevier, Amsterdam, pp. 61–93, 1986.

Kadlec, R.H. and R.L. Knight. Treatment Wetlands. Lewis Publishers, Boca Raton, FL, 1996, p. 893.

Meiorin, E.C. *Constructed Wetlands for Wastewater Treatment: Municipal, Industrial and Agricultural*, Hammer, D.A. Ed., Proceedings from the First International Conference on Constructed Wetlands for Wastewater Treatment. 13–17 July, 1988, Chattanooga, TN. Lewis Publishers, Boca Raton, FL, 1989, pp. 667–685.

Mendelssohn, J.A. and M.T. Postek. Elemental Analysis of Deposits on the Roots of *Spartina alterniflora* Loisel. *Am. J. Bot.*, 69, pp. 904–912, 1982.

Mungur, A.S., R.B.E. Shutes, D.M. Revitt, and M.A. House. A Constructed Wetland for the Treatment of Highway Runoff in the United Kingdom, in International Association on Water Quality (IAWQ); Specialist Group on the Use of Macrophytes in *Water Pollution Control Newsletter* 11, pp. 7–12, 1994.

Mungur, A.S., R.B.E. Shutes, D.M. Revitt, and M.A. House. An Assessment of Metal Removal from Highway Runoff by a Natural Wetland. *Wat. Sci. Tech.*, 32(3), pp. 169–175, 1995.

Revitt, D.M. and G.M. Morrison. Metal Speciation Variations with Separate Stormwater Systems. *Environ. Technol. Lett.*, 8, p. 373, 1987.

Rexnord, Inc. Sources and Migration of Highway Runoff Pollutants. Rpt. RE-84/059, U.S. Federal Highway Administration, VA, 1984.

Simpson, R.L., R.E. Good, R. Walker, and B.R. Frasco. The Role of Delaware River Freshwater Tidal Wetlands in the Retention of Nutrients and Heavy Metals. *J. Environ. Qual.*, 12, pp. 41–48, 1983.

Sinicrope, T.L., R. Langis, R.M. Gesberg, M.J. Busanardo, and J.B. Zedler. Metal Removal by Wetland Mesocosms Subjected to Different Hydroperiods. *Ecol. Eng.*, 1(4), pp. 309–322, 1992.

Taylor, G.J. and A.A. Crowder. 1983. Uptake and Accumulation of Copper, Nickel and Iron by *Typha latifolia* Grown in Solution Culture. *Can. J. Botany*, 61(7), pp. 1825–1830, 1992.

Ye, Z.H., A.J.M. Baker, and M.H. Wong. Heavy Metal Tolerance, Uptake and Accumulation in Populations of *Typha latifolia* L. and *Phragmites australis* (Cav.) Trin. ex Steudal, in Proceedings of the 4th International Conference on Wetland Systems for Water Pollution Control. 6–10 November 1994, Guangzhou, China, IAWQ, 1994, pp. 297–306.

Zhang, T.T., J.B. Ellis, D.M. Revitt, and R.B.E. Shutes. Metal Uptake and Associated Pollution Control by *Typha latifolia* in Urban Wetlands, in *Constructed Wetlands in Water Pollution Control*, Cooper, P.F. and B.C. Findlater, Eds., Pergamon Press, Oxford, 1990, pp. 451–459.

Chapter 33

Vermitechnology for the Remediation of Degraded Soils in Western Siberia

N.F. Protopopov, B.R. Striganova, and J.A. Manakov

INTRODUCTION

Western Siberia, which is rich in different mineral resource deposits, has become an area of large mining and processing complexes. The problems of remediating degraded soils related to industrial expansion are topical. The remediation process strategy is based on the technical and biological steps of works. Usage of biological methods provides more complete and rapid restoration of ecological functions in technogenic landscapes. Vermitechnology (earthworm technology) has a high potential in biological remediation.

VERMITECHNOLOGY AS AN ECOLOGICAL ENGINEERING OPTION

The earthworms are known as accelerators of the conversion of organic wastes into stable soil conditioners, which improve the soil. Many researchers have investigated some aspects of earthworm breeding, the use of earthworms for improving the soil fertility and soil physical conditions, vermitechnological stabilization of organic wastes, etc. (Watanabe and Tsukamoto, 1976; Edwards and Lofty, 1977; Mitchell et al., 1977; Kaplan et al., 1980; Haimi and Huhta, 1987). The destructive impact of earthworms on some human and plant pathogens has also been found (Brown and Mitchell, 1981; Toyota and Kimura, 1994).

The special properties of earthworms to decrease heavy metals content and other toxic components in digested wastes and soils by incorporation into body tissues and into humic substances complexes are used in some organic waste utilization technologies.

The most popular species of earthworms using in vermitechnology is *Eisenia foetida* spp. (*Oligochaeta, Lumbricidae*).

The application of vermitechnology for remediation of degraded soils in an industrial scale under extreme conditions of Western Siberia requires effective large-scale production of earthworm populations and vermicompost. The investigations related to the development of the intensive technologies for breeding earthworms and the production of vermicompost have been carried out since 1989 in Tomsk (Novikov et al., 1994; Protopopov, 1996, 1996a, 1996b; Protopopov et al., 1996). There are short humid summers and long winters in Western Siberia. The conventional approaches were founded to be ineffective for large-scale vermiculture in Siberia. New approaches to vermiproduction have been developed. There are three main ways to achieve success of vermitechnology:

(a) using and preparing optimal substrates for vermiculture
(b) development of effective technologies for vermiculture
(c) selection of more highly productive earthworms.

Some examples of Western Siberian experiences include following:

- using fresh wastes as substrates
- accelerated methods of substrate pretreatment
- decreasing expenditure
- low energy consumption vermitechnologies
- turn negative climatic conditions into positive ones
- using accessible resources, equipments and materials
- integration of vermitechnology with other technologies for complete treatment and utilization of organic wastes
- using different vermitechnologies and techniques depends on main production trends and aims:
  - stabilization of wastes (vermistabilization)
  - biomass production (vermicultivation)
  - production of fertilizers (vermicomposting)
- standardization of products (vermicompost)
- observance and usage of economic and law mechanisms.
- New developments in year-round vermiproduction have brought applications to other fields. In particular, these new developments have helped in restoring degraded soils which are related to oil production, coal mining, organic municipal and agricultural waste dumps.

DEGRADED SOILS IN OIL AND GAS FIELDS OF WESTERN SIBERIA

The intensive exploration of oil and gas deposits in Siberia began in the 1960s. Oil and gas deposits are mainly in the Tyumen region and particularly in the Tomsk, Novosibirsk, and Krasnoyarsk areas. Presently, the oil and gas fields cover an area of about 1 m km^2 (Sedykh, 1995). The main types of degraded lands include oil and/or high mineralized water spills near pipelines and pads of wells, pits, or ground pools for the storage of well-drilling liquid and solid wastes, soil and vegetation cover disturbances, and sites around flares.

There are some effective in situ and ex situ technologies for the treatment of oil-contaminated soils. For example, since the beginning of the 1980s the *ex situ* technologies for contaminated soil treatment have been applied in the Netherlands (Honders, 1995). The soil treatment is based on heat (thermal desorption/incineration), liquid (classification/flotation) and biological (landfarming) technologies.

Most of the Western Siberian oil fields are situated in natural wetlands (marshes and meadows). More than 80% of the soils consist of semihydromorphic and hydromorphic soils of nonagricultural use. This is the reason why most conventional soil remediation systems and technologies are ineffective and unusable.

The in situ technologies for soil treatment emphasize biological methods. The conventional technologies of recultivating the oil contaminated soils and pits which are widely used in Western Siberia are based on the mechanical harvesting of free oil, addition of commercial mineral fertilizer and/or organic fertilizer (slurry, etc.) and/or oil-digesting microorganism cultures, loosening, covering by pure mineral soils, planting cereal and/or legume cultures.

Most of the oil fields are situated far from urban areas, rural habitats, and traffic systems. High cost and ecological safety requirements have made the transportation of unconditioned organic fertilizers expensive and hazardous. The application of commercial mineral fertilizers is also expensive. Surplus supply of both may increase the corrosion risk for well and pipeline systems.

The vermirecultivation method for the pits of the well drilling wastes and the oil-contaminated soils was developed by Protopopov (1995b). The method is based on the addition of pure vermicompost and/or vermicompost with the inoculated commercial oil-digesting microorganism culture (named NBD-Vermic). This is applied to contaminated sites with other technical recultivation operations, and with the planting of native plants which are common to the natural landscapes and surrounding areas. The level of vermicompost application depends on the extent of soil contamination and goals of the treatment.

The application of vermicompost, as an accelerator of soil self-remediation, is economically and ecologically feasible, because it is a highly effective stabilized fertilizer. This recycled bioproduct is odorless and lightly granulated. Also, it is easily and safely transportable. The small quantity of vermicompost required for treating sites decreases the corrosion risk for operational equipment and construction.

The use of native plants corresponds to the requirements of the landscape conservation. Particularly, application of vermicompost excludes the contamination by trace elements which may occur when applying some mineral fertilizers.

PHYTOVERMIREMEDIATION OF SOILS DEGRADED BY COAL MINING

Kuzbas is the largest coal region in Russia. At present the area disturbed by mining works is about 1,000 km^2, and more than 8 km^3 of rocks are dumped on the surface. These technogenic formations are one of the serious factors that contributes to environmental pollution in the region. The primary phytocenoses formation in Kuzbas dumps occurs slowly over a 20–30 year period. Rock dumps (sandstones, siltstones, argillites) have a low fertility potential, and they have especially low levels of nitrogen content (Gadzhiev et al., 1992; *Restoration of Technogenic Landscapes...*, 1977).

Investigations and practical works on the remediation of coal mining dumps have been carried out by Kuzbas Botanical Gardens (Kemerovo, Russia) since 1987. The investigations have essentially focused on creating culturphytocenosis in rock dumps. Plant growth for agriculture profits on rocks dumps has been investigated. In this case new approaches are evaluated. One of these is the application of vermitechnology which might be implemented by:

(i) introduction of earthworms into wet sites of dumps with/without addition of organic wastes (sewage sludge, slurry, etc.).
(ii) addition of vermicomposts with/without the sowing of plants.

The phytovermiremediation technique is based on the second step (Protopopov et al., 1996). The technique is based on adding vermicompost to smoothed rock dumps and the planting of a grass-mixture (one- and two-year cereals, legume cultures).

VERMISTABILIZATION OF ORGANIC WASTE DUMPS

The organic agricultural and municipal waste dumps are widely spread throughout Western Siberia. These pose a real risk to environmental and human health.

There are two main ways for vermitechnology applications to stabilize or remediate organic waste dumps:

(a) introduction of earthworms into sites of organic waste dumps
(b) utilization of organic wastes "on-site."

Introduced earthworms stabilize the wastes, turn them into soil aggregates, and decrease pathogenic microorganisms populations.

The utilization of organic wastes "on-site" is a preventative approach. Some vermitechnologies for the intensive utilization of sewage sludge and pig farming effluents under extreme conditions of Western Siberia were developed (Novikov, 1993; Protopopov and Tuchak, 1994; Protopopov, 1995a, 1996a, 1996b; Protopopov and Tuchak, 1996).

All methods were developed as year-round cycle technology. The methods are divided into "inside" and "outside" technologies.

The "inside" technology which is based on container-technology is more preferable for vermicultivation and vermicomposting (Novikov, 1993). For large-scale vermistabilization of sewage sludge and animal wastes, the "outside" approach is more acceptable.

The winter methods of vermicultivation (Protopopov and Tuchak, 1994) are based on vermistabilizing organic wastes in beds year-round. The winter method was used in the "Vermibloc" technique for surplus sewage sludge utilization in a conventional waste water treatment plant (Protopopov, 1995a; Protopopov, 1996a).

Due to the application of vermitechnology since 1993 in Kyslovka CWS (waste water treatment complex), the dumping of sludge from sludge-ponds was not needed (Protopopov and Tuchak, 1996).

One of the attractive peculiarities of vermitechnology is that it is based on using the products of treated wastes (vermicompost) with a purely biological or ecological approach. The investigations of the vermitechnology for remediation of degraded lands in Western Siberia are still in progress.

ACKNOWLEDGMENTS

We are grateful to the John D. and Catherine T. MacArthur Foundation for its financial support in the conference participation (Grant #: 96-45001A-FSU), and to Mrs. Lyudmila Chikina and Mr. Steve Vrooman for proofreading the paper.

REFERENCES

Brown, B.A. and M.J. Mitchell. Role of the Earthworm, *Eisenia foetida*, in Affecting Survival of *Salmonella enteritidis ser. typhimurium*. *Pedobiologia*, 22, pp. 434–438, 1981.

Edwards, C.A. and J.R. Lofty. *Biology of Earthworms*. Halsted Press, New York, 1977.

Gadzhiev, I.M., V.M. Kurachev, F.K. Ragim-zade, et al. *Ecology and Recultivation of Technogenic Landscapes*. Nauka, Siberian Branch, Novosibirsk, 1992, p. 305, (in Russian).

Haimi, J. and V. Huhta. Comparison of Composts Produced from Identical Wastes by "Vermistabilization" and Conventional Composting. *Pedobiologia*, 30(2), pp. 137–144, 1987.

Honders, A. Treatment of Oil-Bearing Soils in Western Siberia—Solutions Derived from Dutch Experiences, in *Workshop on Decontamination and Rehabilitation of Polluted Areas in Western Siberia, Novosibirsk, Russia, October 24–25, 1995*, EU/ THERMI, Tyumen, 1995.

Kaplan, D.L., R. Hartenstein, and E.F. Neuhauser. Physicochemical Requirements in the Environment of the Earthworm *Eisenia foetida. Soil Biol. Biochem.*, 12, pp. 347–352, 1980.

Mitchell, M.J., R.M. Mulligan, R. Hartenstein, and E.F. Neuhauser. Conversion of Sludges into "Topsoils" by Earthworms. *Compost Science*, 18(4), pp. 28–32, 1977.

Novikov, J.M. Vermicultivation and Vermicompost Production and Indicator of Vermicompost maturity. Russian Patent Application No. 93058080/ 15(058032), (in Russian), 1993.

Protopopov, N.F. All-Year-Round Method of Sewage Sludge and Wood-Waste Utilization by Earthworms and Vermicompost Produced from Sewage Sludge and Coniferous Wood-Wastes. Russian Patent Application No. 95115438/13(026354), (in Russian), 1995.

Protopopov, N.F. Method of Vermirecultivating the Pits of the Wells Drillings Wastes and the Oil-Contaminated Soils and Composition for Treatment of Oil-Contaminated Soils. Russian Patent Application No. 95109604/15(016527), (in Russian), 1995a.

Protopopov, N.F. New Vermitechnology Approach for Sewage Sludge Utilization in Northern and Temperate Climatês All Year Round. *Environ. Res. Forum*, 5–6: 413–416, 1996.

Protopopov, N.F. Using Earthworms for Utilization of Sewage Sludge. Communication presented in Computer Conference on Waste Water Treatment for United Nations Conference HABITAT II, Istanbul, June, 1996. Swedish Royal Institute of Technology, GAP, March–April, 1996, INTERNET. 1996a.

Protopopov, N.F., J.A. Manakov, S.A. Skoblikov, and L.G. Kolesnichenko. Method of Phytovermirecultivating Rock Dumps of Coal Mining. Russian Patent Application No. 96120971/20(027604), (in Russian), 1996.

Protopopov, N.F. and S.G. Tuchak. Application of Vermitechnology in Combined Wetland Systems, in *5th International Conference on Constructed Wetland Systems for Water Pollution Control: Preprints, Vienna, 1996, IAWQ*, IWGA, Vienna, Vol. 2, p-r 25-1-4, 1996.

Protopopov, N.F. and V.N. Tuchak. Winter Method of Vermicultivation. Russian Patent Application No. 94041156/15(040480), (in Russian), 1994.

Protopopov, N.F., V.N. Tuchak, and B.R. Striganova. Utilization of Effluents from Pig Breeding Plants by Vermitechnology Methods under Extreme Conditions of Siberia, in *International Conference on Ecological Engineering, October 7–10, 1996, Beijing, PRC*. IEES, CAS, 1996.

Restoration of Technogenic Landscapes in Siberia (Theory and Technology). Nauka, Siberian Branch, Novosibirsk, p. 160, (in Russian), 1977.

Sedykh, P. Ecological Statement of Western Siberia. Main Contaminated Lands, Main Sites Which Are Under Conditions of Petrochemical Pollution, in *Workshop on Decontamination and Rehabilitation of Polluted Areas in Western Siberia, Novosibirsk, Russia, October 24–25, 1995*, EU/THERMI, Tyumen, (in Russian), 1995.

Toyota, K. and M. Kimura. Earthworms Disseminate a Soil-Borne Plant Pathogen, *Fusarium oxysporum f. sp. raphani. Biol. Fertil. Soils*, 18, pp. 32–36, 1994.

Watanabe, H. and J. Tsukamoto. 1976. Seasonal Change in Size Class and Stage Structure of Lumbricid *Eisenia foetida* Population in a Field Compost and Its Practical Application as the Decomposer of Organic Waste Matter. *Rev. Ecol. Biol. Sol.*, 13, pp. 141–146, 1994.

Chapter 34

Utilization of Pig-on-Litter Compost and Anaerobically Digested Sewage Sludge for the Growth of Edible Crops: Rate of Application and Effects of Heavy Metals

C.M.L. Lai and R.Y.H. Cheung

INTRODUCTION

According to Oswell and Rootham (1992), sludge production in Hong Kong from sewage treatment work amounts to 15,400 t of dry solids per year. With tighter environmental control and standards enactment, combined with general economic growth in the future, sludge production within the territory by 2000 would increase to between 25,000 and 400,000 t yr^{-1} (d.s.).

In terms of sewage sludge, instead of regarding disposal as the first option of waste management strategy, much research has been done on sludge to develop by-products either to recycle to the land or for the recovery of usable materials and production of energy, such as oil, as reported by Bridle and Hertle, 1988.

Application of organic wastes such as sewage sludge to land has long been practiced for nutrient recycling, in particular organic matter, N and P (Sommers, 1977; Suss, 1979). Also, land application of sludge is an inexpensive and attractive way of disposal. The European Community Urban Waste Water Treatment Directive (91/271/EEC) had already mandated that sea disposal of sewage sludge will end by 1998. This will make land application of such waste an even more logical and popular means of waste disposal. In all cases, no matter whether disposal to agricultural land, forest, disturbed land, or specified disposal sites, further sludge stabilization and treatment occur at the receiving site due to exposure to sunlight and dehydration. Furthermore, through the action of soil microorganisms and pathogens, many toxic organic-borne chemicals in sludge would be eventually destroyed. Organic matter in sludge improves soil texture and structure, which increases water retention and supports microbial growth. It also facilitates nutrient transport (Garcia-Delgado et al., 1994).

Through a slow but continuous mineralization process, organic form nutrients are converted into phyto-available forms. These simple inorganic nitrogenous and phosphorous compounds facilitate overall plant growth in the waste-amended soils. Organic waste can, therefore, be a partial substitute for the expensive inorganic fertilizers.

Dowdy et al. (1978) reported that the increase of crop yield by sludge application often exceeded that of a well managed fertilized control.

By design, sewage treatment processes lead to the high removal efficiency of more than 90% of solids from the water phase. So the treated effluent often fulfills the water quality standards for discharge. However, the sludge itself may contain environmental contaminants, such as heavy metals from domestic, commercial, and industrial activities. They are usually tightly bonded in the sludge matrix and concentrated in the solids generated from sewage treatment facilities. Heavy metals (Cd, Cu, Cr, Ni, Pb, and Zn) in wastewater sludge may persist in soil for very long periods of time (McGrath, 1987).

Disposal of organic wastes with environmental contaminants to the receiving environment has, therefore, raised concerns. Phytotoxic heavy metals (such as Cu, Ni, and Zn) can retard plant growth. With plants grown on contaminated soils, uptake may occur via root system and/or dust may be blown onto leaves or stems. Direct uptake of pollutants through roots and leaves may lead to lethal and sublethal effects on the health of flora (Pett and Eduljee, 1994). Heavy metals will potentially affect not only plants directly grown on waste-amended soil, but also *secondary consumers*, such as animals and finally humans. For example, zootoxic metals (such as Cd, Cr, and Pb) can be bioconcentrated in the vegetation (grown from the waste-amended soil), making it unsuitable for human consumption.

For the application of sludge to land, management practices should consider concentrations of organic and inorganic chemicals in sludge, maximum allowable accumulation levels in soils, accumulation rates, local soil conditions (pH, organic matter content, cation exchange capacity), climatic conditions, and topography as provided in EC Directive, 1986. Limits are set for total zinc, nickel, cadmium, mercury, and lead, with chromium being considered in the EC directive. If good management of sludge land application (adaptation of new developments, monitoring of sludge characterization, estimation of effect on crops and consumers) were practiced, there would be a cost-effective method of sludge reutilization with minimal risk to health.

The aim of this chapter is to study the effect of anaerobically digested sewage sludge application to a sandy soil, in comparison with POL-compost, on crop yield as well as the supply of phytonutrients (N, P, K) and the impact of heavy metals and bioaccumulation pattern in various plant tissues.

MATERIALS AND METHODS

A representative sample of anaerobically digested sewage sludge was collected from the Tai Po Sewage Treatment Work in Hong Kong, which was classified as secondary wastewater treatment of mesophilic anaerobically digestion with dewatering. 'Pig-on-Litter' compost was collected from the Hong Kong Government Experimental animal farm in Ta Kwa Ling, Hong Kong.

The freshwater sand, sewage sludge, and POL-compost were analyzed for their nutrient profile. Total nitrogen was measured by the Kjeldahl method (H_2O_2 and H_2SO_4 acid digestion), according to Allen, 1989. Soil inorganic nitrate was measured in 25:1, 10-min extraction with deionized and double distilled water, using copperized cadmium reduction, (Willis and Gentry, 1987), FIA, Lachat QuickChem. Sample digest and extract for total and extractable phosphorus were prepared (4: 5 H_2O_2 and H_2SO_4 acid digestion and 40:1, 1-h extraction, 2.5% acetic acid extraction), by FIA Lachat Quick Chem, Molybdenum Blue Method (Golterman et al., 1978; Mackereth et al., 1978). Total carbon was determined by CHN analyzer.

Total and extractable metals (K, Cd, Cr, Cu, Ni, Pb, and Zn) were prepared by 1:1 H_2O_2 and conc. HNO_3 digestion (Jones et al., 1991) and 25:1, 1-h extraction with 1M ammonium acetate extraction, respectively. After sample preparation, the sample was then filtered through Whatman® No. 44 filter paper. The filtrates were analyzed for metals by Atomic Absorption Spectrophotometry, AA 6501S, Shimadzu.

In order to study the biological response of crops, a pot trial experiment was conducted in the Botanical Experimental Facility in the City University of Hong Kong. Two locally available edible leafy and root crops; namely, Chinese White Cabbage (*Brassica chinensis*) and Chinese Radish (*Raphanus sativus*) were selected. Each organic waste was mixed separately to form a series of pots with different dosages (0, 0.75, 1.25, 2.5, 5, and 10% w/w) with freshwater sand without supplementary fertilizer application as growth media (6 kg pot^{-1}) and was placed into 20-cm plastic pots. They were left for seven days for equilibrium. Ten pregerminated seedlings (seven days old) in Meliculite were transplanted into each waste-amended pot.

Each treatment was triplicated, which then made a total of {3*2*[(2*5)+1]}= 63 pots. Control pots received neither sewage sludge nor POL-compost.

All pots were watered regularly during their growing period. Plants were thinned to reduce competition at day 35. Four *B. chinensis* and two *R. sativus* individual plants were retained for further investigation, respectively.

B. chinensis and *R. sativus* were grown for a total of 90 days and 120 days, respectively. All remaining plants were taken out from their pots and the harvested tissues were washed under running tap water to remove soil particles and then deionized water. After washing, the vegetation samples were allowed to drain dry and then separated into aerial and root parts, then oven-dried (60°C for seven days) for dry weight determinations. The metal contents (Cd, Cr, Cu, Ni, Pb, and Zn) of various tissue parts were determined after grinding (passing a 0.5-mm screen) and conc. HNO_3 digestion, as described for the organic wastes.

After harvesting, top soil samples (5 cm) were collected from each pot. After air-drying and sieving, a fraction of smaller than 2 mm was collected for nutrient and metal analysis as the procedure for the original organic wastes.

All essential plastics and glassware used in the experiment were soaked at 10% HNO_3 for 24 h and then rinsed repeatedly in deionized and bidistilled water to prevent any metal contamination.

Standard reference material (NIST 1515 and 1547) was analyzed along with the test sample for quality control purposes.

RESULTS AND DISCUSSION

Nutrients in Solid Wastes

In order to study the ease of nutrient availability, extractable NO_3^-, PO_4^{3-}, and K^+ were determined using different extractants. POL-compost showed a higher content of phytoavailable nutrient than that of sewage sludge. It was due to the chemical fixation of nutrients such as binding in the sludge matrix during wastewater treatment processes, while only simple physical adsorption of nutrient on sawdust surface occurred in POL-compost.

However, in terms of total phytonutrients (N, P, K), the situation was different and not so overwhelming; sewage sludge showed nearly three times more than that of POL-compost in the case of total nitrogen. For phosphorus, both sewage sludge and POL showed comparable content, while POL-compost provided much higher potassium than that of sewage sludge (Table 34.1).

Table 34.1. Chemical Characteristics of Freshwater Sand, POL-Compost and Sewage Sludge.[a]

Parameters		Freshwater Sand	POL-Compost	Sewage Sludge
Total carbon %		0.23	31.301	30.43
Total nitrogen %		0.03	1.76	4.57
Extractable NO_3^- ($\mu g\ g^{-1}$)		0.03	103.10	42.66
C/N ratio		8.41	17.79	6.66
Total P %		0.01	1.72	1.94
Extractable P ($\mu g\ g^{-1}$)		16.92	2114.60	909.71
K ($\mu g\ g^{-1}$)	Total	2240.56	23318.60	3438.11
	Extractable	261.45	23317.89	1745.56
Cd ($\mu g\ g^{-1}$)	Total	0.03	4.29	2.05
	Extractable	0.03	0.34	0.26
Cr ($\mu g\ g^{-1}$)	Total	76.63	18.10	60.97
	Extractable	0.32	0.78	0.71
Cu ($\mu g\ g^{-1}$)	Total	5.90	1282.53	1557.31
	Extractable	0.14	14.47	54.80
Ni ($\mu g\ g^{-1}$)	Total	43.29	10.13	54.59
	Extractable	0.509	1.198	5.516
Pb ($\mu g\ g^{-1}$)	Total	14.747	26.181	158.552
	Extractable	0.028	0.908	0.784
Zn ($\mu g\ g^{-1}$)	Total	34.181	1000.23	2045.17
	Extractable	0.166	8.8715	825.964

[a] Each value is a mean value of 3 replicates.

Heavy Metals in Solid Wastes

On the other hand, anaerobically digested sewage sludge generally contained higher levels of heavy metals when compared with POL-compost (except Cd). Zn and Cu were the most evident and the second most dominant metals in both sludge and POL-compost ($>1,000$ mg kg^{-1} dry weight basis). The high Cu and Zn content in sewage sludge was of the same order of magnitude as in previous studies in the U.S. and Europe (Baker et al., 1985). Then, the concentration was followed by Pb> Cr, Ni>Cd in sewage sludge, and Pb, Cr, Ni>Cd in POL-compost (Table 34.1).

As the Tai Po Sewage Treatment Work received a mixture of domestic sewage and industrial effluent, substantial content of electroplating-contaminants, Cu, Pb, and Zn was expected. Also, the intensive use of household plumbing in industries explained the high Cu and Zn content.

The high Cu and Zn content in POL-compost was due to the incorporation of Cu-containing compound, such as $CuSO_4$, in pig feed to increase feed conversion efficiency and growth rates, and Zn amendment for Cu toxicity antagonism (Chaney, 1973).

It was interesting to note that Cd was the least dominant metal in sludge and, in fact, it is a relatively *Cd-free sludge*. POL-compost contained even more Cd than sludge.

Fraction of Exchangeable Heavy Metals in Solid Wastes

Most of the heavy metals in the wastes were not bioavailable, as the percentage of exchangeable metals in wastes accounted for not more than 13% of total heavy metal con-

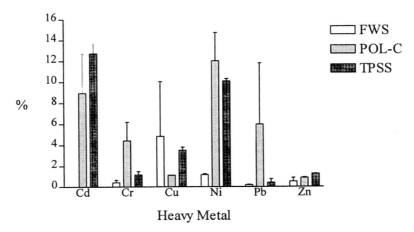

FIGURE 34.1. Exchangeable heavy metals (%) in freshwater sand (FWS), POL computer (POL-C) and sewage sludge (TPSS).

tent. A relatively higher percentage of Cd and Ni (Figure 34.1) existed in the exchangeable phase of the solid wastes. Only very small fractions of Cu and Zn existed in the exchangeable phase, although their total content in POL-compost and sewage sludge was rather high. Similar results were obtained by Pun et al. (1995).

Biological Response

In terms of sewage sludge application, *B. chinensis* showed a gradual increase in biomass with waste amendment up to 2.5% (w/w) (Figure 34.2a). There was no further significant difference with further sludge amendment above this level. For *R. sativus*, there is a maximum biomass with sludge amendment up to 1.25% (w/w) (Figure 34.2b). Again, further addition made no difference, although there was a slight drop in biomass at 10% (w/w) amendment. The presence of heavy metals probably limits the growth and neutralizes the effect of extra nutrients.

However, there is a different picture for POL-compost amendment (Figure 34.2a). Biomass of *B. chinensis* increased with POL-compost application dosage up to 2.5% (w/w). Further dosage increase made no difference. This could be explained by the fact that total nitrogen content in POL-compost (1.8%) was around one-third of sludge (4.6%). Application of POL-compost at 2.5% (w/w), double the dose of sewage sludge at 1.25% (w/w), had met the optimal nutrient requirement for *B. chinensis*. Further increase in POL-compost and sewage sludge dosage above 2.5% (w/w) and 1.25% (w/w), respectively, cannot further increase biomass production of the plant species.

On the other hand, the biomass of *R. sativus* was proportional to application rates used in this study (i.e., up to 10% (w/w)) (Figure 34.2b). In the case of *R. sativus* as one of the root crops, the improvement of the soil physical quality, by the addition of POL-compost, facilitated better root development. The humus nature of refuse compost, which reduced the pore space of the sandy soil, resulted in an improved water-holding and nutrient-holding capacity (Hortenstine and Rothwell, 1968) and rendered the mass more permeable to air (Hearman, 1977).

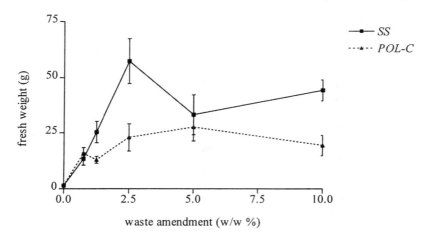

FIGURE 34.2a. Fresh weight (g) of *B. chinensis* grown from waste-amended soil.

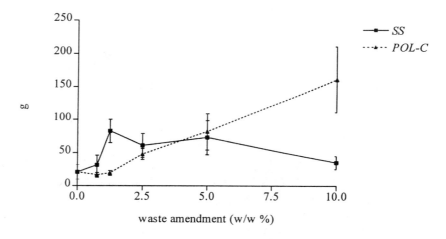

FIGURE 34.2b. Fresh weight (g) of *R. sativus* grown from waste-amended soil.

Biomass Distribution

When the biomass distribution of *R. sativus* (leaf dry wt./root dry wt.) was investigated, it was found that ratio increased with sewage sludge application dosage (Table 34.2). This was due to the dominance of heavy metals, especially Cu, Zn, which inhibited root growth. For sewage sludge, there is a reduction of root biomass beyond the 1.25% (w/w) amendment. Although there was an overall stimulation of leaf growth, the rate of increase was reduced after 2.5% (w/w) amendment. It was observed that roots are more sensitive to heavy metal toxicity than shoots in terms of inhibition, as illustrated in the case of sewage sludge (Figure 34.3). Previous researchers such as Wong and Bradshaw (1982) and Cheung et al. (1989) also reported similar phenomena. While there is no observable trend for POL-compost amendment, there was a parallel growth stimulation of both leaf and root which resulted in a more and less constant ratio (Table 34.2).

Table 34.2. Biomass Distribution of *R. sativus* Grown from Waste-Amended Soil.

Waste Amendment Dosage (w/w %)	Biomass Ratio (Leaf/Root)	
	Sewage Sludge	POL-Compost
0.00	0.75	0.75
0.75	1.09	0.50
1.25	0.82	0.85
2.50	1.53	0.55
5.00	2.22	1.13
10.00	3.32	0.79

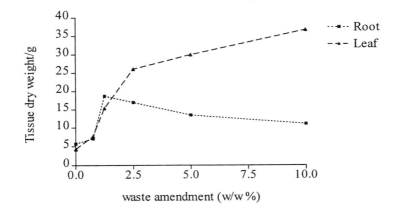

FIGURE 34.3. Dry weight of *R. sativus* grown from sewage sludge-amended medium.

Metal Toxicity and Metal Tolerance

Soil microbiological activity is much more sensitive to sludge Cu input than higher plants. An increase in heavy metals, such as Cd, Cu, Pb, and Zn, affected the decomposition rates of organic matter (Coughtrey et al., 1979) and would result in subsequent nitrification. Buresh and Maragtian (1976) reported that trace amounts of Cu can act as a catalyst to reduce the energy barrier for chemical reduction of nitrate to nitrite. Chang and Broadbent (1982) reported that an increase in extractable Cu in soil can decrease soil enzymatic activities and inhibit nitrification. Overall, increase of Cu in soil can reduce nitrogen availability.

According to Mellor and Malley, 1948, the order of stability of metallic organic complexes are ranked in the following order, Cu>Mn>Ni>Pb>Cd>Zn>Cr>Hg>Al>Fe. The metal toxicity in terms of root inhibition followed the same order as complex stability.

Overall, the growth of *B. chinensis* was better on sewage sludge than on POL-compost in terms of same-weight amendment (Figure 34.2a). The improved growth of *B. chinensis* was due to the higher nitrogen content and continuous mineralization in sewage sludge, which is essential for plant growth. Although there is a high metal content in sludge, the *Brassicaceae* family is well-known for their metal tolerance and uptake ability. For example, *B. juncea*, a high biomass crop plant, has been identified as a potentially useful plant for phytoremediation because of its high metal bioconcentration coefficient of shoot and root (Banuelos and Meek, 1990, Salt et al., 1995).

Heavy Metals in Crops

Figure 34.4 illustrates an obvious and direct relationship between the levels of Cu in the tissues of the harvested crops and the levels of Cu in the waste-amended media. Results of linear regression are presented in Table 34.3. It was observed that there was a linear relationship between the Cu content in *R. sativus* (especially root portion) and the level of waste amendment, as reflected by their high regression coefficient.

Several previous researchers have published data on the increased uptake by vegetable crops and timothy (*Phleum pratense*) on sewage sludge-amended soil that showed high levels of extractable Cu (Davies, 1980; Kabata-Pendias and Pendias, 1992; Adediran and Kramer, 1987; Harter, 1986).

For Cd, Cr, and Ni (Figure 34.5a, b, c), it was found that the levels of metals did not increase when the application level increased. This is particularly obvious for *R. sativus*. There were maximum values of heavy metals at application dosage of 5% (w/w), while any further increase in dosage resulted in a decline in the heavy metal content of the crop.

It was observed in this study that there was no measurable amount of lead in the aerial parts (leaf) of *B. chinensis* or the root and leaf of *R. sativus*. Again, there were no significant differences in the Pb-content of roots and leaves of *R. sativus* in this study, as reported by Gaweda (1991). Previous researchers also reported very limited translocation of Pb to aerial parts (including leaf), (Broyer et al., 1972; Kabata-Pendias 1977) and reproductive and storage organs (Berthet et al., 1984; Foroughi et al., 1983). Also, organic matter in soils would reduce the uptake of Pb by radish *Raphanus raphanistrum* L. *subvar. radicula* Pers. The strong barrier presence in plants further prevents the absorption of excessive amounts of Pb. That lead toxicity was not clearly observed in this study may be due to the dominance of Zn in pot media, which would reduce Pb toxicity in radish root (Gaweda, 1991).

It was clearly indicated in the results that *B. chinensis* has a higher Cr content than *R. sativus*. It is known that heavy metal absorption by plants depends on the species (Sequi and Petruzzelli, 1978; Delcarte et al., 1979; Hani and Gupta, 1983; Hovmand, 1983), because they differ in terms of cation exchange capacities. Metal uptake abilities were determined by their interaction and thermodynamic characteristics (Pearson, 1968 a,b) between special components in protein, such as nitrogen, sulfur, and oxygen.

Bioaccumulation of Heavy Metals

The bioaccumulation pattern of heavy metals into the crops was further illustrated by the bioconcentration factor (BCF) based on the ratio of heavy metal content in the vegetation to the total heavy metal content in the pot media. The BCF values were not found to be significantly correlated with application rate of the organic wastes, so the averaged BCF values were presented in Table 34.4. In general, the BCF values of the leafy crop (*Brassica chinensis*) was higher than that of the root crop (*Rahanus sativus*). Different tissues of *Raphanus sativus* showed different levels of heavy metal uptake. Particularly for sewage sludge-amended pots, levels of metals in leaves were always higher than those in roots.

Groupings of metals according to their mobility from waste-amended soil, reflected by the bioaccumulation, are as follows: Ni, Pb, Cd <Cr, Cu <Zn (Table 34.4).

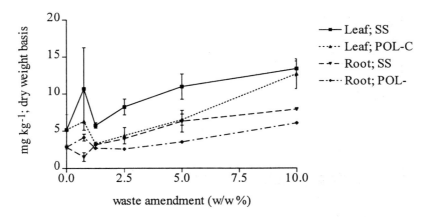

FIGURE 34.4. Copper content of *R. sativus* grown from waste-amended medium.

Table 34.3. Results of Linear Regression Analysis of Cu Content in Vegetation Against Waste-Amendment Dosage of Pot Medium.

Sample	Organic Waste Amended	Regression Coefficient
B. chinensis	SS	0.5904
B. chinensis	POLC	0.1021
R. sativus-Leaf	SS	0.6406
R. sativus-Leaf	POLC	0.8101
R. sativus-Root	SS	0.9023
R. sativus-Root	POLC	0.8346

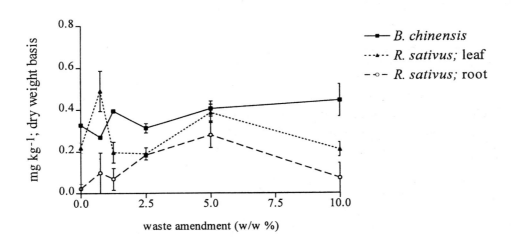

FIGURE 34.5a. Cd content of vegetation grown from sewage sludge-amended medium.

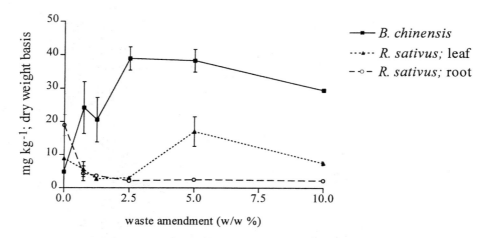

FIGURE 34.5b. Cr content of vegetation grown from sewage sludge-amended medium.

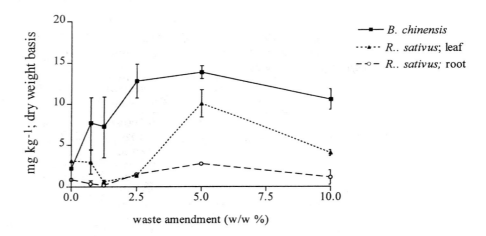

FIGURE 34.5c. Ni content of vegetation grown from sewage sludge-amended medium.

The BCF_{Cu}, (bioconcentration factor for copper) of *B. chinensis* tended to be higher than *R. sativus* leaf, which in turn is higher than that of root. It is the same in the case of BCF_{Zn}. In 1980, Davis and Calton-Smith reported that there was a relative metal accumulation of Cu and Zn in aerial parts of plants.

Ni, Cr, and Cd were accumulated in plants as suggested in the literature (Davis and Stark, 1980). Maize plants accumulate large amounts of Cr, Ni, and Mn when aerobically, anaerobically digested sludge and chicken manure, is added to soil and unamended pots (Hernández et al., 1991).

Ni seems to be the least mobile element in this study. As reflected by the BCF, it was different from previous workers' results, which stated that Ni was considered to be a highly mobile element (Cataldo et al., 1978).

It was observed that mobility of lead to aboveground parts (leaf) was low in cabbage. Zinc was the element most selectively taken up. It was also noted that BCF, in both plant species, always exceeded 1. Channey and Giordano (1977) classified Zn as an element that

Table 34.4. Bioconcentration Factors (BCF) Data for Waste-Amended Pots Based on the Total Heavy Metals in Pot Media [mean (s.d.)].

	Cd	Cr	Cu	Ni	Pb	Zn
(A) Sewage Sludge						
B. chinensis	0.487	0.659	0.709	0.294	0.314	2.921
	(0.267)	(0.505)	(0.681)	(0.198)	(0.142)	(1.601)
R. sativus	ND[a]	0.143	0.876	0.093	0.431	2.145
(Leaf)	(0.105)	(0.924)	(0.081)	(0.401)	(0.509)	
R. sativus	ND	0.104	0.467	0.032	0.098	1.228
(Root)	(0.104)	(0.514)	(0.025)	(0.07)	(0.33)	
(B) POL Compost						
B. chinenisis	0.605	0.538	1.690	0.1965	0.372	3.669
	(0.424)	(0.358)	(1.235)	(0.129)	(0.151)	(1.599)
R. sativus	ND	0.069	0.651	0.027	0.393	2.337
(Leaf)	(0.046)	(0.934)	(0.026)	(0.244)	(0.651)	
R. sativus	ND	0.173	0.375	0.033	0.221	1.434
(Root)	(0.223)	(0.505)	(0.052)	(0.172)	(0.928)	

[a] ND: cannot be determined as the metal level (in pot media) falls below detection limit.

is readily translocated to the plant tops. Webber (1972) also reported that Zn is readily taken up and translocated within wheat grown on Zn-contaminated soils. This reflects extensive uptake of Zn, which is one of the essential elements for the normal physiology of the plants (Table 34.4).

Zinc malate may be involved in zinc transfer and perhaps tolerance in zinc-tolerant species such as *Deschampsia caespitosa* and *Agrostis tenuis* (Brookes et al., 1981).

Only Cr was bioaccumulated in the crops. Results indicated that the edible part of *B. chinensis*, grown with high rates of sewage sludge, exceeded the permissible level of Cr, in Hong Kong, which is 1 ppm (wet weight basis).

Few chromium accumulators have been discovered. Mainly Zimbabwean species have been reported to contain high concentrations of chromium in their ash—up to 5% (Wild, 1974). Also, leaf tissue of wide *Leptospermum scoparium* from a disused chromite mine in New Zealand contained up to 1% of soluble chromium (Lyon et al., 1971) as trioxalato-chromate (III) ion, $[Cr(C_2O_4)_3]^{3-}$ (Lyon et al., 1969). Chromium was, however, transported in the xylem sap as chromate, indicating that metabolism to the complex took place within the leaf tissues.

The influence of transpiration on cation uptake by intact plants has more significant effect on the accumulation of cations by the shoot than root (Bowling and Weatherley, 1965). The continued delivery of ions via the transpiration stream ensures that the leaf is supplied with the mineral nutrients essential to its growth and development. However, the continued supply of ions and other solutes to a mature leaf can lead to excessively high levels within the leaf and may result in some toxicity. Some excess ions are accumulated into the vacuoles of the leaf mesophyll cells (Baker, 1983).

REFERENCES

Adediran, S.A. and J.R. Kramer. Copper Adsorption on Clay, Iron-Manganese Oxide and Organic Fractions Along a Salinity Gradient. *Appl. Geochem.*, 2, pp. 213–216, 1987.

Allen, S.E. Analysis of Vegetation and Other Organic Materials, in *Chemical Analysis of Ecological Materials*, 2nd ed., S.E. Allen, Ed., Blackwell Scientific Publications, London, 1989, p. 59.

Baker, D.A. Uptake of Cation and the Transport Within the Plant, in *Metals and Micronutrients: Uptake and Utilization by Plants*. Robb, D.A. and W.S. Pierpoint, Eds., Academic Press, Inc., London, 1983, pp. 3–19.

Baker, D.E. *Criteria and Recommendations for Land Application of Sludges in the Northeast*. Pennsylvania State University, Agric. Exp. Sta. Bull. 851, 1985.

Banuelos, G.S. and D.W. Meek. Accumulation of Selenium in Plants Grown on Selenium-Treated Soil. *J. Environ. Qual.*, 19, pp. 772–777, 1990.

Berthet, B., C. Amiard-Triquet, C. Metayer, and J.C. Amiard. Etude des voies de transfert du plomb de l'environnment aux vegetaux cultives; application a l'utilisation agricole de boues de station d'epuration. *Water, Air Soil Pollut.*, 21, pp. 447–460, 1984.

Bowling, D.J.F. and P.E. Weatherley. The Relationships Between Transpiration and Potassium Uptake in *Ricinus communis*. *J. Exptl. Botany*, 16, pp. 732–741, 1965.

Bridle, T.R. and C.K. Hertle. Oil from Sludge—A Cost-Effective Sludge Management System. *J. Aust. Water Wastewater Assoc.*, 15(3), 1988.

Brookes, A., J.C. Collins, and D.A. Thurman. The Mechanisms of Zinc Tolerance in Grasses. *J. Plant Nutrition*, 3, pp. 695–705, 1981.

Broyer, T.C., C.M. Johnson, and R.E. Paull. Some Aspects of Lead in Plant Nutrition, *Plant Soil*, 2, pp. 301–313, 1972.

Buresh, R.J. and J.T. Maragtian. Chemical Reduction of Nitrate by Ferrous Iron. *J. Environ. Quality*, 5, pp. 320–325, 1976.

Cataldo, D.A., T.R. Garland, R.E. Wilung, and H. Drucker. Nickel in Plants. *Plant Physiol.*, 62, pp. 566–570, 1978.

Chaney, R.L. *Crop and Food Chain Effects of Toxic Elements in Sludge and Effluents*, Champaign, Urbana, IL. U.S. Environment Protection Agency, U.S.D.A. University workshop report, 1973.

Chang, F.H. and F.G. Broadbent. Influence of Trace Metals on Some Soil Nitrogen Transformations. *J. Environ. Quality*, 11, pp. 1–4, 1982.

Channey, R.L. and P.M. Giordano. *Soils for the Management of Organic Wastes and Waste Waters*, Elliot, L.F. and F.L. Stevenson, Eds., Soil Sci. Soc. Am., Am. Soc. Agron. & Crop Sci. Soc. Am., Madison, WI, 1977, pp 235–279.

Chaussod, R., J.C. Germon, and G. Catroux. Determination de la valeur fertilisante des boues résiduaires. Aptitude à liberer l'azote. Ministère de l'environnement et du cadre de vie. Convention d'étude n. 74050 (in French), 1978.

Cheung, Y.H., M.H. Wong, and N.F.Y. Tam. Root and Shoot Elongation as an Assessment of Heavy Metal Toxicity and 'Zn Equivalent Value' of Edible Crops. *Hydrobiologia*. 188/189, pp. 377–383, 1989.

Coughtrey, P.J., C.H. Jones, M.H. Martin, and S.W. Shales. Litter Accumulation in Woodlands Contaminated by Lead, Zinc, Cadmium and Copper. *Oikos.*, 39, pp. 51–60, 1979.

Davies, B.E. *Applied Soil Trace Elements*. John Wiley & Sons, New York, 1980.

Davis, D.R. and C. Calton-Smith. Crop as Indicators of the Significance of Contamination of Soil by Heavy Metal, Technical Report 140. Water Research Centre, Sterenage, UK, 1980.

Davis, R.D. and J.H. Stark. Effects of Sewage Sludge on the Heavy Metals Content of Soils and Crops. Field trials at Carrington and Royston, in Proceedings of the Second European Symposium Characterization, Treatment and Use of Sewage Sludge, Vienna, 1980.

Delcarte, D., C. Xanthoulis, and R. Impens. Valorisation agricole des boues. Problèmes liés a la présence de métaux lourds, in Proceedings of the First European Symposium C.E.E. Treatment and Use of Sewage Sludge, Cadarache, France (in French), 1979.

Dowdy, R.H., W.E. Larson, J.M. Titrud, and J.J. Latterell. Growth and Metal Uptake of Snap Beans Grown on Sewage Sludge Amended Soil. A Four-Year Study. *Environ. Qual.*, 7, pp. 252–257, 1978.

Foroughi, M.V., D. Fritz, K. Teicher and F. Venter. Die Wirkung einiger Schwermetalle auf Gemüsepflnzen-Gegenüberstellung der Ergebnisse aus Wasserkultur- und Substratversuchen. Landwirtshafliche For-schung, *Sonderheft.*, 39, pp. 426–433, 1983.

García-Delgado, R.A., F. García-Herrnzo, C. Gómez-Lahoz, and J.M. Rodrígnez-Maroto. Heavy Metals and Disposal Alternatives for an Anaerobic Sewage Sludge. *J. Environ. Sci. Health*, A29(7), pp. 1335–1347, 1994.

Gaweda, M. The Uptake of Lead of Spinach *Spinacia oleracea L.* and Radish *Raphanus raphanistrum L. subvar. radicula pers.* as Affected by Organic Matter in Soil. *Acta Physiologiae Plantarum*, 13 (3), pp. 167–174, 1991.

Golterman H.L., R.S. Clymo, and M.A.M. Ohnstad. *Methods of Physical and Chemical Analysis of Freshwater*, Blackwell, Oxford, 1978.

Hani, H. and S. Gupta. Second Report on the Standalized Cd-Pot Experiment 1980–81, in Concerted Action Treatment and Use of Sewage Sludge, Cost 68 Ter, 1983.

Harter, R.D. *Adsorption Phenomena*. Van Nostrand Reinhold, New York, 1986.

Hearman, J.D. Composting—An Approach to Using Sewage Waste. *Compost Sci.*, 18(1), pp. 28–29, 1977.

Hernàndez, T., J.I. Moreno, and F. Costa. 1991. Influence of Sewage Sludge Application on Crop Yields and Heavy Metal Availability. *Soil Sci. Plant Nutrition*, 37, pp. 201–210, 1977.

Hortenstine, C.C. and D.F. Rothwell. 1968. Garbage Compost as a Source of Plant Nutrients for Oats and Radishes. *Compost Sci.*, 9(2), pp. 23–25, 1977.

Hovmand, M.F. Cycling of Pb, Cd, Zn and Ni in Danish Agriculture, in C.E.C. The Influence of Sewage Sludge Application on Physical and Biochemical Properties of Soils, Cartroux, L'H. and D. Suess, Eds., Reidel Publishing Company, Dordrecht and London, 1983.

Jones, J., Jr., B. Benton, B. Wolf, and H.A. Mills. *Plant Analysis Handbook. A Practical Sampling, Preparation, Analysis, and Interpretation Guide*. Micro-Macro Publishing, Inc., Athens, GA, 1991, p. 213.

Kabata-Pendias, A. 1977. Wplyw olowiu w pozywce wondnej na sklad chemiczny stoklosy (*Bromus unioloides* L.). *Roczniki Nauk Rolniczych*, Seria A, 102(2), pp. 29–38, 1991.

Kabata-Pendias, A. and H. Pendias. *Trace Elements in Soils and Plants*, 2nd ed., CRC Press, Boca Raton, FL, 1992, p. 365.

Lyon, G.L., P.J. Peterson, and R.R. Brooks. Chromium-51 Distribution in Tissues and Extracts of *Leptospermum scoparium. Planta (Berl.)*, 88, pp. 282–287, 1969.

Lyon, G.L., P.J. Peterson, R.R. Brooks, and G.W. Butler. Calcium, Magnesium and Trace Elements in a New Zealand Serpentine Flora. *J. Ecol.*, 59, pp. 421–429, 1971

Mackereth, F.J.H., J. Heron, and J.F. Talling. *Water Analysis*. Freshwater Biological Association. Sci. Publ. No. 36, 1978.

McGrath, S.P. Long-Term Studies of Metal Transfers Following Application of Sewage Sludge, in *Pollutant Transport and Fate in Ecosystems*. Coughtrey, P.J. and M.H. Martin, Eds., Unsworth. Blackwell Scientific, Oxford, 1987, pp. 301–317.

Mellor, C.P. and L. Malley. Order of Stability of Metal Complexes, *Nature*, London. clxi. 436, 1948.

Oswell, M.A. and R.C. Rootham. Strategic Studies on Sludge Disposal for Hong Kong, in *Sludge Treatment and Disposal: Current Practice and Developments*, Holmes, P.R., Ed., The Institution of Water and Environmental Management, Hong Kong, 1992, pp. 7–16.

Pearson, R. Hard and Soft Acid and Bases. Part I: Fundamental Principles. *J. Chem. Ed.* 45, pp. 581–587, 1968a.

Pearson, R. Hard and Soft Acid and Bases. Part II: Underlying Theories. *J. Chem. Ed.*, 45, pp. 643–648, 1968b.

Pett, J. and G. Eduljee. Flora and Fauna, in *Environmental Impact Assessment for Waste Treatment and Disposal Facilities*, John Wiley, Chichester, 1994, pp. 125–140.

Pun, K.C., R.Y.H. Cheung, and M.H. Wong. Characterization of Sewage Sludge and Toxicity Evaluation with Microalgae. *Marine Pollut. Bull.*, 31, pp. 394–401, 1995.

Salt, D.E., R.C. Prince, I.J. Pickering, and I. Raskin. Mechanisms of Cd Mobility and Accumulation in Indian Mustard, *Plant Physiol.*, 109, pp. 1427–1433, 1995.

Sequi, P. and G. Petruzzelli. Il riciclaggio dei rigiuti. Pericolo per l'inquinamento del terreno. *Ital. Agric. Ann.*, 115, n. 10 Oct (in Italian), 1978.

Sommers, L.E. Chemical Composition of Sewage Sludge and Analysis of Their Potential Use as Fertilizer. *J. Environ. Qual.*, 6, pp. 225–232, 1977.

Suss, A. Nitrogen Availability in Sewage Sludge, in *Converted Action E.E.C. Treatment and Use of Sewage Sludge, Cost Project* 68 bis, Dijon, 1979.

The European Community Urban Waste Water Treatment Directive (91/271/EEC).

Webber, J. Effects of Toxic Metals in Sewage on Crops. *Water Pollut. Control*, 71 (3), p. 404, 1972.

Wild, H. *Kirkia*. 9, pp. 209–232, 1974.

Willis, R.B. and C.E. Gentry. *Commun. Soil Sci. Plant Anal.*, 18, p. 625, 1987.

Wong, M.H. and A.D. Bradshaw. A Comparison of Toxicity of Heavy Metals by Using Root Elongation of Ryegrass, *Lolium perenne. New Phytol.*, 91, pp. 255–261, 1982.